Citrus
Second Edition

CROP PRODUCTION SCIENCE IN HORTICULTURE SERIES

Series Editor: Jeff Atherton, Professor of Tropical Horticulture, University of the West Indies, Barbados

This series examines economically important horticultural crops selected from the major production systems in temperate, subtropical and tropical climatic areas. Systems represented range from open field and plantation sites to protected plastic and glass houses, growing rooms and laboratories. Emphasis is placed on the scientific principles underlying crop production practices rather than on providing empirical recipes for uncritical acceptance. Scientific understanding provides the key to both reasoned choice of practice and the solution of future problems.

Students and staff at universities and colleges throughout the world involved in courses in horticulture, as well as in agriculture, plant science, food science and applied biology at degree, diploma or certificate level, will welcome this series as a succinct and readable source of information. The books will also be invaluable to progressive growers, advisers and end-product users requiring an authoritative, but brief, scientific introduction to particular crops or systems. Keen gardeners wishing to understand the scientific basis of recommended practices will also find the series very useful.

The authors are all internationally renowned experts with extensive experience of their subjects. Each volume follows a common format covering all aspects of production, from background physiology and breeding, to propagation and planting, through husbandry and crop protection, to harvesting, handling and storage. Selective references are included to direct the reader to further information on specific topics.

Titles Available:
1. **Ornamental Bulbs, Corms and Tubers** A.R. Rees
2. **Citrus** F.S. Davies and L.G. Albrigo
3. **Onions and Other Vegetable Alliums** J.L. Brewster
4. **Ornamental Bedding Plants** A.M. Armitage
5. **Bananas and Plantains** J.C. Robinson
6. **Cucurbits** R.W. Robinson and D.S. Decker-Walters
7. **Tropical Fruits** H.Y. Nakasone and R.E. Paull
8. **Coffee, Cocoa and Tea** K.C. Willson
9. **Lettuce, Endive and Chicory** E.J. Ryder
10. **Carrots and Related Vegetable** *Umbelliferae* V.E. Rubatzky, C.F. Quiros and P.W. Simon
11. **Strawberries** J.F. Hancock
12. **Peppers: Vegetable and Spice Capsicums** P.W. Bosland and E.J. Votava
13. **Tomatoes** E. Heuvelink
14. **Vegetable Brassicas and Related Crucifers** G. Dixon
15. **Onions and Other Vegetable Alliums, 2nd Edition** J.L. Brewster
16. **Grapes** G.L. Creasy and L.L. Creasy
17. **Tropical Root and Tuber Crops: Cassava, Sweet Potato, Yams and Aroids** V. Lebot
18. **Olives** I. Therios
19. **Bananas and Plantains, 2nd Edition** J.C. Robinson and V. Galán Saúco
20. **Tropical Fruits, 2nd Edition Volume 1** R.E. Paull and O. Duarte
21. **Blueberries** J. Retamales and J.F. Hancock
22. **Peppers: Vegetable and Spice Capsicums, 2nd Edition** P.W. Bosland and E.J. Votava
23. **Raspberries** R.C. Funt
24. **Tropical Fruits, 2nd Edition Volume 2** R.E. Paull and O. Duarte
25. **Peas and Beans** A. Biddle
26. **Blackberries and Their Hybrids** H.K. Hall and R.C. Funt
27. **Tomatoes, 2nd Edition** E. Heuvelink
28. **Grapes, 2nd Edition** G.L. Creasy and L.L. Creasy
29. **Citrus, 2nd Edition** L.G. Albrigo, L.L. Stelinski and L.W. Timmer

CITRUS

SECOND EDITION

L. Gene Albrigo

Professor Emeritus of Horticulture, Citrus Research and Education Center, University of Florida, Lake Alfred, Florida, USA

Lukasz L. Stelinski

Professor of Entomology, CREC, UF

Lavern W. Timmer

Professor Emeritus of Plant Pathology, CREC, UF

CABI is a trading name of CAB International

CABI
Nosworthy Way
Wallingford
Oxfordshire OX10 8DE
UK

CABI
745 Atlantic Avenue
8th Floor
Boston, MA 02111
USA

Tel: +44 (0)1491 832111
Fax: +44 (0)1491 833508
E-mail: info@cabi.org
Website: www.cabi.org

Tel: +1 (617)682-9015
E-mail: cabi-nao@cabi.org

© L. Gene Albrigo, Lukasz L. Stelinski and Lavern W. Timmer 2019. All rights reserved. No part of this publication may be reproduced in any form or by any means, electronically, mechanically, by photocopying, recording or otherwise, without the prior permission of the copyright owners.

A catalogue record for this book is available from the British Library, London, UK.

Library of Congress Cataloging-in-Publication Data

Names: Albrigo, Leo Gene, 1940- author. | Stelinski, Lukasz L., author. | Timmer, L. W. (Lavern Wayne), 1941- author.
Title: Citrus / L. Gene Albrigo, Lukasz L. Stelinski, Lavern W. Timmer.
Description: 2nd edition. | Boston, MA : CABI, 2019. | Includes bibliographical references and index. | Summary: "Citrus, 2nd Edition covers the biology and cultivation of citrus and considers the citrus industry from an international perspective. Fruits including oranges, mandarins, lemons, limes, grapefruit and hybrids such as tangelos are covered and the fundamental topics of taxonomy, cultivars, breeding, husbandry, weeds, pests and diseases are also addressed in detail. This updated new edition includes coverage of new cultivars, advances in the molecular approaches to taxonomic studies, new findings in the physiological understanding of key citrus fruits, recent research into environmental factors affecting external and internal fruit quality and expanded coverage of pests and diseases. The fusion of scientific coverage and practical management make the text suitable for a range of horticulturalists including breeders, growers, researchers and academics"-- Provided by publisher.
Identifiers: LCCN 2019016379 (print) | LCCN 2019980150 (ebook) | ISBN 9781845938154 (paperback) | ISBN 9781780642727 (pdf)
Subjects: LCSH: Citrus. | Citrus fruit industry.
Classification: LCC SB369 .A572 2019 (print) | LCC SB369 (ebook) | DDC 634/.304--dc23
LC record available at https://lccn.loc.gov/2019016379
LC ebook record available at https://lccn.loc.gov/2019980150

ISBN-13: 9781845938154 (Paperback)
            9781780642727 (ePDF)
            9781789242096 (ePub)

Commissioning Editor: Rebecca Stubbs
Editorial Assistant: Tabitha Jay
Production Editor: James Bishop

Typeset by SPi, Pondicherry, India
Printed and bound in the UK by Bell & Bain Ltd, Glasgow

Contents

	Preface	VII
1	**History, Distribution and Uses of Citrus Fruit**	**1**
	History	1
	Distribution and Production	4
	Uses of Citrus Fruits	11
2	**Taxonomy, Cultivars and Breeding**	**13**
	Taxonomy	13
	Citrus Genetics and Breeding	45
3	**Rootstocks**	**54**
	Description of Traditional Citrus Rootstocks	56
	Morphology and Anatomy of Citrus Rootstocks	71
	Physiology of Citrus Rootstocks	72
	Mycorrhizae	76
	Recent Advances in Rootstock Development	77
4	**Environmental Constraints on Growth, Development and Physiology of Citrus**	**78**
	Tropical Regions	78
	Subtropical Regions	80
	Environmental Constraints on Vegetative Growth and Development	82
	Environmental Constraints on Flowering and Fruiting	92
	Environmental Constraints on Fruit Yields	106
	Environmental Factors Affecting Fruit Growth, Development and Quality	109

5	**PLANT HUSBANDRY**	**117**
	Nursery Operations	117
	Establishing the Orchard	127
	Water Management	136
	Freeze-hardiness and Freeze Protection	146
	Mineral Nutrition	153
	Tree Size Control	168
	Growth Regulators	171
	Weed Control	173
	Plant Husbandry Summary	182
6	**ARTHROPOD PESTS**	**183**
	Asian Citrus Psyllid	184
	Citrus Leafminer	186
	Root Weevil Complex	187
	Mites	190
	Secondary Insect Pests	193
	Pest Management	204
7	**DISEASES**	**214**
	Tree Declines and Disorders	214
	Soil-borne Diseases	229
	Diseases of Fruit and Foliage	232
	Postharvest Decays	239
	Control Measures	241
8	**FRUIT QUALITY, HARVESTING AND POSTHARVEST TECHNOLOGY**	**246**
	Characteristics of Citrus Fruit	246
	Fresh Fruit	252
	Postharvest Technology	260
REFERENCES		**273**
INDEX		**305**

Preface

Citrus fruits have been cultivated and enjoyed for over 4000 years. Moreover, they are grown in nearly every country within the 40° north–south latitude. This worldwide dissemination has been associated with many of the great explorations and conflicts in history, including the conquests of Alexander the Great, the spread of Muhammadanism and the explorations of Columbus, who brought citrus to the New World.

The intent of this second edition of *Citrus*, like the original by F.S. Davies and L.G. Albrigo, is to provide the reader with an updated overview of citriculture from a worldwide perspective. As a practical matter, individual cultural programmes for every citrus-growing region cannot be discussed, but instead current theories and technological advances in citriculture are emphasized, citing specific examples of how and where they are used. Many current references and reviews on various aspects of citriculture are included for persons desiring more detail on a specific topic. However, due to space limitations the literature review provided is by no means exhaustive.

The text begins with a discussion of major production areas with figures and current trends (Chapter 1). The confusing and controversial taxonomic situation for *Citrus* and related genera is then discussed, emphasizing molecular biology (biotechnological) advances that are clarifying the genetic relationships between various citrus species. This is followed by a discussion of the major commercially important citrus species and cultivars and traditional and current techniques in citrus breeding (Chapter 2). In Chapter 3 we cover the importance of rootstocks in citriculture and discuss the major rootstocks, their advantages and disadvantages. In Chapter 4, the role of climatic factors in worldwide citrus production is emphasized. Climate has a pronounced effect on citrus yields, growth, economic returns and fruit quality. Plant husbandry, including nursery practices, irrigation, fertilization, freeze protection, pruning, growth regulator use and weed control is covered in Chapter 5. In Chapter 6 the major pests and recent changes in their distribution are covered. Diseases of citrus, emphasizing major problems and control measures are the topic of Chapter 7. The final chapter (8) deals with postharvest quality, harvesting and handling of citrus fruits, including the importance of biotic and abiotic problems, as well as packinghouse and processing techniques. The text should provide the reader with the most important basic information on citriculture, although certainly not all aspects have been addressed.

Several reviewers have made extremely helpful and constructive suggestions on improving the text. While the text has undergone numerous revisions, we take full responsibility for any errors that may occur. We would also like to thank the series editors for their helpful suggestions, editorial assistance and patience throughout the project. Dr Fred Davies' efforts on the first edition should be recognized and much of that information is timeless and carried into this edition. Finally, the authors would like to thank Katherine Snyder for her efforts in developing some and assembling all the figures for this book. She was particularly helpful in improving some of the images used in the figures.

L. Gene Albrigo
Lukasz L. Stelinski
L. W. Timmer

1

History, Distribution and Uses of Citrus Fruit

HISTORY

The major area of origin of *Citrus* species, particularly edible fruits, is believed to be southeastern Asia, including from eastern Arabia east to the Philippines and from the Himalayas south to Indonesia or Australia (Fig. 1.1). Within this large region, northeastern India and northern Burma were believed to be the centre of origin, but a 1990 evaluation suggested that Yunnan Province in south-central China may be as important due to the diversity of species found and the system of rivers that could have provided dispersal to the south (Gmitter and Hu, 1990). More recently Wu *et al.* (2018) have identified the slopes of the Himalayan mountains as the likely area of origin for much of citrus. Within the area, more tropical types, such as limes, probably evolved in more tropical areas and more subtropical fruit types, such as oranges and some mandarins, in the more northern zones or more mountainous, cooler, regions. However, polyembryony stabilized cultivars and therefore 'species' so that citrus types may have been moved further than expected.

Extensive movement of the various types of citrus probably occurred within the general area of citrus origin from before recorded history. Many types of citrus are believed to have moved west before Christ to various Arabian areas, such as Oman, Persia, Media (Iran) and even Palestine (Tolkowsky, 1938). Major types of edible citrus from these areas include citron, sour orange, lime, lemon, sweet orange, shaddock (pummelo), mandarin and kumquat. Cultivar characteristics, genetics and taxonomy of the major edible citrus and the complex nature of named citrus species are presented in Chapter 2.

Citrons (*Citrus medica* L.) originated in the region from south China to India. The citron was found in Media when Alexander of Macedonia entered Asia (about 330 BC) and was subsequently introduced into the Mediterranean region. Other citrus types were introduced to Italy during the early Roman Empire (27 BC–AD 284), but they are believed to have been destroyed at the end of this era. Controversy exists about whether citron is mentioned in the

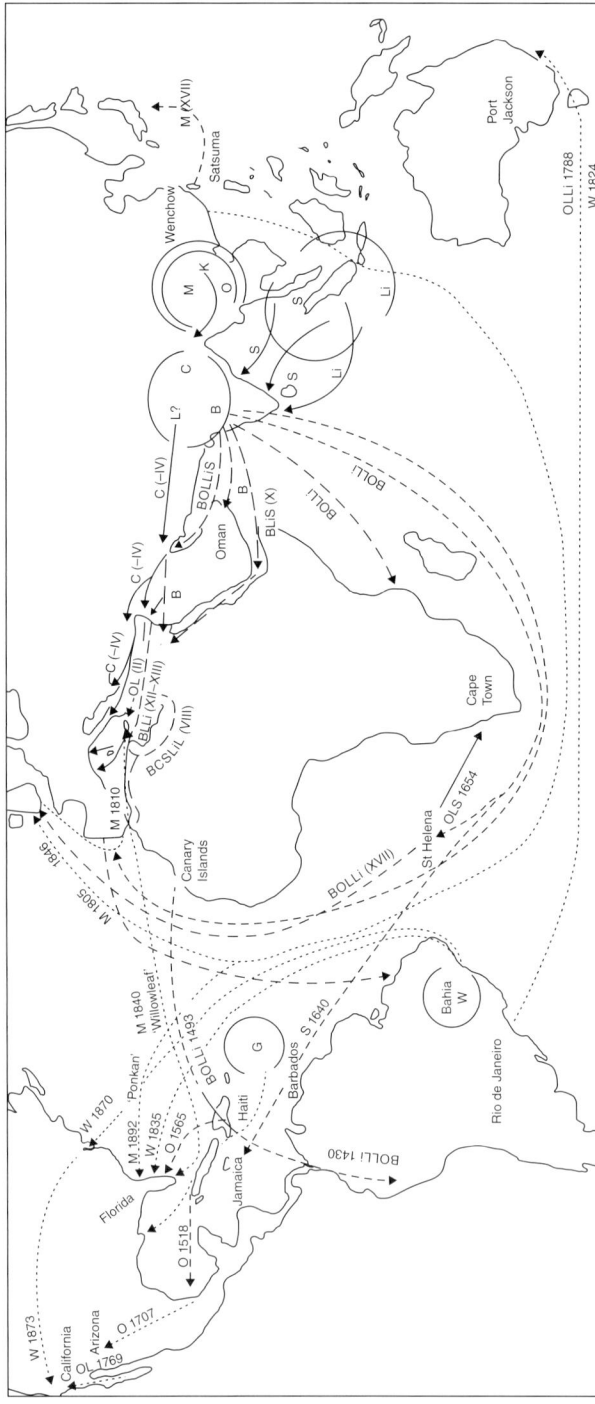

Fig. 1.1. The areas of origin of major citrus species and their paths of distribution (Chapot, 1975). B, bigarade (*Citrus aurantium*); C, citron (*Citrus medica*); G, grapefruit (*Citrus paradisi*); K, kumquat (*Fortunella margarita*); L, lemon (*Citrus limon*); Li, lime (*Citrus aurantifolia*); M, mandarin (*Citrus reticulata*); S, shaddock (*Citrus grandis*); W, 'Washington' navel; ⎯ BC; ⎯ ⎯, AD 1–700; ⎯ ⎯ ⎯, AD 700–1492 (711: Arab occupation of Spain); ⎯ ⎯ ⎯ ⎯, 1493–1700 (1493 second expedition of Christopher Columbus); ⎯ ⎯ ⎯ ⎯ ⎯, after AD 1700 (first appearance of grapefruit). Centuries are given in roman numerals (minus sign indicates BC); years are given in arabic numerals.

Bible, but it is clear that the Jewish religion was using it in their ceremonies by AD 50–150 (Webber et al., 1967).

Limes (*Citrus aurantifolia* Swingle) apparently originated in the east Indian archipelago. They were probably brought across the Sea of Oman by Arabian sailors and subsequently transported to Egypt and Europe.

Lemons (*Citrus limon* Burmann) are of unknown origin, possibly a hybrid between citron and lime, creating an intermediate species (Chapot, 1975; Barrett and Rhodes, 1976). Citron is believed to be a more primitive species and limes and lemons are at least closely related (Barrett and Rhodes, 1976). Lemons are known to have been spread to North Africa and Spain about AD 1150 in connection with expansion of the Arabian Empire.

The area of origin of the sour orange (*Citrus aurantium* L.) is believed to be southeastern Asia, possibly India. Sour orange was introduced progressively westwards in the early centuries AD associated with Arab conquests until it reached North Africa and Spain in around AD 700.

Sweet oranges (*Citrus sinensis* [L.] Osbeck) originated in southern China and possibly as far south as Indonesia (Webber et al., 1967). The sweet orange may have travelled a similar route as the citron and been introduced to Europe by the Romans. Greenhouses were developed during this time to protect the cold tender plants in pots in winter in the gardens of influential Roman families and were called orangeries, implying that oranges were a primary plant maintained by the Romans (Tolkowsky, 1938). Early introductions were apparently lost after the fall of the Roman Empire and oranges were reintroduced around 1425 through the Genoese trade routes (Webber et al., 1967). More selections were no doubt introduced during the extensive trading of the Venezia era, between the 15th and 17th centuries AD. The Portuguese brought superior selections of sweet oranges from China in about AD 1500. The great families of the Italian City States maintained large collections of lemons, oranges, etc. (paintings and descriptions of the cultivar collections of citrus and other fruits of the Medici family are on display in museums in Florence, Italy).

The 'Washington' navel orange originated in Bahia, Brazil and is probably a mutation of 'Seleta' sweet orange. It was introduced to Australia (1824), Florida (1835) and California (1870) through Washington, DC, where it apparently received its current name (Fig. 1.1). 'Washington' navel and many cultivars arising from mutations of it have been distributed worldwide (Davies, 1986a).

The shaddock or pummelo (*Citrus grandis* [L.] Osbeck) originated in the Malaysia and Indian archipelagos and is widely distributed in the Fiji Islands. Hybrids of shaddock were found by crusaders in Palestine by AD 900 and were distributed to Europe and then to the Caribbean, apparently by an East Indian ship captain named Shaddock (Webber et al., 1967).

The origin of grapefruit appears to be different in that it developed as a mutation or hybrid of the shaddock in the West Indies, perhaps Barbados. The binomial *Citrus paradisi* Macf. was assigned to this species (of questionable validity, see Chapter 2). Grapefruit were introduced into Florida, still a major producer worldwide – although production has been greatly reduced by the disease huanglongbing (HLB) (see Chapter 7) – from the Caribbean in about 1809 by Don Phillippe, probably from seed collected in Jamaica.

The area of origin for mandarins (*Citrus reticulata* Blanco) was probably Indo-China and south China with traders carrying selections to eastern India. Traditional production areas of this species have been in Asia. Mandarins were transported from Asia to Europe much later than other citrus; the 'Willowleaf' (*Citrus deliciosa* Tenole) was introduced from China after 1805 to become the major species in the Mediterranean region and *C. reticulata* was introduced even later.

Movement of citrus to Africa from India probably occurred during AD 700–1400 and various citrus, particularly limes and oranges, were introduced to the Americas by the Mediterranean explorers and settlers settled in Hispaniola in the Caribbean (Spanish) and Bahia, Brazil (Portuguese). The movement of citrus throughout the Americas was aided by the development of missions by the Roman Catholic Church that established plantings of various fruit including citrus.

Kumquats (*Fortunella margarita* [Lour.] Swingle) from southern China, other minor use citrus types and trifoliate orange (*Poncirus trifoliata* [L.] Raf.) from central and northern China, for freeze-hardy rootstocks, are also important species. Other related species are being employed in new biotechnology-oriented breeding programmes (see Chapter 2). The origin concept of the various 'citrus species' is probably valid as locations from which stable hybrids evolved that we refer to as species. The genetic relationships in *Citrus* are discussed in Chapter 2 also.

Some unusual citrus relatives appear to have originated in Australia – the xerophytic *Eremocitrus* and *Microcitrus*, and from Africa – *Aegle*, *Aeglopsis*, *Afraegle* and *Balsamocitrus* (Khan, 2007).

DISTRIBUTION AND PRODUCTION

Major Production Areas

Citrus is grown primarily between the latitudes 40°N to 40°S. More northern and southern locations of commercial production exist where temperatures are moderated by large bodies of water or ocean winds. World changes in citrus production from the early 2000s (2000–2004) to recent times (2013–2014) are presented from UNFAO data (http://www.fao.org/economic/est/est-commodities/citrus-fruit/en/) in Table 1.1. During this 10-year period there has been about a 23% increase in production overall. Most of this increase has been due to increases in mandarin production, although all types of citrus showed an increase in production, even though São Paulo, Brazil and Florida, USA are having reduced yields of oranges and grapefruit in Florida because of HLB disease. Further, most of the production increase was in the northern hemisphere. At least 32 countries produce more than 300,000 million tonnes of citrus annually. China, Brazil and the USA, in that order, are the major producers

Table 1.1. Change in world citrus production from 2003–2004 and 2004–2005, two season average to 2012–2013 and 2013–2014. All values are expressed in 1000 × tonnes (http://www.fao.org/economic/est/est-commodities/citrus-fruit/en/).

Location/years	Total	Oranges	Mandarins	Lemons/limes	Grapefruit
World – 2003–2004 & 2004–2005	97,823	62,699	18,271	11,537	5,316
Northern hemisphere	69,501	40,245	16,109	8,486	4,662
Southern hemisphere	28,322	22,454	2,163	3,051	654
World – 2012–2013 & 2013–2014	120,219	68,060	30,302	13,058	7,430
Northern hemisphere	92,603	46,549	28,117	10,131	6,704
Southern hemisphere	27,616	21,512	2,186	2,927	726

(28 to 10 million tonnes in 2012–2013 and 2013–2014), but Mexico, India and Spain were major producers also (6 to 7.5 million tonnes annually) in this same period. Several Mediterranean countries, Peru, Argentina, Iran, Japan, South Africa, Pakistan, Vietnam, Australia and the Republic of Korea are also currently major commercial production regions of citrus in the world. Of the two largest citrus producers, Brazil is the largest producer of processed citrus (oranges), and China is the largest producer of fresh citrus, primarily mandarins, navels and pummelos. The USA is the third largest producer of citrus and is the largest producer of grapefruit. From 1989–1990 until 2004–2005, comprehensive FAO citrus data indicated that world production of citrus increased about 44%, with a balanced increase in both southern and northern hemisphere countries, but since 2003–2004 citrus production has increased at only about half of this rate (23%) and mostly in the northern hemisphere.

Orange production has had a moderate increase, with Brazil and the USA still the major producers. In addition to oranges for processing, Brazil has a significant production of limes and mandarins used primarily fresh in its domestic market. China, with the largest individual country production increase (40%), is currently number one in overall production. This production increase from 1969 was nearly tenfold and emphasis has shifted to oranges from mandarin-type fruit which used to predominate (Spurling, 1969; Zhang, 1981) but now mandarins are only two and ½ of the orange production. Many of the popular cultivars in China are local selections that are not grown in the western world. These include several cultivars of satsuma mandarin and over 30 orange cultivars, but navel orange cultivars predominate production and are mostly selections obtained from California. For the USA, Florida produces mostly round oranges for processing and grapefruit for fresh and processed use. In California, Arizona and Texas, navels, lemons, seedless mandarins and grapefruit are the major fresh fruit types depending on the state. In spite of decreases in Brazil and Florida, world orange production has still increased in recent years.

Production areas and layout tend to follow topography limitations and the size of overall production (Fig. 1.2). Large farms of more than 1000 ha are common in Brazil (Fig 1.2A), while many commune farms fit together with coordinated production to cover mountainsides in China (Fig. 1.2B) and intermediate size farms may be separated by other crops or sit alone in countries like Spain (Fig. 1.2C). Countries with smallholdings, particularly in Southeast Asia and the Middle East, are usually intercropped with vegetables or other annual crops such as soybean (wet season) and black gram/green gram (winter season) in India or alfalfa in southern Yemen (Fig. 1.2D). Intercropped citrus may still be a viable export crop, like Pakistan's 'Kinnow' production. In most cases of intercropping, not enough clear space is left around the citrus trees as they are growing to bearing age (Fig. 1.2D).

Severe freezes during the 1980s, hurricanes in 2004–2005 and citrus canker caused considerable crop and tree losses in Florida. Since 2003–2004 Florida has seen a reduction in production of more than 70% primarily due

Fig. 1.2. (A) Typical large citrus planting in Brazil; (B) Composite production by many growers in one area in Central China; (C) A smallholding in Spain adjacent to disked plot for an annual crop; (D) A smaller production area of young citrus with intercropped alfalfa in Yemen showing the lack of cleared ground space for the citrus tree.

to HLB disease. Grapefruit production has been the most affected by HLB in Florida. A recent USDA NASS acreage report put Florida citrus acreage at 447,000 acres, less than half of the highest acreage in the late 1970s before freezes, canker and HLB took their toll (https://www.nass.usda.gov/Statistics_by_State/Florida/Publications/Citrus/). While California has land development and available irrigation water issues, it is a major shipper of lemons and navel oranges – 61% of its orange plantings being navel oranges – but seedless mandarin production has expanded significantly in recent years. Spain and Mexico are the next largest producers followed by India, Iran, Italy, Japan, Egypt, Argentina, Turkey and Morocco. The relative ranking of these countries has varied with reported production increasing in India, Mexico, Iran, Argentina, Turkey, Egypt and Spain, while decreasing in Japan. Production in Cuba and other Central American and Caribbean countries increased significantly from 1987 to the early 1990s but has declined since then due to low fruit prices, hurricanes and now diseases, most notably HLB (greening). All citrus production areas in the American continent from Florida to Argentina have been affected by HLB. Brazil and Florida have been the most affected in terms of tree and production losses.

Prior to the large impact of HLB starting in 2011, from 2004–2005 until 2009–2010 a 37% increase in overall production took place worldwide. Most of the increase was in countries not on the top production list, unlike the increase in the previous 10 years. According to FAO reports, countries showing consistent growth over the 20-year period until 2010 included India, Mexico, Turkey, Egypt, Pakistan, South Africa, Indonesia, Syria, Peru and Algeria. An exception in the countries with large production was China which had a 10.1% increase from 2004–2005 to 2009–2010 following an 8.5% increase in the previous 10 years.

Recent Distribution and Utilization of Citrus Fruit

Large citrus industries exist today, primarily where climates are particularly suitable for production and fresh or processed citrus markets could be developed both internally and, more importantly, externally (Albrigo and Behr, 1992). Major fresh fruit exports, according to FAO data, account for about 13% of total world production, but USDA Foreign Agric. Service data from 2009/2010 through 2014/2015 suggests that at least 70% of citrus production is consumed fresh, apparently with the majority being domestic consumption (https://www.fas.usda.gov/Data/Citrus-World-Markets-and-Trade).

As transportation to other western European countries has improved and populations have increased, the Mediterranean countries have developed a strong fresh fruit export market from their traditional internal use of citrus. Mandarins, lemons and oranges are all important in the total fresh export marketing from the region, and Spain has become the major exporting country.

Mandarin cultivars are the leading export because of the development of seedless selections (particularly clementines) and attention to marketing of high-quality fruit. Over the last 15–20 years, mandarin production increased in Spain, but production decreased slightly in recent years, probably due to economics, production costs and a saturated market for easy-peel mandarins. Before the downturn, in the 14 years from 1991 to 2005, exports increased by 76% as the seedless mandarin market was being developed. Consumer preference for easy peeling, seedless fruit, led to improved selections and improved production techniques, partially to meet this demand while maintaining quality. These successful steps strengthened Spain's marketing position. Morocco followed a similar pattern to increase its citrus industry but has also stabilized in recent years. Some of the details of this production are covered in Chapter 5. The Mediterranean region now often overproduces these mandarins, leading to poor price stability. There is little evidence that marketing under the European Community (EC) has improved the balance of supply and demand as while mandarin production has increased, orange production is nearly double mandarin production in the Mediterranean region. In the USA, California has greatly increased its mandarin production with seedless cultivars and small size being used as the major marketing tools.

'Navels' and 'Valencias' are the predominant cultivars in Spain and Greece, but in Italy blood oranges such as 'Tarocco', 'Moro' and 'Sanguinello' constituted about 70% of the annual orange production of 1.8 million tonnes (Russo, 1981). 'Biondo Comune' is by far the most popular non-blood orange (was 0.4 million tonnes annually). Italy may still produce some bergamots (sour orange hybrids) for perfume essence (Barone *et al.*, 1988), but northern Africa and Paraguay have primarily abandoned their sour orange orchards. Morocco and Egypt have significant export marketing with cultivars similar to those of Spain, but again with seedless mandarins predominating, particularly the local cultivar Afourer in Morocco. 'Navels' and 'Valencias' also predominate in western hemisphere production of oranges for fresh fruit markets.

In the USA from the late 1970s until 1991, mandarin-type production and their exports declined from 429,000 tonnes total and 18,000 tonnes annual average exports, 1976–1981, compared to 263,000 tonnes total and 13,000 tonnes annual average fresh exports for 1989–1991. Since that decline, total production has recovered to 393,000 tonnes with exports at 20,000 tonnes by 2005. Harvesting without clipping, because of cost, and limiting handling and packaging methods in the USA contributed to quality problems and a loss of market demand, but available cultivars played a major role in market loss. 'Orlando', 'Murcott', 'Dancy', 'Nova', 'Robinson', 'Sunburst' and 'Fallglo' were the primary cultivars in Florida; however, because of their seediness and in some cases poor keeping and handling characteristics under most growing conditions they were not widely accepted in the world, especially the European markets. Only 'Murcott', 'Sunburst' and 'Fallglo' are major cultivars, with increasing planting of 'Sugarbelle' in Florida – the major planted mandarin.

'Bingo' and 'Tango' are also being planted. Seedless mandarins are a goal of the Florida industry with a number of selections being evaluated. In recent years there has been a resurgence of fresh mandarin production in California. First, Spanish and Moroccan cultivars were introduced. More recently, seedless cultivars have been introduced from California's own breeding programme. Currently, thousands of acres have been planted in the San Joaquin Valley as far north as Fresno, California. A recent report indicates that 38,780 acres of mandarins now exist in California and 2 million trees of 'Tango', irradiated nearly seedless W. Murcott, have been planted (www.google.com/patents/USPP17863).

As California successfully followed the Spanish model, it is now marketing large quantities of seedless mandarins, adding to the over-supply in the market. Because seedless mandarins do not size well (see Chapters 4 and 5), successful marketing has required convincing consumers that small, easy-peeler citrus are ideal for fresh consumption. This has been promoted with names like 'Pixies', 'Cuties', 'Delites' and 'Halos'.

Development of southern hemisphere export citrus industries in Australia and South America somewhat followed the example of South Africa. As the population increased in the late 1800s (due to the discovery of diamonds and gold in South Africa), a local citrus industry was started. In about 1906, refrigerated shipping was developed and the UK became involved in fresh citrus marketing. This stimulated further development for export (Oberholzer, 1969). Currently, 'Valencia' and navel oranges make up the major portion of the total production in South Africa, but seedless mandarin marketing has increased for the off-season to northern hemisphere producing countries. Development of the citrus industry in Australia followed a similar time frame, but after a marketing slow-down in the 1970s, exports have increased to expanding industrialized Asian cities such as Singapore, Bangkok and Hong Kong. The primary production is 'Valencia' and navel oranges. Mandarins ('Ellendale'), grapefruits ('Marsh') and lemons ('Eureka' and 'Villafranca') are produced in near equal amounts and equal about one-sixth of the total orange production of about 600,000 tonnes (Gallasch and Ainsworth, 1988; FAO, 2006). Recent production and planting reflect some growth in the industry.

In South America, Argentina, Chile, Uruguay and Brazil have significant internal marketing of fresh fruit, but rather small fresh citrus export activities (FAO, 2016). Of these, Argentina has the strongest marketing programme, but in addition Peru has developed a strong mandarin export market along with fruit exports in general.

Many tropical countries in the Americas increased commercial plantings in the 1970s and particularly in the 1980s following freezes in Florida, but most of these countries are now facing HLB disease. All tropical and marginally tropical countries of the world produce some citrus, but the major portion of the production is from backyard plantings or small farms and is sold locally. This production is not totally accounted for in available production statistics. Many of the tropical areas have limited production because of severe dry or wet climate cycles and/or disease limitations such as citrus canker, citrus tristeza virus (CTV) or greening (see Chapters 4 and 7). Southeast Asia has been limited in citrus production, primarily to limes and some mandarins, because of these disease constraints. India produces a moderate amount of citrus in spite of these constraints, with a balance of orange, mandarin and lime production. Popular cultivars include 'Mosambi', 'Blood Red', 'Pineapple', 'Hamlin', 'Jaffa', and 'Valencia Late' oranges and 'Nagpur Santra' and 'Coorg' mandarins. Japan, with its geographically isolated location and cool climate, has developed a historically strong citrus industry, but it is based primarily on the satsuma mandarin because of its citrus canker tolerance and freeze-hardiness. Decreasing consumption trends of satsumas in Japan, however, have resulted in a decrease in production (Kitagawa *et al.*, 1988; FAO, 2016).

World exports of fresh lemons and limes remained relatively stable from the late 1970s (961,000 tonnes annual average 1976–1981) to the late 1980s (1,002,000 tonnes annual average 1989–1991). By 2015, fresh lemon/lime production had increased to 13 million tonnes (FAO, 2016). Spain, Italy, Turkey, Greece and Egypt produce significant quantities of lemons primarily from the 'Eureka' and 'Villafranca' cultivars. In the western hemisphere, California and Arizona are primary producers of lemons for domestic and export markets, relying on 'Eureka' and 'Lisbon' lemons as their major cultivars. Southern hemisphere countries with subtropical climates, such as Argentina, Uruguay and Chile, have developed a lemon export industry for the off-season to the northern hemisphere with these same cultivars (FAO, see recent website pages). Argentina also produces 'Genova' lemons in addition to 'Eureka' and 'Villafranca'. Argentina, while strong in fresh export, is the largest processed lemon producing country. Mexico is the largest producer and exporter of limes, primarily 'Mexican', in the world. Persian limes, grown on Mexico's east coast, are more valuable than Mexican limes, grown on the west coast. Limes are also a major citrus type grown in most tropical countries in the Americas and Southeast Asia.

Grapefruit exports increased primarily due to increased consumption in Japan, western Europe, the Pacific Rim countries (eastern Asia), and eastern Europe, until the recent economic downturn in 1992. Major exporters were the USA (Florida), Cuba, Israel, Argentina, South Africa and Honduras (Albrigo and Behr, 1992). In recent years, production has shifted away from 'Marsh' (white-fleshed) to pink or red cultivars, which are more popular with consumers. The standard 'Ruby Red' was replaced by cultivars with deeper red flesh colouration, such as 'Henderson' (Flame), 'Ray Ruby', 'Rio Red' and 'Star Ruby'. Recent problems with citrus canker and particularly HLB are reducing grapefruit production in Florida, USA, which has been the major grapefruit producer in the world.

The major citrus-producing countries are also the major fresh fruit exporters, although noticeable exceptions are Brazil (primarily São Paulo state) and to a lesser extent the USA (Florida), where industries are based more on citrus production for processing. The exceptionally large industries in these two locations are the result of their climatic and industrial capacity to produce high-quality, processed, frozen concentrate orange juice (FCOJ) or not from concentrate (NFC). While only 21% of all citrus worldwide is processed, 81% of this is oranges with 66% of the orange processing occurring in São Paulo, Brazil and Florida, USA. Florida produces juice primarily from the early and late season cultivars 'Hamlin' and 'Valencia' (FAO, 2016), while in Brazil 'Hamlin', 'Pera', 'Valencia' and 'Natal' are the primary cultivars used (Amaro, 1984). 'Pera' accounts for nearly 50% and 'Natal' 25% of the trees in São Paulo State.

The development of a large processing industry also occurred because of the invention of FCOJ in Florida in 1948, the selection of orange cultivars with high yields, juice content and percentage soluble solids and the continued development of sophisticated industrial infrastructure to meet the world demand for this product (Albrigo and Behr, 1992). In the 1980s, many smaller producing countries in the Americas began to develop or expand their orange production with the addition of processing facilities for FCOJ. Most lowland tropical locations produce fruit with lower percentage soluble solids, very low acidity and poorer colour than that produced in humid subtropical regions such as Florida and São Paulo (see Chapters 4 and 8). Spain and Italy are producing more orange juice than in the past (FAO, 2016). Argentina processes a large part of its citrus production, particularly lemons, and about a third of its oranges (FAO, 2015). The lemons have an especially high yield and quality of oil. Major importing countries of fresh citrus fruit and citrus products include most EU countries, Sweden, Switzerland, Japan, the USA, Canada and South Korea. Significant imports are also purchased by Saudi Arabia, Singapore, Israel, Australia and several Pacific Rim countries besides South Korea and Malaysia. Major importing countries that are also producing countries are the USA, Japan, Australia and Israel. Several eastern European countries imported significant amounts of citrus, although these imports significantly decreased after 1991 (FAO, 1991) due to political and economic restructuring of these countries. In recent years (2010–2015), Russia has been a strong importer of all types of citrus (FAO, 2016).

USES OF CITRUS FRUITS

Fresh citrus has been appreciated to the extent that it was carried from its area of origin and cultivated from before the 5th century BC. Early sailors recognized that citrus fruits were important for prevention of scurvy which results from a lack of vitamin C in the diet. Hence, the origin of the term 'limey' for British sailors. The Italians grew citrus in glasshouses (orangeries), perhaps from

Roman times (Webber *et al.*, 1967). From the area of origin, citrus was collected, distributed and appreciated by the Arabians, Italians, Spanish and Portuguese. Export of fresh citrus expanded as shipping became available. Sweet oranges, lemons, grapefruit and most recently mandarins, mostly seedless, are exported considerable distances by ship (e.g. Florida to Japan, South America to the EU, South Africa to the EU and most recently the Mediterranean Basin to North America). Generally, mandarin fruits are more easily damaged than oranges or grapefruit during storage and shipping, but with the greater demand for seedless, easy-peeling mandarins, shipping limitations have largely been overcome (see Chapter 8).

Processing of citrus fruit may have begun with the use of citrus peel oils in perfumes from selected cultivars of mandarins by the Chinese in AD 300 and later the use of some sour orange hybrids called bergamots for this same purpose starting in the early 1700s. The first edible processing product may have been orange marmalade prepared from sour orange or candied peel made from citrons. Kumquats and calamondins have been sweet-pickled for many years. Modern processing began with the use of canned sections prepared by hand. Later, machines were developed to peel and separate the sections. Mandarin sections are the main canned section product today. Processing greatly expanded with the invention (1948) and promotion (1950s) of FCOJ. By-products from juice extraction are important in soft drink production (oils and juice for flavours and juice pulp for cloud), pectin production (peel and rag extracts) and cattle feed production (dried peel pellets). Lightly pasteurized NFC is now a popular product with more advanced handling procedures. Further details of processing and product handling can be found in Chapter 8.

Citrus production worldwide continues to increase and will probably do so in the future if demand strengthens and severe diseases are managed (see Chapter 7). Estimates in 1993 suggested that citrus production would exceed 80 million tonnes by the year 2000, with major increases occurring in China, Mexico and the USA. While China more than met these expectations, the USA was not able to, for the reasons mentioned. Worldwide, production did exceed 80 million tonnes by almost 18 million by 2005, more than 2.6% per year over a 15-year period. For the past 15 years, demand, particularly for processed juice, has stagnated or declined in US and European markets (Ferdman, 2014). This, coupled with disease and commercial development losses of citrus land, has slowed citrus expansion in some countries like the USA and Brazil so that while demand for juice has decreased significantly so has processed juice by about 200 million boxes (40.82 kg, 8.2 million metric tonnes). Lack of available water may be the next limitation to expansion and maintenance of citrus-producing land. Serious water restrictions are currently causing reductions in citrus acreage in California, USA. Nevertheless, many new or expanded markets have become available since the late 1980s including former Eastern-bloc countries and parts of Asia that have improved economies which have allowed expansion of citrus markets, particularly for fresh citrus products.

2

TAXONOMY, CULTIVARS AND BREEDING

TAXONOMY

Commercial citrus and related genera are primarily evergreen species of subtropical and tropical origins belonging to the order Geraniales and the family Rutaceae. Rutaceae is 1 of 12 families in the suborder Geraniineae. Species within the Rutaceae generally have 4 important characteristics: (i) the presence of oil glands; (ii) the ovary is raised on a floral (nectary) disc; (iii) pellucid dots are present in the leaves; and (iv) fruit have axile placentation (Swingle and Reece, 1967). The family is further subdivided into 6 subfamilies including the Aurantioideae to which true citrus and related genera belong. Plants within the Aurantioideae are unusual because the fruit are hesperidium berries (a single enlarged ovary surrounded by a leathery peel) and contain specialized structures, the juice vesicles (sacs). Furthermore, many species contain polyembryonic seeds which may contain both zygotic and nucellar embryos.

Historically, the taxonomic situation of tribes, subtribes, genera and species within the Aurantioideae is controversial, complex and confusing. Citrus and many related genera hybridize readily and have done so in the wild for centuries (Swingle and Reece, 1967). Therefore, there is no clear reproductive separation among species. Moreover, many species reproduce via nucellar embryony. Often embryos arise from nucellar tissue and are true-to-type to maternal tissue. Nucellar embryony permits the continued asexual existence of a species or a hybrid as an undifferentiating clone except by mutation.

Evolution of Taxonomic Systems for Rutaceae-Aurantioideae-Citrus

The original taxonomic systems of Hooker (1875) and Engler (1896 [1931]) are artificial systems based on morphological characteristics and the putative origin of a species. Hooker originally proposed that the Aurantioideae consisted of 13 genera with only 4 species in the genus. Engler later revised this to

6 species and finally to 11 species in 1931. These taxonomists did not strongly consider the existence and importance of nucellar embryony when developing their classifications, nor did they have access to biotechnological approaches to taxonomy.

During the mid-1900s Swingle (1948) and later Swingle and Reece (1967) developed a taxonomic system based on 2 tribes, the Clauseneae and the Citreae (Table 2.1). This system classifies citrus and its relatives based on several morphological characteristics, and it is useful and functional from a practical standpoint. The Citreae were further divided into 3 subtribes, the Triphasiinae, Balsamocitrineae and the Citrinae which contains primitive citrus relatives and the true citrus group of 6 genera including *Citrus*, *Poncirus*, *Eremocitrus*, *Microcitrus*, *Fortunella* and *Clymenia*. Swingle's system further divides *Citrus* into 16 species. A description of each of the 6 'true citrus' genera is given later based on the taxonomic system of Swingle (1948).

The genus *Fortunella* (kumquat), named after English plantsman Robert Fortune, includes 4 species of small trees and shrubs. Leaves are unifoliate and have distinctive silver colouration on their underside. Flowers are borne singly or in clusters in the leaf axils; there are typically 5 sepals, 5 white petals, 16–20 stamens and 3–7 carpels. Kumquats flower much later in the season than other true *Citrus* species and are quite freeze-hardy. Fruit are small, ranging in shape from round ('Meiwa' and 'Marumi') to ovate ('Nagami'). Fruit are served fresh or candied and differ from other true species in that the entire fruit including the peel may be eaten.

Poncirus consists of 2 trifoliolate species, *Poncirus trifoliata* and *Poncirus polyandra*, and is characterized by small trees, usually with trifoliate, deciduous leaves. The 3 leaflets are of similar size and oval. *Poncirus* is the only true citrus species that is deciduous and thus is extremely freeze-hardy when fully acclimated. It is native to northern China. There are 2 distinct types of twigs, those with internodes longer than the petiole and those consisting of foliated spurs which develop from dormant buds and have short internodes. Unlike other true *Citrus* species, bud scales are very pronounced and pubescent, producing long, sharp thorns in the leaf axils throughout the tree's lifetime. Flowers have 5 sepals, 5 petals and numerous stamens that curve outward from the base of the ovary. The fruit consists of 8–13 fused carpels; it is pubescent, has a very bitter taste due to the presence of ponciridin and is not palatable. Seeds are plump with a smooth exterior and often contain many embryos. *Poncirus* trees are used as rootstocks but also produce an excellent protective hedge due to their extreme thorniness. Commercial citrus cultivars can be very small to quite large depending on the P. trifoliate selection used as a rootstock.

Eremocitrus is characterized by a single species of xerophytic trees native only to Australia and has long been geographically isolated from other true *Citrus* genera except *Microcitrus*. The growth habit is spreading due to presence of long, drooping branches. *Eremocitrus* trees are well-adapted to their xerophytic environment. Leaves are unifoliate, greyish green, thick, leathery and lanceolate. Stomata are sunken probably as an adaptive mechanism to reduce

Table 2.1. List of tribes, subtribes, subtribal groups and genera of the orange subfamily, Aurantioideae (Swingle 1948; Swingle and Reece 1967).

Tribe I. *Clauseneae*: Very remote and remote citroid fruit trees (3 subtribes, 5 genera, 79 species, 20 varieties)
    Subtribe A. *Micromelinae*: Very remote citroid fruit trees (1 genus, 9 species, 4 varieties)
        1. *Micromelum*: (9 species, 4 varieties)
    Subtribe B. *Clausenmae*: Remote citroid fruit trees (3 genera, 69 species, 16 varieties)
        2. *Clycosrhis*: (35 species)
        3. *Clausena*: (23 species, 4 varieties)
    Subtribe C. *Merrillinae*: Large-fruited remote citroid fruit trees (1 genus, 1 species)
        4. *Merrillia*: (1 species)
Tribe II. *Citreae*: Citrus and citroid fruit trees (3 subtribes, 9 subtribal groups, 28 genera, 124 species, 18 varieties)
    Subtribe A. *Triphasiinae*: Minor citroid fruit trees (3 subtribal groups, 8 genera, 46 species, 3 varieties)
        A. *Wenzeha* group (4 genera, 15 species, 1 variety)
            5. *Wenzelia*: (9 species, 1 variety)
            6. *Monanthocitrus*: (1 species)
            7. *Oxanthera*: (4 species)
            8. *Merope*: (1 species)
        B. *Triphasia* group (2 genera, 4 species)
            9. *Triphasia*: (3 species)
            10. *Pamburus*: (1 species)
        C. *Luvunga* group (2 genera, 27 species, 2 varieties)
            11. *Luvunga*: (12 species)
            12. *Paramignya*: (15 species, 2 varieties)
    Subtribe B. *Citrinae*: Citrus fruit trees (3 subtribal groups, 13 genera, 65 species, 15 varieties)
        A. Primitive citrus fruit trees (5 genera, 14 species)
            13. *Severinia*: (6 species)
            14. *Pleiospermium*: (5 species)
            15. *Burkillanthus*: (1 species)
            16. *Limnocitrus*: (1 species)
            17. *Hesperthusa*: (1 species)
        B. Near-citrus fruit trees (2 genera, 22 species, 4 varieties)
            18. *Citropsis*: (11 species, 1 variety)
            19. *Atalantia*: 11 species, 3 varieties)
        C. True citrus fruit trees (6 genera, 29 species, 11 varieties)
            20. *Fortunella*: (4 species, 1 variety)
            21. *Eremocitrus*: (1 species)
            22. *Poncirus*: (1 species, 1 variety)
            23. *Clymenia*: (1 species)
            24. *Microcitrus*: (6 species, 1 variety)
            25. *Citrus*: (16 species, 8 varieties)

Continued

Table 2.1. Continued.

>   Subtribe C. *Balsamocitrinae*: Hard-shelled citroid fruit trees (3 subtribal groups, 7 genera, 13 species)
>   A. *Tabog* group (1 genus, 1 species)
>   26. *Swinglea*: (1 species)
>   B. Bael-fruit group (4 genera, 8 species)
>   27. *Aegle*: (1 species)
>   28. *Afraegle*: (4 species)
>   29. *Aeglopsis*: (2 species)
>   30. *Balsamocitrus*: (1 species)
>   C. Wood-apple group (2 genera, 4 species)
>   31. *Feronia*: (1 species)
>   32. *Feroniella*: (3 species)
> Totals: 2 tribes, 6 subtribes, 9 subtribal groups, 33 genera, 203 species, 38 varieties.

transpiration under arid growing conditions. Moreover, *Eremocitrus* appears to be freeze-hardy (Barrett, 1985). Flowers are borne singly or in small groups in the leaf axils, have 3 to 5 sepals and petals, and 15–20 unfused stamens. The ovary consists of 3 to 5 fused carpels with 2 ovules per carpel. The fruit are oval to pyriform in shape with a 'typical' fleshy peel found in all true citrus species. Seeds are small with hard, wrinkled seed coats. *Eremocitrus* is not of commercial importance but has been used in citrus breeding programmes to enhance drought tolerance and freeze-hardiness of the progeny. It has also been used as an interstock.

Microcitrus is indigenous to the northeastern rain forest of Australia and has also been geographically isolated from other true *Citrus* genera for millions of years. The genus consists of 6 species of moderate-sized trees. Leaves are unifoliate dimorphic, changing from greatly reduced cataphylls on seedlings to small elliptical to ovate leathery leaves on mature plants. The petiole is small and the upper leaf surface is pubescent. New foliage often has a characteristic purplish tint. Flowers are borne in leaf axils; they are small with 4 to 5 sepals and petals, 10–20 fused stamens and 4 to 8 carpels containing 4 to 8 ovules per carpel. Petals are white, also having purplish colouration at their base. Fruit are small and roundish containing considerable amounts of acidic oil droplets. Seeds are small and ovate in shape. Some selections of *Microcitrus australasica* are resistant to burrowing nematode (*Radophilus citrophilus*) and *Phytophthora* fungi (Barrett, 1985). Trees are not nearly as freeze-hardy as *Eremocitrus*. The prefix 'micro' very accurately describes the general size of the leaves, flowers and fruit relative to other true citrus genera. *Microcitrus* has not been of much commercial importance but has been used as an interstock and in rootstock breeding on a limited basis.

Clymenia was also geographically isolated from other true *Citrus* genera millions of years ago, being found in Papua New Guinea and nearby islets, and differs in a number of ways from the other genera in the true *Citrus* group. It is characterized by 2 species of small trees with unifoliate acuminate leaves tapering into a very short petiole. Branches are thornless and subangular when young, changing to cylindrical as they mature. Flowers are borne singly in leaf axils on short, stout pedicels; they have 5 sepals and petals which contain numerous oil glands. There are 10–20 times more stamens than petals which are fused at the base, but free at the top. The ovary consists of 14–16 fused carpels containing many ovules per carpel. The style is shorter and wider than that of other true *Citrus* species and the stigma is large and flattened. The fruit of *Clymenia* are ovoid, thin-peeled, contain numerous oil glands and many small seeds. *Clymenia* is certainly the most unusual and unique of the true *Citrus* group but is not of commercial importance. It has been used as an interstock on occasion. Recent studies suggest that this may be a hybrid between *Fortunella* and *Citrus* (Berhow et al., 2000).

Based on the taxonomic scheme of Swingle the genus *Citrus* consists of 16 species of moderate to large evergreen trees. Tree shape varies from upright in some mandarins to spreading in grapefruit and satsuma mandarins. Branches are angular when young, changing to cylindrical as they mature. Thorns are numerous in leaf axils of young trees or some species like lemons and limes; however, they become less prominent as the trees mature. Leaves are unifoliate with laminae ranging in size from very large (grapefruit and pummelos), to moderate (oranges and lemons) to small (most mandarins). Petiole size also varies with species, usually in the same manner as leaf size. Petioles of some species have prominent wings, as for grapefruit. Flowers are borne singly or in groups in leaf axils and may be either perfect or staminate. Flowers generally have 4 to 5 sepals, 4 to 8 petals, 20–40 fused stamens and 8–15 fused carpels usually containing 4–8 ovules in seedy cultivars. The style is longer than the ovary for perfect flowers, but tends to be shorter for staminate-only flowers. The fruit consists of a single ovary of 8–15 fused carpels (segments) surrounded by a leathery peel. Fruit shape varies from spheroid (oranges) to oblate (grapefruit and mandarins) to prolate (lemons and some lime cultivars). Juice vesicles have distinct stalks which attach to the segment walls. Segments are separated from the outer peel by white endocarp tissue called the albedo. The outer peel, flavedo, contains numerous oil glands and varies in colour from green-yellow (lemons, limes, grapefruit) to orange and reddish orange (oranges) to deep orange to red-orange (mandarins). Seeds are obovoid to roundish in shape and contain 1 to many embryos. Cotyledon colour ranges from white (oranges and grapefruit) to green (mandarins).

A contemporary of Swingle, T. Tanaka, believed that the differences he observed among many citrus species warranted further division of the genus into 162 species with 13 primary elements (Tanaka, 1977). Critics of Tanaka believed that he included many hybrids that do not warrant species status, while others believed that Swingle combined too many dissimilar plants into single species. Taxonomists at Cornell University favoured a system of identifying some species in the genus using nothomorphs to emphasize the hybrid but sometimes unknown parentage of some citrus species (Bailey and Bailey, 1978). For example, Swingle considered grapefruit (*Citrus paradisi*) a true species; whereas the Cornell system considers it a natural hybrid thus designated *Citrus x paradisi*. R.W. Hodgson (1967) revised the taxonomic systems of Swingle and Tanaka and with his own observations suggested the use of a system having 36 species. Several other taxonomic systems have been proposed separating *Citrus* into 6–42 species. Spiegal-Roy and Goldschmidt (1996) suggested a modified list of species (Table 2.2) but did not reduce the true *Citrus* to the current level of 3 or 4 true species that are shown at the bottom of Table 2.1.

Table 2.2. List of species of citrus[1] in the subgenus *Citrus* or close relatives, conventional name, suggested origin[2] and species concept[3] (modified from Spiegal-Roy and Goldschmidt, 1996).

Species	Year named	Conventional name	Assumed native habitat	Species concept
Subgenus *Citrus*				
C. medica L.	1753	Citron	India	True species
C. aurantium L. (*C.reticulata* x *C.grandis*)	1753	Sour orange	China	Hybrid origin
C. sinensis Osbeck.	1757	Sweet orange	China	Hybrid origin
C. maxima Osbeck.	1765	Pummelo	China	True species
C. limon (L.) Burm.f.	1766	Lemon	India	Trihybrid
C. reticulata Blanco	1837	Mandarin	SE Asia	True species
C. aurantifolia Christm.	1913	Common lime	SE Asia	Trihybrid
C.paradisi Macf.	1930	Grapefruit	Barbados	Hybrid origin
C.tachibana Tan.	1924	Tachibana	Japan	[4]
C.indica Tan.	1931	Indian wild orange	India	[4]
Fortunella spp.[5]		Kumquat	SE China	True species
C.madurensis Lour.[5]		Calamondin	Phillipines	Hybrid origin
Subgenus *Papeda*				
C. hystrix D.C.	1813	Mauritius papeda	S.E. Asia	[4]
C.macroptera Mont.	1860	Melanesian papeda	S.E. Asia	[4]
Poncirus trifoliata		Bitter orange	China	True species

[1]According to Swingle (1948).
[2]Based on Cooper and Chapot (1977).
[3]Based on Barrett and Rhodes (1976).
[4]Not investigated.
[5]A true genus and its hybrid, which is classified as a species of *Citrus*, Spiegal-Roy and Goldschmidt (1996).

Some taxonomists have advocated combining all *Citrus* into a single citrus species because nearly all species readily hybridize. Most taxonomists and certainly most horticulturists view such a combination as impractical and not very utilitarian.

The confusion about many *Citrus* species comes about because the free exchange of genes has been limited by widespread apomixis – replacement of the normal sexual seed reproduction by asexual seed reproduction without fertilization. This suppresses and replaces the zygotic embryos with nucellar ones. Apomictically produced offspring are genetically identical to the parent plant. In flowering plants, the term 'apomixis' is commonly used in a restricted sense to mean agamospermy, i.e. asexual reproduction through seeds. Adventive embryony, also called nucellar embryony, describes a reproductive system where there could be an embryo sac (gametophyte) in the ovule, but the embryos do not arise from the cells of the gametophyte; they arise from cells of nucellus or the integument. Nucellar embryony is important in several species of citrus, including sweet orange and grapefruit (Frost and Soost, 1968), in maintaining these hybrid species. Most cultivars of lemon produce both zygotic as well as nucellar seedlings (Ray, 2002). Efforts to produce vigorous lemon cultivars from intercrossing 'Lisbon' and 'Eureka', the true Sicilian

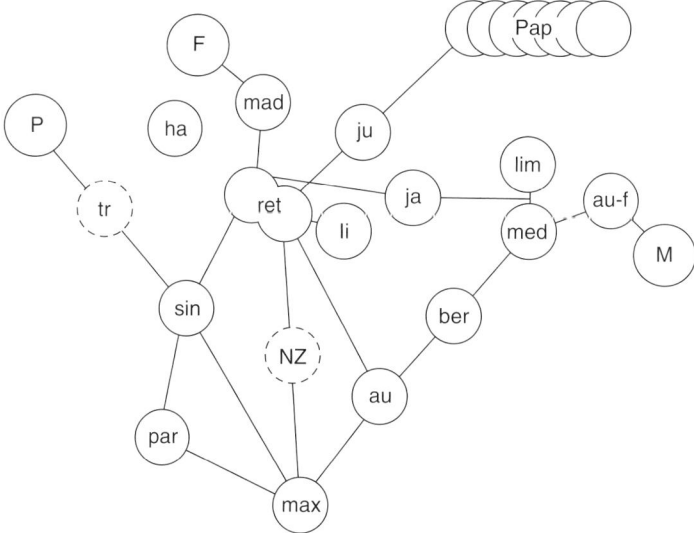

Fig. 2.1. Citrus biotype interrelationships. P = *Poncirus*; F = *Fortunella*; M = *Microcitrus*; Pap = *Papedas*; ha = *C. halimii*; mad = *C. madurensis* = calamondin; ret = *C. reticulata* = mandarin; sin = *C. sinensis* = sweet orange; max = *C. maxima* = pummelo, shaddock; tr = 'Troyer' citrange, NZ = 'New Zealand' goldfruit; li = *C. limoma* = rangpur; ja = *C. jambhiri* = rough lemon; ju = *C. junos* = yuzu; au = *C. aurantium* = sour orange; ber = *C. bergamia* = bergamot; lim = *C. limon* = lemon; med = *C. medica* = citron; au-f = *C. aurantifolia* = lime (Scora, 1988).

lemon cultivars grown in the USA (Davies and Albrigo, 1994), have not been successful (Soost and Roose, 1996).

Only 3 citrus species are true botanical species that reproduce true to type zygotic seedlings. These include the commercially important citrus species, C. *maxima* L. (pummelo), C. *medica* L. (citron) and C. *reticulata* L. (mandarin). Of these, only mandarin readily produces both zygotic and nucellar seedlings (Scora, 1988). In breeding programmes, mandarins and pummelos were often used as the female parent, which led to true zygotic seedlings. This greatly limited the complete use of the citrus genome in breeding. As an example, most useful orange and grapefruit cultivars originated as chance hybrid seedlings in the area of origin which, due to good fruit quality, were protected and found to produce identical plants from seed. These were eventually moved around and traded, and mutations from them led to most of the sweet orange cultivars we have today.

With careful evaluation of leaf oils, pollen surface architecture, biochemical reactions such as oxidative browning and other stable phenotypic characteristics, first Barrett and Rhodes (1976) and then Scora (1988) suggested that the commonly used citrus were comprised of 3 to 4 true species and all others were hybrids that derived from these species (Fig. 2.1). C. *halimii* was included as a fourth true species originating in the tropical highlands of Thailand/west Malaysia since this species does not appear to be in the parentage of our modern citrus cultivars. Further support for this classification came from molecular biology studies using isozyme analyses and restriction fragment length polymorphisms (RFLP) (Roose,1988). Isozymes are enzymes with similar biochemical functions in *in vitro* assays but have differing structures. Since their structure is specified by genes, isozyme analysis may indicate genetic relatedness among species.

This earlier work was confirmed by molecular biology methods which led to more genomic techniques and eventually citrus genomic elucidation (Talon and Gmitter, 2008; Gmitter *et al.*, 2012). Some references to using different genetic marker methods and general evolution of the confirmation of the true species and the sources of hybridization leading to all the other species are: Nicolosi *et al.* (2000), Abkenar *et al.* (2004), Uzun *et al.* (2009), Bayer *et al.* (2009) and Amar *et al.* (2011). Basically, all of these studies have supported the conclusions of Barrett and Rhodes (1976) and Scora (1988) that there are only 3 or 4 true species of citrus. Further, they have elucidated the parents that resulted in the species hybrids as listed in the previously outlined taxonomic hierarchy. Wu *et al.* (2018) have presented the most comprehensive evaluation of how the citrus species evolved starting some 8 million years ago on the slopes of the Himalayan mountains. It will be interesting to see how this proposal plays out in the future. Figure 2.2 shows the common and distinct DNA bands that can distinquish parents and hybrids that have been used to identify the few true citrus species.

Despite the strong biochemical and molecular evidence for only 3 major affinity groups, there are also practical considerations in any taxonomic

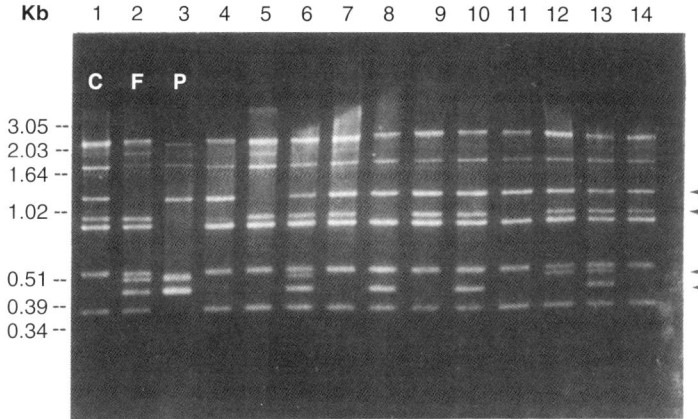

Fig. 2.2. Agarose gel profiles of random amplified DNA products (RADPs) from original parents. C. grandis (C) and P. trifoliata (P), F, hybrid (F) and some *Citrus* × *Poncirus* backcrosses to *Citrus* (lanes 4–14). Arrowheads indicate polymorphic lanes. Kb = kilobases (G.A. Moore, University of Florida, Gainesville, USA – Sahin-Cevik and Moore, 2011).

system. From a commercial and horticultural standpoint, the major species are most conveniently separated based on the system of Hodgson (1967). Hodgson's system, which is predicated on Swingle's, establishes 8 commercially important species within *Citrus*, the mandarins (*C. reticulata*), sweet oranges (*C. sinensis*), pummelo (*C. grandis*), grapefruit (*C. paradisi*), lemons (*C. limon*), limes (*C. aurantifolia*), citron (*C. medica*) and kumquats (*Fortunella* sp.). Kumquats are grown to a limited extent and the shaddock or pummelo is of economic importance in much of Southeast Asia and China but not as much on a worldwide basis. These groups are readily identifiable to citriculturists, citrus brokers and consumers compared to the complex system of Tanaka discussed earlier.

Major Taxonomic Groups within Citrus

On a worldwide basis, these 8 citrus groups are horticulturally and commercially of economic significance and their characteristics and some important cultivars are discussed in the following sections.

Sweet oranges

Sweet orange (*C. sinensis* [L.] Osb.) possibly originated in northeastern India and central China. Sweet orange is the most widely distributed and has the greatest production of all commercial citrus species (see Chapter 1). Its moderate freeze-hardiness and the adaptability of a wide range of cultivars to many climatic conditions make it very adaptable to many growing regions.

Sweet orange fruit are in general low to moderate in acids and moderate to high in per cent soluble solids. Therefore, the flavour of fresh fruit and juice products appeals to a broad range of people worldwide. Sweet oranges may be separated into 4 groups based on fruit morphological characteristics, chemical constituents and convenience in eating: (i) the common or round oranges; (ii) the navel oranges; (iii) the pigmented (blood) oranges; and (iv) the acidless oranges. The round oranges are the most important commercially and represent a major portion of sweet oranges grown worldwide. Navel oranges are the second most widely planted group, while blood orange plantings are limited primarily to areas with Mediterranean-type climates. Acidless oranges are planted primarily for backyard use and are not of importance commercially.

Sweet oranges are also grouped based on season of maturity, namely early, mid or late season. In a subtropical climate, early season cultivars generally reach maturity within 6–9 months after full bloom, mid-season 9–12 months and late season greater than 12 months. These maturation rates, of course, depend on growing conditions (see Chapter 4) and the factors chosen to define fruit maturity (see Chapter 8).

Sweet oranges may also be classified based on seediness. Cultivars range in seediness from commercially seedless (0–8 seeds), to moderately seedy (9–15 seeds), to very seedy (>15 seeds). Although many seedy cultivars are used for local consumption, most important commercial cultivars such as 'Valencia', 'Pera' and navel oranges are commercially seedless. The characteristics of the major round orange cultivars are discussed in the following sections. The cultivars described below by no means represent the large number of local selections that are planted worldwide; however, this group represents a very high percentage of the total production worldwide.

Round oranges
The 'Hamlin' orange originated as a chance seedling near DeLand, Florida, in 1879. Yields for mature 'Hamlin' orchards may average 60–80 tonnes ha^{-1} with some orchards producing over 100 tonnes ha^{-1}. For this reason and early maturation before likely freezes in Florida, 'Hamlin' continues to be planted, particularly in colder areas of Florida and some areas of São Paulo state in Brazil. Some selections of 'Hamlin' in Brazil, however, are extremely prone to premature fruit drop. 'Hamlin' is used primarily for processing into juice, but also is marketed as an early season fresh fruit. 'Hamlin' often has heavy early fruit drop in Florida where the crop load is very high but generally fruit hold on the tree well and may be harvested for processing as late as February or March. In Florida fruit drop has become much more severe with the introduction of huanglongbing (HLB) disease (see Chapter 7).

'Hamlin' trees grow somewhat upright and symmetrical, and are moderately freeze-hardy. Fruit mature early in the season (September–December, 6–9 months after bloom in the northern hemisphere) and thus are rarely damaged by freezes or can be rapidly harvested after an early freeze. Fruit are spherical and generally smaller than other sweet orange cultivars. In some seasons fruit size may be limiting for the fresh market. The peel is smooth and thin, and when harvested early is susceptible to rind tearing (plugging). Peel colour is generally poor, especially in growing regions with high average annual temperatures. Fruit are susceptible to splitting in some seasons due to the thin peel.

'Valencia' was identified and named in Portugal prior to 1865, but most likely is of Chinese origin. Trees from the original selection were shipped to California in 1876 and Florida in 1877 and since then have been widely distributed throughout the world's citrus-growing regions. The original selection in Florida was named 'Hart's Tardiff' due to the late maturity of the fruit. However, the name was later changed to 'Valencia' because the fruits resembled a similar cultivar growing in Valencia, Spain.

'Valencia' is the most important late-season sweet orange in the world. Trees are similar in appearance to most other sweet orange cultivars, with yields varying from moderate (40–50 tonnes ha^{-1}) to high (>60 tonnes ha^{-1}) in some growing regions. Fruit usually matures from February to October in the northern hemisphere and August to April in the southern hemisphere, and holds quite well on the tree without serious loss of fruit quality. Fruit held late, however, may re-green toward the stem-end on the tree in some areas, making them less acceptable for the fresh market. Alternate bearing may be induced, particularly in areas where fruit are held late (over 18 months from bloom), due to a reduction in vegetative growth, flower bud production and fruit set for the subsequent season. Some alternate bearing almost always occurs (see Chapter 4).

'Valencia' fruit are of medium size, spherical to oblong, and commercially seedless. Fruit quality is excellent, primarily due to the development of deep orange peel and juice colour. 'Valencia' juice is in demand by processors for blending with lower quality juices because of its high levels of total soluble solids (TSS) and colour. Several cultivars from 'Valencia' are prized on the fresh fruit market. These include 'Olinda', 'Frost', 'Campbell', 'Midnight' and 'Delta' among others. These cultivars differ primarily in fruit shape, quality and peel thickness as well as in date of maturity. However, all fruit are very similar morphologically and all mature in late season.

Because of their late maturity, 'Valencia' trees have some unusual problems when grown in relatively freeze-prone subtropical regions like Florida or California (USA). Fruit must remain on the tree throughout the coldest periods of the season. As a result, fruit losses due to freezing may be more severe than those of other, earlier-maturing cultivars. In addition, two crops are usually present on the tree at the same time in the spring: the mature crop which was set in the previous year and the current year's crop. The presence of two crops causes difficulties in hedging and topping operations because some of the current or future crop is removed. Although removal of a portion of the crop is disconcerting to growers, this practice stimulates new vegetative growth and ultimately improves yields and fruit quality. 'Valencia' fruit are also subject to excessive fruit drop in some seasons and postharvest losses in some areas due to fruit creasing or splitting (see Chapter 8). Some somoclonal selections of 'Valencia' mature much earlier, and may replace 'Hamlin' and other early maturing cultivars (Grosser et al., 2007b). Some of these selections are available in Florida for grower trials. 'Natal' is also a late-season sweet orange grown in Brazil, primarily for processing. The tree is very similar morphologically to 'Valencia'. Fruit are also similar to those of 'Valencia' and the cultivar has similar cropping problems due to the late-season maturation.

'Pera' is an important and extensively planted mid- to late-season sweet orange grown for the processing and fresh markets in Brazil (Saunt, 1990). Although 'Pera' is not widely disseminated worldwide, its extensive use in Brazil, the world's largest orange producer, makes it of relatively great economic importance. 'Pera' trees are morphologically similar to 'Valencia' and other sweet orange cultivars, being upright, vigorous and densely foliated. Fruit are medium-sized, ovate and contain 5–10 seeds. Fruit quality is moderate, superior to that of 'Hamlin' but inferior to that of 'Valencia', due to lower juice colour score and TSS. Under Brazil's climate, 'Pera' trees produce multiple blooms and crops during the season. The multiple crops make it difficult to harvest at the correct maturity for best processed quality, but multiple crops allows multiple harvests for fresh fruit marketing.

'Shamouti' is a mid-season sweet orange which originated near Jaffa, Israel, in 1844 as a bud mutation of 'Beladi' orange (Saunt, 1990). Sometimes it is referred to as the 'Jaffa' orange which had been grown to a limited extent in Florida, although the two cultivars are not synonymous. 'Shamouti' trees have a slightly different leaf and tree morphology, having broader leaf laminae and more upright growth than other sweet orange cultivars. The fruit are ovate and the peel often is quite rough and thick when grown under Mediterranean-type climates. Fruit are commercially seedless with moderate soluble solids, low acidity and good colour, making the fruit excellent for the fresh market. Yields are moderate to poor in most citrus-growing regions, although in some areas (southern South Africa) 'Shamouti' trees are quite productive.

The 'Pineapple' orange was an important mid-season cultivar grown primarily in Florida with some plantings in Brazil and the Republic of South Africa. Trees are moderately vigorous and productive; however, preharvest fruit drop often causes yield reductions and coupled with a tendency toward alternate bearing has led to disfavour of this cultivar. Fruit are also seedy which is disadvantageous for the fresh market, but internal quality is excellent. 'Pineapple' juice is deep orange in colour and TSS are high, making this cultivar desirable for processing. 'Parson Brown' is another mid-season cultivar.

Some newer cultivars such as 'Early Gold', 'Westin', 'Vernia' and others from Brazil and other locations have been evaluated in Florida (Castle and

Baldwin, 2011). 'Vernia' seems to be the most promising of these, but new releases from the University of Florida breeding programme are being sent for grower evaluation through a 'fast track' programme. The most interesting appear to be early maturing somatic hybrids of 'Valencia' with 'Valquarius' being 4 to 8 weeks earlier than standard 'Valencia'.

Navel oranges

Navel orange fruit differ from those of most other citrus cultivars due to the presence of a distinctive secondary or even tertiary or quaternary fruit (navel) at the stylar end of the fruit. Tree morphology and leaf characteristics, however, are similar to those of other sweet orange cultivars. Fruit of most navel orange cultivars are seedless due to complete pollen and partial ovule sterility, are generally larger than those of other sweet orange cultivars and are grown primarily for the fresh market (Davies, 1986a). The presence of limonin, a compound that when oxidized imparts bitterness to juice, generally limits the use of navel oranges for processing. Methods have been developed to remove limonin from juice. The juice first undergoes ultrafiltration to remove pulp which interferes with limonin removal. The juice then passes through polystyrene resin exchange columns which remove a high percentage of the limonin (Wethern, 1991). However, this process is relatively slow and moderately costly, and therefore little processing occurs. Navels are juiced fresh in restaurants, juice bars, etc., particularly in Spain.

Navel oranges are more susceptible to environmental stresses than other sweet orange cultivars, probably because seedless fruit do not compete well for nutrients (Davies, 1986a). Insufficient soil moisture during initial fruit set (bloom) and physiological (June or November) drop periods, and high temperatures during these same time periods cause significant yield reductions. Moreover, navel orange trees are susceptible to fruit drop during the summer and autumn in areas like Florida that may further reduce yield by as much as 30% (Lima and Davies, 1984). Production of navel oranges is especially poor in climates with very high average day- and night-time temperatures (see Chapter 4). Highest quality fruit are produced in Mediterranean-type climates such as Spain and California coastal regions.

'Washington' navel is by far the most widely planted and commercially important navel orange cultivar (Davies, 1986a). Most other cultivars, with the exception of the Australian group, have originated from 'Washington' either as limb sports or as nucellar seedling variants. Washington navel was originally the 'Bahia' navel and was selected because of its superior productivity and fruit quality compared with so-called Australian types. 'Washington' navel is commercially important in California, Australia, Florida, Spain, Morocco and South Africa and is planted in most citrus-growing regions of the world. The trend recently, however, is to replace old-line 'Washington' navel trees with improved local or nucellar selections.

'Atwood', 'Fisher' and 'Newhall' originated in California (the California budwood certification programme), and are among the earliest maturing navel orange cultivars. 'Newhall' is widely planted in central China. Tree and fruit characteristics of these selections for the most part are similar to those of 'Washington' navel with the exception of 'Newhall' which is a more oblong fruit (Davies, 1986a).

'Navelina' and 'Navelate' are the most important navel cultivars in Spain, a major producer of navel oranges. 'Navelina' is a limb sport of 'Washington' that originated in Riverside, California, in about 1910. As the name implies, it is usually a smaller tree than 'Washington', is earlier maturing, and fruit hold well on the tree. Fruit tend to be more oblong than other navel cultivars except 'Newhall'. 'Navelate' is a late maturing selection that produces a very vigorous tree. It originated as a sport of 'Washington' in Alcanar, Spain, in 1948.

'Summerfield' navel was widely planted in Florida because it is earlier maturing and appears to be better adapted to Florida's humid subtropical climate. However, currently nucellar selections such as F-56-11 are the most widely disseminated cultivars and a number of other nucellar selections are promising because yields are superior to those of the 'Washington' navel. 'Marrs' orange, which lacks a distinct secondary fruit, is a limb sport of 'Washington' navel. It is an early-maturing, productive, low-acid selection widely planted in Texas. 'Marrs' is also lower in limonin than other navel orange selections.

'Baianinha', the most widely planted navel selection in Brazil, originated as a limb sport of 'Bahia'. It is generally less vigorous and has smaller primary and secondary fruit than 'Washington'. Thus, it is less prone to physiological fruit drop than 'Washington' or 'Summerfield' navel under humid subtropical climates. Secondary fruit size appears to be inversely related to incidence of fruit drop (Davies, 1986a). 'Baianinha' also appears to be well adapted to hot, arid growing conditions in northern South Africa.

'Leng' is a major navel orange cultivar in Australia. It is also a limb sport of 'Washington', found near Mildura, Australia, in 1934. 'Leng' fruit are generally smaller and have a thinner peel than 'Washington' fruit (Davies, 1986a). However, 'Leng' is more susceptible to peel disorders and does not ship as well as 'Washington'. 'Lane Late', an important late-season cultivar, was selected at Curlwaa, New South Wales, Australia, in 1963. It is similar in fruit size to 'Washington' but may be more subject to re-greening and granulation late in the season. It is the most popular late navel cultivar in California.

'Palmer' is a nucellar seedling of 'Washington' that originated near Brenthoek, South Africa, in the 1930s. It has been the most widely planted cultivar in South Africa since the 1970s. 'Palmer' is a vigorous, productive cultivar with fruit that holds well on the tree. 'Robyn' is a promising selection in cooler locations of South Africa where it matures later than 'Palmer' or 'Washington'.

The 'Cara Cara' navel was discovered in Venezuela. It develops red flesh (lycopene based), even in lowland tropical climates, unlike blood oranges. It has been planted to a limited extent in California, Florida and South America.

Pigmented oranges
Pigmented (blood) oranges are of commercial importance in several Mediterranean countries including Italy, Spain, Morocco, Algeria and Tunisia, but are not widely grown outside this region. When grown under Mediterranean-type climates with hot days and, most importantly, cool nights, fruit develop a very deep red flesh colour which may also occur in the peel. Fruit appearance is very appealing and internal quality is usually excellent. However, flesh colour does not attain the same redness in subtropical or tropical areas with high night-time temperatures. The red colour is due to anthocyanin pigment, which is also the primary pigment in blueberries and apples, rather than due to lycopene or carotenoid pigments, which predominate in red grapefruit and oranges, respectively. The most important cultivars of blood oranges include 'Tarocco', 'Sanguinello', 'Sanguinelli', 'Moro' and 'Maltaise Sanguine' (Saunt, 1990).

Mandarins
The mandarin group comprises numerous purported species as well as intergeneric and interspecific hybrids which possess several unique characteristics, including easy peeling (usually) and the requirement for cross-pollination for some cultivars to achieve commercially acceptable yields. The term 'mandarin' is used throughout most of the major citrus-producing regions including Japan, China, Spain and Italy. The term 'tangerine' is used to refer to many mandarin-type citrus in the USA and refers to more deeply pigmented mandarins in Australia and China. Mandarins are referred to as soft citrus in South Africa. They are produced primarily for the fresh fruit market and for canned segments (Japan, Spain and China), although the deeply coloured juice may also be blended with orange or other juices to improve their colour, or mandarin juice may be sold as single strength. With the knowledge that most commercial citrus are hybrids, the term mandarin now is more commonly used.

Satsuma mandarin group
The progenitor of the satsuma group of mandarins (*C. unshiu* Marc.) probably originated in China but was transported centuries ago to Japan, where it has become the major type of citrus planted. The satsuma itself possibly originated on Nagashima Island, Japan, from seeds brought from China (Saunt, 1990). There are over 100 cultivars of satsuma differing from each other primarily in time of maturity but also in fruit shape and internal quality (Saunt, 1990). The satsuma is well adapted to cool subtropical regions of Japan, Spain, central China and southern South Africa, and has a low heat unit requirement for fruit maturity. Leaves have large laminae and reduced petioles. The tree has a

spreading, drooping growth habit distinctively different from other mandarin types. Satsuma foliage and wood are the most freeze-hardy of all commercially grown citrus cultivars, withstanding minimum wood temperatures of −9°C when fully acclimated (Yelenosky, 1985) (see Chapter 5).

Satsuma fruit are moderately large compared with several other mandarins and are oblate to obovate in shape depending on growing conditions. Fruit grown under humid subtropical conditions tend to be large, obovate and have a coarse exterior, often with internal colour changes occurring before peel colour changes. Fruit produced under cool conditions tend to be small, oblate and develop a deep orange peel colour. Satsuma fruit have a moderately hollow central axis and are seedless in most instances due to ovule sterility.

The tendency in recent years in Japan, China and Spain has been to select or breed for earlier maturing satsuma cultivars to expand the harvest season since satsumas, like most mandarins, store poorly on the tree due to segment drying and thus have a limited harvest period. The Japanese have selected for earlier maturing satsumas such as 'Miyagoma wase' and 'Okitsu wase' (wase = early maturing) among others which may be marketable as early as September in the northern hemisphere when field grown (Saunt, 1990). Moreover, when grown under artificial conditions in the greenhouse, fruit may be harvested as early as May. Similarly, the Chinese have selected early maturing cultivars 'Xinjin', 'Gongchuan' and 'Nangan', also to extend the harvest and marketing period. The 'Wenzhou' satsuma ('Mikan') of China and the 'Owari' of Japan are early to mid-season cultivars, maturing from November to December for 'Wenzhou' and October for 'Owari'. 'Owari' followed by 'Okitsu' are the most widely planted satsumas in Spain, although 'Clausellina', an early maturing bud mutation of 'Owari', is also planted to a limited extent. All these are relatively early maturing in Spain.

Common mandarin group

The common mandarin group (*C. reticulata* Blanco) represents an assemblage of cultivars. Further, some cultivars in this group, namely 'Murcott' (Honey) and 'Ellendale' tangor, are naturally occurring hybrids which, based on fruit characteristics, are most suitably classified to this group (Saunt, 1990). Common mandarins differ morphologically from satsumas in having a more upright growth habit and in general small flowers and fruit. Laminae are also generally smaller than those of satsuma, and petiole size is also reduced. Fruit, while having typical mandarin characteristics of a hollow central axis, easy segmentation and green cotyledons, are more difficult to peel than satsumas or fruit in the Mediterranean group. However, the peel is more readily separated than that of sweet oranges. The firmer, more adherent peel also improves handling, storage and shipping characteristics compared with satsuma mandarins. The exceptions are 'Ponkan' and 'Dancy' which, when grown under humid subtropical or tropical conditions, may produce puffy, difficult-to-ship fruit.

The 'Clementine' mandarin probably originated in China and is similar in characteristics to the 'Canton' mandarin of China. It was selected by Father Clement Rodier in Oran, Morocco, in the 1890s and clementine types have become the most widely planted and economically important of all mandarins on a worldwide basis (Saunt, 1990). It is particularly well-adapted and widely planted in North Africa and throughout much of the Mediterranean region and parts of South Africa. 'Clementine' trees are densely foliated, moderately large and have consistently high yields. Fruit quality is excellent, but attainment of minimum marketable size is sometimes a problem (see Chapters 4 and 5). Fruit are usually self-incompatible and seedless (with the exception of 'Monreal') if grown without pollinators and thus are highly prized in most markets. Clementine cultivars are not as well adapted to humid subtropical or tropical growing regions as to Mediterranean climates. Several cultivars of clementine are available, differing in time of harvest, yield and fruit size. 'Fina' was the mainstay of the Spanish industry, producing a high quality mid-season fruit, although fruit size is less than adequate in some years. As with satsuma, earlier cultivars have been selected to extend the harvest season. 'Marisol' and 'Oroval' are earlier maturing selections of 'Fina'; 'Clemenules' is a mid-season maturing cultivar which holds well on the tree. 'Oronules', 'Arrufatina' and 'Orogrande' are early maturing as are four newer clementine selections. Four hybrids selected for late maturation are also marketed by Anecoop (https://anecoop.com/en/product-range).

'Dancy' tangerine was discovered near Orange Mills, Florida, in 1857 by Colonel Dancy, and was the most widely planted true tangerine cultivar in Florida. This cultivar is similar to several tangerines grown under other names in other parts of the world and is representative of mandarin issues that have seen these older cultivars become disfavoured for commercial use. 'Dancy' comes true-to-type from seed because of a high rate of nucellar embryony (nearly 100%) and its period of juvenility is much less than that of sweet oranges (4–5 vs 8–13 years under subtropical conditions). 'Dancy', like many seedy tangerines, tends toward alternate bearing and has typical brittle wood, a combination often resulting in limb breakage. Fruit size of 'Dancy' is often too small for commercial use in 'on' years. Rough lemon rootstock was commonly used in the past to improve size but led to early segment drying. 'Dancy' has a very high heat requirement and is well suited to hot, humid areas. Adequate orange peel colour develops even in tropical areas, but is greatly affected by light; the best colour develops in full sunlight.

Poor peel colour is often a problem in densely foliated orchards. 'Dancy' may be spot-picked for both size and colour, although this practice is uncommon in the USA because it is too costly. Most 'Dancy' fruit are shipped fresh in November and December in the northern hemisphere. 'Dancy' fruit hold poorly on the tree, tending to dry out and become ricey; thus it is necessary to market 'Dancy' soon after it meets maturity standards (see Chapter 8 for maturity standards). 'Dancy' trees are quite freeze-hardy but the fruit, like all citrus fruit, are not.

Moreover, the 'Dancy' tree regrows slowly following freeze-injury. 'Dancy' is not used as a pollinizer for tangerine hybrids because it blooms quite late and in some years ('off' years) does not bloom at all. It is susceptible to *Alternaria* brown spot on fruit and foliage. There have been few new plantings of 'Dancy', because of its many problems and the introduction of mandarin hybrids; however, 'Dancy' was a very profitable cultivar when managed, handled and marketed properly. Most plantings in Florida are now gone.

The 'Ponkan', also called the 'Nagpur Santra' in India or 'Warnurco' tangerine, and 'Honey' orange (not to be confused with 'Honey' of California or Honey ['Murcott'] in Florida) is of limited importance worldwide, but is still popular in some regions of China. 'Ponkan' fruit are low in juice acidity, but high in quality and early maturing. Fruit are easily damaged during harvesting and packing due to the inherent puffiness of the peel. Tree growth is very upright, and, like other mandarins, 'Ponkan' has a strong tendency towards alternate bearing and limb breakage is a problem in years of heavy crop load. Cultivars are still being evaluated in China as this fruit is still appreciated for local consumption (Liu and Deng, 2007).

Mediterranean 'Willowleaf' mandarin

The Mediterranean mandarin (*Citrus deliciosa* Ten.) also originated in China. It is often called the 'Willowleaf' mandarin because its lanceolate leaves resemble those of a willow tree. The tree is compact, densely foliated and, like most mandarins, is relatively freeze-hardy. The fruit is seedy, of small to moderate size and oblate to necked in shape. Fruit mature in mid-season and flavour is moderate to good. The popularity of the Mediterranean mandarin has declined because the tree has a strong tendency towards alternate bearing and the peel tends to become loose and puffy leading to extensive peel damage during shipping. Improved cultivars of 'Clementine', satsuma and citrus hybrids are planted in lieu of 'Willowleaf' in most citrus regions. The mandarin hybrid 'Kinnow' does have a willowleaf parent and is the primary export cultivar from Pakistan.

Naturally occurring mandarin hybrids

'Temple' orange (or 'Temple' mandarin as it is sometimes called) is a natural hybrid of tangerine and sweet orange (tangor) which originated in Jamaica in the late 1800s. Budwood of 'Temple' was brought to Florida in 1885. 'Temple' is a mid-season (January–March in the northern hemisphere) mandarin-type and in Florida has been mostly replaced by 'Murcott'. It is also grown to a limited extent in other areas such as South Africa. 'Temple' orange has distinctive tree and fruit characteristics that attest to its hybrid origin. The tree has a dense, spreading growth habit unlike that of most true tangerines. Leaves are lanceolate with thin petioles characteristic of true mandarins. Fruit are of excellent quality, possessing high TSS and deep orange juice colour. Fruit, flattened at the stylar end, are obovate to slightly subglobose in shape. Fruit are easier to peel and segment than sweet oranges but more difficult to peel than regular mandarins. 'Temple' fruit contain 20–30 seeds with white rather than the bright green cotyledons typical of mandarins. 'Sue Linda' is a commercially seedless selection from Florida. Some commercially seedless 'Temple' trees are also grown in South Africa.

Because 'Temple' produces only zygotic seedlings, no nucellar virus-free selections were available. Consequently, most old-line 'Temple' trees carry exocortis, xyloporosis and citrus tristeza virus (CTV). Moreover, 'Temple' trees tend to produce a number of growth flushes throughout the season, predisposing them to aphid feeding and potential transfer of CTV and now HLB. The production of many growth flushes also makes 'Temple' trees less hardy to freezes than other mandarin types during most seasons. Virus-free 'Temple' selections were produced by shoot-tip grafting but the cultivar is still not very popular with growers. 'Temple' is highly susceptible to the scab fungus.

'Murcott', also known as 'Honey', 'Honey Murcott' or 'Honey' orange, is a mandarin hybrid which originated either from an abandoned US Department of Agriculture (USDA) nursery some time around 1916, or as a chance seedling near Miami, Florida. It was first propagated by Charles Murcott Smith. 'Murcott' is probably a tangor (tangerine × sweet orange), rather than a true tangerine. 'Murcott' has many tree and fruit characteristics typical of a true mandarin. Trees are moderately freeze-hardy and vigorous and have a very distinctive upright growth habit. Leaves are lanceolate, but not as much as those of mandarins like 'Clementine'. Fruit are borne in terminal clusters causing typical lodging of the branches, as in other mandarin cultivars. The fruit are moderately sized, seedy and oblate with a semi-hollow central axis, also characteristics of true mandarins. However, the fruit are not as easily peeled as other mandarins and the cotyledons are white, not green. The peel is not loose or puffy and fruit ship well. The combination of fruit and tree characteristics supports the contention that 'Murcott' is a mandarin hybrid.

'Murcott' fruit attain commercial maturity from January to March in the northern hemisphere making them the latest maturing of all commercially important mandarin cultivars. Late maturation has the advantage of placing fruit on the market at times of high prices, but also requires that fruit survive the coldest months of the year, risking freeze-damage in areas such as Florida. Most 'Murcott' fruit must be harvested for the fresh fruit market because their juice is high in limonin, making it less desirable for processing. Internal fruit quality for fresh use is among the best of any mandarin or mandarin hybrid. Peel and juice attain a deep reddish orange colour at optimum maturity and flavour is excellent.

'Murcott' mandarin is very prone to alternate bearing. Large crops of late-maturing fruit, in particular, may induce severe alternate bearing with considerably reduced cropping the next season. Moreover, 'Murcott' trees in some instances will set such heavy crops that root growth is severely impaired. Heavy crops can actually kill the tree because carbohydrate reserves in the roots become depleted (Smith, 1976). This problem, known as 'Murcott' collapse, can be controlled by judicious pruning or by crop thinning to balance the crop and lessen the stress on the tree. Fruit splitting and drop reduce yield significantly in some seasons. Higher N and K levels are recommended for 'Murcott' than for sweet oranges, grapefruit or other mandarin cultivars, but don't appear to correct 'Murcott' collapse (see Chapter 5).

'Afourer' originated in Morocco and is described in the University of California Citrus Variety Collection (http://www.citrusvariety.ucr.edu/citrus/wmurcott.html). The fruit matures late, has good colour and size and is low seeded when grown in solid plantings. It is very popular in Morocco, as 'Afourer', and in many mandarin producing countries it may be marketed as 'W. Murcott' or other variations of this name. The cultivar is probably not closely related to 'Murcott'.

Man-made mandarin hybrids (tangelos)

Tangelos are interspecific hybrids of *C. reticulata* (mandarin or tangerine) and *C. paradisi* (grapefruit or pummelo). The most important commercial tangelo cultivars, 'Orlando' and 'Minneola', arose from a cross between 'Dancy' tangerine and 'Duncan' grapefruit made by Webber and Swingle in Florida in 1897. Tangelo fruit and tree characteristics are varied and diverse with some selections being similar to grapefruit and others more similar to true mandarins. In general, tangelos are quite vigorous, freeze-hardy and produce weakly parthenocarpic fruit requiring cross-pollination or gibberellic acid (GA_3) sprays to achieve adequate fruit set and yields. Leaf morphology ranges from broad and cupped to lanceolate. Petiole size also ranges from narrow to broadly winged.

'Orlando' tangelo produces a vigorous, large tree with distinctively broad, cupped leaves with moderate-sized petioles. The leaves and wood are some of the most freeze-hardy of all commercially important citrus cultivars with the exception of satsumas. Fruit are oblate to subglobose, ranging from seedless where no cross-pollination occurs to seedy (10–20 seeds) when sufficient numbers of pollinizers (trees) and pollinators (bees) are present. There is generally a positive correlation between fruit size and seed number. Fruit reach maturity in December to January in the northern hemisphere, attaining moderate to bright orange peel colour. 'Orlando' juice has a flavour intermediate between that of mandarin and grapefruit, with a noticeable distinct grapefruit after-flavour.

Cross-pollination of 'Orlando' with a compatible cultivar like 'Temple' orange or 'Robinson' tangerine is recommended for optimum production. 'Minneola', its sister cultivar, is not cross-compatible with 'Orlando' and should not be used for cross-pollination. Fruit may also be produced parthenocarpically for fresh fruit markets where seedlessness is desirable. 'Orlando' requires higher rates of N fertilization than round oranges and frequently shows N deficiency symptoms during winter in subtropical areas, known as winter chlorosis. Fruit and foliage are susceptible to the *Alternaria* brown spot fungus.

'Minneola' tangelo ('Honey Bell'), although a sister seedling to 'Orlando', is distinctively different from it in tree and fruit characteristics. Trees are very vigorous, large and spreading with large pointed laminae and moderate-sized petioles. 'Minneola' trees are nearly as freeze-hardy as 'Orlando', but are generally less productive in areas with high springtime temperatures. Fruit are large, obovate and have a pronounced neck at the stem end. Fruit, peel and juice colour are deep reddish orange at maturity and juice flavour is excellent. The peel is moderately adherent and finely pebbled. Fruit of 'Minneola' reach commercial maturity between January and March in the northern hemisphere and July to August in the southern hemisphere. Seed number ranges from 0 to 10–20 depending on the degree of cross-pollination. 'Minneola', like 'Orlando', requires cross-pollination with, for example, 'Temple' orange or 'Robinson' tangerine to achieve adequate yields and optimum fruit size. However, fruit can be produced parthenocarpically by planting large solid blocks and using GA_3 sprays at bloom and postbloom for fresh fruit markets where seedlessness is desirable. Both 'Orlando' and 'Minneola' set low crops parthenocarpically, even with GA_3 sprays, and 'Robinson' has many adverse tree characteristics, is not grown in large amounts in Florida and therefore is not a viable pollinator for these cultivars. Both 'Orlando' and 'Minneola' have declined in hectarage.

Other man-made mandarin hybrids
The high incidence of nucellar embryony in citrus has caused considerable problems for fruit breeders wishing to develop new citrus cultivars. Consequently, breeders found it necessary to use seed parents that produced only zygotic progeny in many of their crosses. In 1942, Gardner and Bellows at the USDA research laboratory in Orlando, Florida, crossed 'Orlando' tangelo with 'Clementine' mandarin (zygotic progeny only). The resulting seedlings were planted and screened for desirable characteristics. Four selections were named and released in 1959: 'Robinson', 'Lee', 'Osceola' and 'Nova'. A cross between 'Minneola' tangelo and 'Clementine' was also made in 1942 leading to the release of 'Page' orange in 1963. 'Sunburst' tangerine, resulting from a cross between 'Robinson' and 'Osceola', was released by the USDA in 1982. 'Fallglo', a 'Bower' × 'Temple' cross made in 1962 by P.C. Reese, was selected for further testing in 1972 and released for propagation in 1987. 'Ambersweet', a 1989 release, resulted from a cross between a mandarin hybrid ('Clementine' × 'Orlando') and a mid-season sweet orange (Accession # 15-3) made in 1963 by C.J. Hearn and P.C. Reese in Florida. A seedling from this cross was selected in 1972 for the 1989 release. These hybrids are grown to a limited extent worldwide and are produced primarily for the fresh fruit market.

'Robinson' trees are moderately vigorous and similar in morphology to true mandarins. 'Robinson' requires cross-pollination with compatible cultivars such as 'Temple' or 'Orlando' to produce adequate yields under most growing conditions, although fruit are also produced parthenocarpically. Sometimes the fruit load becomes excessive, causing limb breakage. Leaves and wood of

'Robinson' are very freeze-hardy, but tree vigour is usually decreased due to excessive cropping or 'Robinson' dieback, and freeze-damage often occurs. Fruit of 'Robinson' mature from October to December in the northern hemisphere. Fruit tend to be small to moderate in size and oblate in shape. External and internal colour attain a deep, rich orange hue, although in some years changes in peel colour lag behind internal quality changes causing difficulty in judging maturity. The fruit has typical mandarin characteristics of easily separable peel and segments and a hollow, open centre. The fruit quality of 'Robinson' is excellent. Seed number varies from 0 to more than 20 per fruit depending on the degree of cross-pollination. In general fruit with more seeds are larger than those with 0 to a few seeds. Seeds produce zygotic progeny only. 'Robinson' is prone to limb dieback which reduces yields. Fruit are susceptible to splitting and plugging because of excessively thin peels in some years. Small fruit size is also a problem in some seasons. Fruit generally do not hold well on the tree nor ship well.

'Sunburst' tree and fruit characteristics are similar to those of 'Nova' and 'Robinson'. Trees are moderately freeze-hardy and tolerant of citrus snow scale, but the foliage and stems are very susceptible to rust mite damage and some other problems under Florida conditions (Albrigo *et al.*, 1987). Fruit are oblate, highly coloured, and thin-peeled; they mature after 'Robinson' (October–November) in the northern hemisphere but before 'Dancy'. As with other mandarin hybrids, seed number varies with degree of cross-pollination. Generally, external and internal colour develop together. Sunburst fruit ship better than fruit of 'Robinson' or 'Dancy'. 'Sunburst' requires pollination with cross-compatible cultivars like 'Temple', 'Orlando' or 'Nova' to produce an acceptable crop. Seeds of 'Sunburst' are polyembryonic.

'Fallglo' is an early-season cultivar which is harvested between late October and November in the northern hemisphere. Growth and leaf characteristics are similar to 'Temple' orange, being densely foliated, moderately vigorous and upright but somewhat spreading. Leaves are lanceolate with reduced petioles reminiscent of 'Temple' but are resistant to sour orange scab. 'Fallglo' trees are less freeze-hardy than those of 'Orlando' or 'Sunburst' and may be less suitable for planting in areas where freezes are likely to occur. The oblate fruit of 'Fallglo' are relatively large with a small but distinctive navel and contain 30–40 seeds depending on degree of cross-pollination. Seeds of 'Fallglo' produce only zygotic progeny. The peel is smooth, contains prominent oil glands and is easily removed. Peel and juice colour are deep orange and fruit quality is comparable with that of 'Temple'. 'Fallglo' is very sensitive to developing peel pitting, which appears to be related to relative humidity changes during harvesting and shipping (Alferez *et al.*, 2004). Unlike many other mandarin hybrids, 'Fallglo' seems not to require cross-pollination to achieve optimum production.

'Ambersweet' is an early-season (mid-October to December) hybrid. Tree growth is somewhat upright and moderately vigorous. Trees have intermediate freeze-hardiness. Fruit are moderately large, convex-shaped and tapered at the stem end, somewhat similar to 'Temple' orange. Occasionally fruit produce a small navel. Juice colour is dark orange under growing conditions typical of northern Florida. Fruit range from seedless to seedy (15 or more) depending on degree of cross-pollination, the flowers have little pollen and poor pollen germination (Albrigo et al., 2001) and some degree of cross-pollination or GA_3 near the end of induction to reduce flowering is necessary to obtain adequate yields. Like 'Robinson', seeds produce zygotic progeny only and most have white cotyledons, with a few green ones. Juice taste and quality are similar to that of round oranges except with better colour. Regulations were changed to allow full use of this cultivar in frozen concentrated orange juice, but its poor yield made it unpopular with growers (see Chapter 8).

Some recent additions to the US mandarin cultivars are 'Tango', irradiated seedless 'W. Murcott' (www.citrusvariety.ucr.edu/citrus/tango.html, by Roose and Williams), 'Sugar Belle' from the University of Florida (http://idtools.org/id/citrus/citrusid/factsheet.php?name=Sugar+Belle) and 'Bingo' (http://patents.google.com/patent/USPP27778P3/en, by F. Gmitter). 'Sugar Belle' has good tree and fruit characteristics, is seedless and appears to have some HLB tolerance. It is being widely planted in Florida. 'Bingo' is relatively new and early plantings have experienced some significant losses to an undetermined dieback.

Seedless mandarin types continue to be in high demand and many breeding programmes around the world have seedless mandarin cultivars as a high priority. Modern breeding techniques are resulting in many selections for evaluation. Citrus industries should expect more selections to be available in the future.

Grapefruit
Grapefruit (*C. paradisi* Macf.), as noted earlier, is probably a hybrid of pummelo and sweet orange (Scora, 1988). Grapefruit is one of the few citrus types that originated in the New World, most probably as a natural hybrid in the West Indies (Scora et al., 1982). The pummelo parentage is quite apparent in similarities between fruit and leaf morphology. Grapefruit, unlike pummelo, produces nucellar and zygotic embryos, rather than zygotic embryos only.

Grapefruit production and distribution are much more limited than that of sweet oranges or mandarins. The high heat requirement limits production of high quality fruit to tropical and hot, humid subtropical regions. In contrast, grapefruit grown in Mediterranean climates are usually high in juice acidity and have a thicker peel and lower juice content than those grown in more tropical or humid subtropical regions. The high juice acid and moderately low TSS levels for grapefruit in general produce fruit less palatable than oranges to many people. The fruit are not often peeled and eaten out-of-hand as are other citrus fruits. Grapefruit consumption is greatest in North America, Europe and Japan; yet, pummelos, with lower acidity, are far more popular in China and most of southern Asia.

Grapefruit trees are very vigorous, having a spreading-type growth habit reminiscent of their pummelo heritage. The structural framework is very sturdy and capable of supporting the greatest crop loads of any commercial citrus type. Leaves are unifoliately compound and much larger than those of sweet orange, mandarins, lemons or limes, and are similar in appearance, although smaller, to those of pummelo. The lamina is ovate with entire to slightly serrated margins. The petiole is large and winged, often with a distinct cordate shape. Grapefruit generally possess a 2/5 phyllotaxis, the same as pummelo, in contrast to the 3/8 phyllotaxis of sweet orange. Grapefruit flowers are perfect and complete, having the same basic structure as that of other citrus species. Flowers are larger than those of other commercially important cultivars with the exception of pummelo. Flowers and thus fruit are often borne in clusters, hence the origin of its name, although fruit are also borne singly, particularly in the interior portions of the tree.

Fruit of grapefruit are the largest of any commercial citrus cultivar, again with the exception of pummelo. Fruit diameter ranges from 8 to 14 cm at maturity depending on cultivar, rootstock, crop load and growing conditions. Grapefruit may remain on the tree for protracted periods and are harvested from September to July in the northern hemisphere. However, some fruit drop may occur which reduces yields, and fruit held too late may become granulated or seeds may germinate within the fruit on the tree (vivipary), thus lowering marketability. Until these problems develop, grapefruit quality generally improves if it remains on the tree and juice acid levels decrease.

Fruit vary from quite seedy to commercially seedless; the latter are preferred on the fresh market. Seeds are large, plump and polyembryonic, with cream-coloured to white cotyledons upon germination.

White-fleshed grapefruit
'Duncan' grapefruit was the major seedy grapefruit cultivar in the USA; however, production has declined and likely no new plantings have been made either in the USA or worldwide. 'Duncan' probably originated as a seedling of the original grapefruit seedlings brought to Florida by Don Phillipe. It was first discovered in the orchard of A.L. Duncan near Safety Harbor in 1875, but was not named until 1892.

'Duncan' is a typical, vigorous grapefruit cultivar. Consequently, trees are generally set at wider spacing than those used for sweet oranges or mandarins. Fruit are produced terminally in clusters and tend to be larger and higher in total soluble solids than those of 'Marsh'. 'Duncan' was used primarily for processing into sections, juice vesicle slurry and juice, since seediness (50–60) limits its demand as fresh fruit.

'Marsh' grapefruit is the most widely planted white-fleshed grapefruit cultivar in the world. It originated as a chance seedling around 1860 near Lakeland, Florida, but was not propagated commercially by C.M. Marsh, a commercial nurseryman, until the late 1880s. 'Marsh' trees are very similar in vegetative characteristics to other grapefruit cultivars, but fruit size and quality are generally not as good as those of 'Duncan'. The major reasons for the popularity of 'Marsh' are its commercial seedlessness, excellent taste and the fact that it holds well on the tree without significant loss of fruit quality, thus extending the harvest season if desired. Fruit are now used more heavily for processing, but it was grown primarily for the fresh market until red fleshed cultivars became more popular.

Red-fleshed grapefruit
All currently popular pink- and red-fleshed grapefruit cultivars have arisen as mutations of white-, pink- or red-fleshed cultivars. The general trend since the early 1900s has been to select and propagate cultivars with progressively redder flesh and peel colours. This selection process has occurred due to increasing consumer demand, and consequently higher prices for redder-fleshed cultivars. In general, these mutations do not differ in internal quality except in the degree of redness, with the exception of 'Star Ruby' which appears to have a more coarse-textured flesh than other cultivars, but Burns and Albrigo (unpublished) could not chemically or morphologically characterize this apparent coarseness. The deeper red colour, however, increases the 'eye appeal' to the consumer and thus demand.

There are two major cultivar developmental lines giving rise to the most important pink- and red-fleshed grapefruit cultivars (Fig. 2.3). The first is derived from 'Walters' (a white-fleshed cultivar) and includes 'Foster', 'Hudson' and 'Star Ruby', and the second line derives from 'Thompson' ('Pink Marsh'), which includes 'Redblush', 'Ruby Red', 'Henninger', 'Ray Ruby', 'Rio Red', 'Flame' ('Henderson') and 'Burgundy' (Saunt, 1990).

'Foster', the first of the pink-fleshed mutations, occurred as a limb sport of 'Walters' in the Atwood orchard near Manatee, Florida, in 1907. The pulp is light pink but the juice of 'Foster' is nearly colourless. 'Foster' was not widely planted primarily because it is seedy, but it has been the source of other more important red-fleshed cultivars. 'Hudson' originated as a bud-sport of 'Foster' in the 1930s in San Benito, Texas. 'Hudson' has deeper red flesh and peel colour than 'Foster' but is also seedy. In 1959, R.A. Hensz of Texas A & M University irradiated seeds of 'Hudson', planted them and subsequently evaluated the seedlings. In 1970 he selected a very deep red-fleshed seedling which he named and released as 'Star Ruby'. 'Star Ruby' has the deepest red peel, flesh and juice colour of any currently available red grapefruit cultivar. The red colouration is even found in the bark of young branches. The leaf colour of 'Star Ruby' differs from that of other red-fleshed cultivars, becoming chlorotic, almost white in some orchards. 'Star Ruby' leaves also become chlorotic in response to some herbicide applications and trees are quite susceptible to *Phytophthora* foot rot. However, despite these problems, yields, fruit quality and economic returns have been very good.

Another line of red-fleshed grapefruit originated from 'Marsh', beginning with 'Thompson' ('Pink Marsh') in 1913 from a limb sport in the orchard of

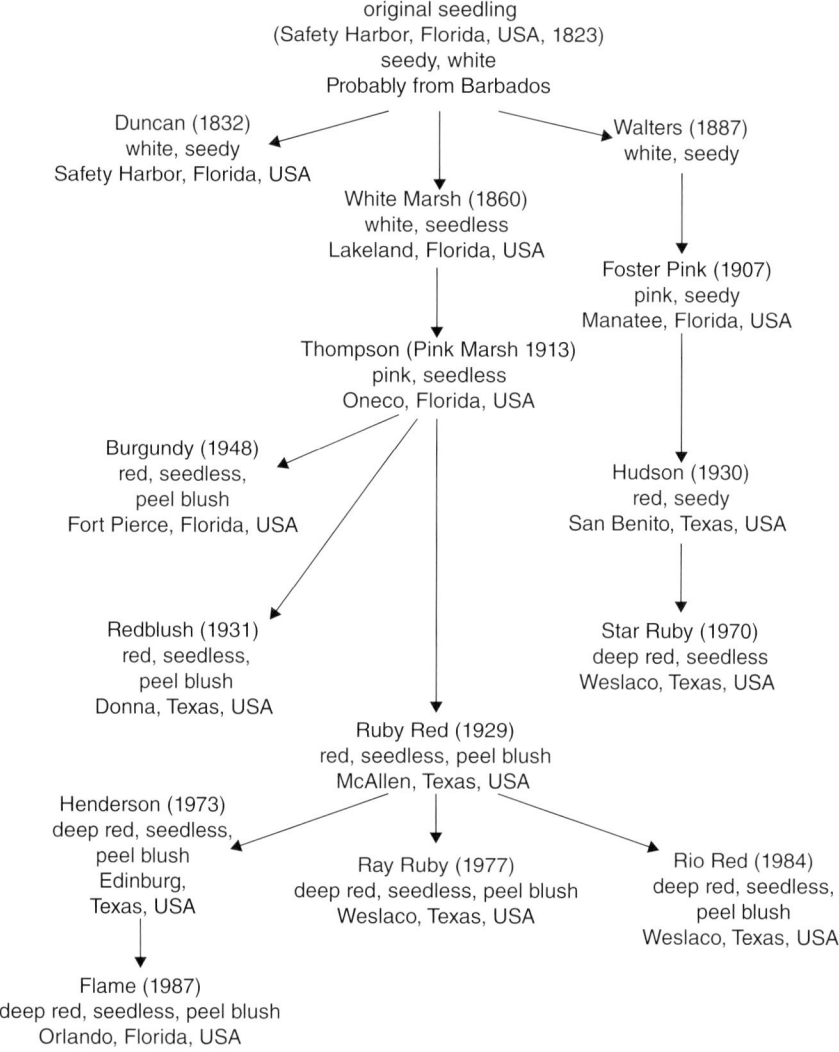

Fig. 2.3. Origin of the most commercially important grapefruit cultivars. Numbers in parentheses indicate date of discovery or release (adapted from Saunt 1990).

W.B. Thompson in Oneco, Florida. 'Thompson' has light pink flesh and peel colour even under optimum conditions, but has the advantage over 'Foster' in being seedless. Tree characteristics of 'Thompson' are identical to those of 'Marsh'. Several of the red-fleshed cultivars are mutations of 'Thompson'. 'Ruby Red' ('Ruby') was discovered as a limb sport of 'Thompson' by A.E. Henninger near McAllen, Texas, in 1929. In 1931 another limb sport of 'Thompson' was selected by J.B. Webb of Donna, Texas and propagated as 'Redblush' ('Webb Redblush'). These cultivars are nearly identical in fruit and tree characteristics. In both, fruit are more deeply red-pigmented than 'Pink Marsh' and moreover a distinct red peel blush is present.

'Ruby Red' has been the source of three more deeply coloured cultivars in Texas: 'Henderson', 'Ray Ruby' and 'Rio Red'. 'Henderson' was first observed as a limb sport of 'Everhard' grapefruit in 1973 in the Henderson orchard near Edinburg, Texas. It has deeper red flesh and peel colour than 'Ruby Red'. 'Ray Ruby' is also a mutation of 'Ruby Red' with peel colour similar to that of 'Henderson' but with a deeper red flesh colour. 'Rio Red' originated as a seedling of 'Ruby Red', planted in 1953 and selected because of favourable yield and fruit characteristics in 1959. Budwood of this selection was irradiated and propagated on sour orange rootstock. A mutation was selected in 1976 for its deep red coloration, named and released as 'Rio Red' in 1984. Peel colour is similar to that of 'Ray Ruby', but the flesh colour is redder.

A fourth red-fleshed cultivar, 'Flame', was released for propagation in 1987 by the USDA in Florida. 'Flame' originated from seeds of 'Henderson' planted in 1973. Peel colour is similar to that of 'Ray Ruby' but internal colour is nearly comparable to that of 'Star Ruby' when grown under Texas conditions. In some areas of Florida, trees have shown leaf chlorosis symptoms similar to those of 'Star Ruby', but internal colour seems to be more stable than that of 'Rio Red' under Florida conditions.

An important characteristic of 'Ray Ruby', 'Rio Red', 'Flame' and 'Star Ruby' not found in 'Ruby Red' is that the deep red internal colour persists much later in the harvest season. This extends the marketing period. Variations in colour may occur depending on climatic conditions, with the best red colour developing in warm to hot, lowland tropical regions because the colour is from lycopene, the red pigment in tomatoes. Currently 'Ray Ruby' and 'Flame' are the most popular new cultivars along with 'Star Ruby'.

A major shortcoming of grapefruit is an apparent susceptibility to HLB of all commercial cultivars. Some tolerance has been noted in non-commercial material (Stover *et al.*, 2013) and a breeding selection with lower furanocoumarins and acidity has been identified (Chen *et al.*, 2011). No selections that have HLB tolerance or low acidity appear to be near commercialization. The University of California Riverside citrus breeding programme did release two pummelo X grapefruit hybrids that are low in acidity (https://citrusvariety.ucr.edu/citrus/oroblanco.html and https://citrusvariety.ucr.edu/citrus/melogold.html).

Limes
Lime (*C. aurantifolia* L.) trees most probably originated in tropical areas along the Malay archipelago and as a result of this heritage are the most freeze-sensitive of all commercial citrus species (Yelenosky, 1985). Thus, their distribution is limited to the tropics and warm, humid subtropical regions where minimum temperatures remain above −2 to −3°C. The two major groups include the acid

and acidless limes of which the acid limes are of commercial importance. The acid limes are subdivided into Citrus × latifolia ('Persian', 'Tahiti' or 'Bearss') and *Citrus aurantifolia* ('Mexican', 'Key' or 'West Indian' limes, probably all of southeastern Asian origin).

Lime trees are extremely vigorous, possessing an upright to very spreading growth habit. Trees are often quite thorny. Leaf morphology differs between the two major groups. The lamina of the 'Tahiti' lime is large and elliptical to ovate with noticeable serrations near the apex. The 'West Indian' lime lamina is much smaller and ovate to nearly round in shape. The lamina is distinctly serrated around the apical margin. Petioles of both types are very reduced and nearly nonexistent. Lime flowers are perfect and complete, having the same general characteristics as other citrus species in most cases. However, occasionally some 'juvenile' flowers are produced which have reduced or nonexistent gynoecia. Flowers of the 'Tahiti' lime, like lemon, are purple-petalled, while those of 'West Indian' lime are white-petalled and smaller than those of 'Tahiti'. Petals of 'West Indian' lime are often extrorse, curving downward from the central axis. Flowering generally occurs in two major peaks seasonally, but also may occur continuously at reduced intensity, producing several crops per year. This flowering pattern is primarily drought–rain driven (see Chapter 4).

Fruit of 'Tahiti' limes are very similar morphologically to lemons (prolate spheroid to spherical); those of 'West Indian' lime are spherical and much smaller than those of 'Tahiti' lime. 'Tahiti' lime rarely produces seeds due to near complete ovule sterility. 'West Indian' lime is moderately seedy, having 10–15 seeds per fruit.

Lemons
Lemon (*C. limon* Burm. f.) trees probably originated in the eastern Himalayan regions of India, although this conjecture is based only on observations of wild species growing in this region. It is generally accepted that lemon is a hybrid closely related to citron (Scora, 1988). Distribution and major production areas of commercial lemons are limited primarily to semiarid to arid subtropical regions, having minimum temperatures greater than −4°C. Lemon trees are more sensitive to low temperatures than other commercial cultivars with the exception of limes (Yelenosky, 1985) and are not well-adapted to humid subtropical or tropical regions due to susceptibility to fungal and algal diseases. In addition, fruit quality, especially peel texture, which becomes coarse, is generally poorer in humid regions than in Mediterranean-type climates.

The lemon tree is extremely vigorous, having an upright, rank and unwieldy growth habit when juvenile which continues when grown in tropical areas – another reason for poor adaptability to tropical climates. As trees mature or when they are grown in Mediterranean climates the growth habit becomes more spreading and manageable. Trees are often quite thorny, although this varies with cultivar, growing conditions and tree age. Generally, young trees are very thorny and leaf morphology is quite variable depending on tree vigour. Laminae are large and ovate with pronounced serrations along the apical margins of leaves developing on vigorous shoots. Laminae become ovate to lanceolate again with serrated margins as shoots mature. Newly developing leaves are purple but as the laminae mature they become green. Petioles are reduced, and nearly nonexistent in some instances.

Lemon flowers are perfect and complete, having the same general characteristics as other commercial citrus species. Flower petals, like newly expanding leaves, are purplish white. Flowers are smaller than those of grapefruit but similar in size to orange flowers. They are typically borne in clusters. Lemon trees have two major flowering periods in Mediterranean climates but tend to have some continuous flowering throughout the year in cool, coastal regions such as those of California. Fruit are produced throughout the year in most major growing regions, but depending on cultivar and environmental factors a large part of the crop is usually harvested in either the summer, autumn or winter. Fruit shape varies from nearly spherical in some cultivars to the more common prolate spheroid of most commercial selections. Fruit have a characteristic apical mammila (nipple) at the stylar end. Lemon fruit are high in total acidity (TA) (5–8%) and low in TSS (7–9%) relative to any other commercially important citrus cultivars except acid limes. Cultivars range from moderately seedy to seedless. Seeds are small with smooth seed coats and pointed micropylar ends.

There are three major groups of lemons not including those used as rootstocks: the *Femminello*, Verna (Berna) and Sicilian types. The *Femminello* and Verna types are primarily grown in North Africa and Europe and Sicilian types in the USA and South Africa (Saunt, 1990). The *Femminello* group, *Femminello comune*, *Femminello siracusano* and *Femminello St Teresa*, represent the most widely grown lemon types in Italy. There are 36 accessions of Sicilian lemon, belonging to three different cultivars '(Femminello', 'Monachello', and 'Lunario'). Since the cultivars are everbearing, several crops can be harvested throughout the year. The Primofiore crop is harvested from September to November; the Limoni crop from December to May; the Bianchetti crop from April to June and the Verdelli crop from June to October (Saunt, 1990). Tree morphology is characteristic of most lemons in being upright and vigorous. Yields are moderate to high and are quite consistent. Femminello accessions are moderately seedy and fruit shape prolate spherical with a moderately sized apical mammila. Femminello fruit have moderately high acidity and lower juice content. They are marketed fresh or for their peel oil content.

'Verna' is the major cultivar of Spain and is very similar morphologically to 'Lisbon' lemon. Unlike the *Femminello* group, 'Verna' trees produce a major crop (February–August) and a second lesser crop (Verdelli, August–October) in the northern hemisphere. Fruit are more elongated and have a more pronounced apical mammila than *Femminello* but are seedless.

The most important cultivar in the Sicilian group is 'Eureka', which is also important in California and Australia. It originated as a seedling from Sicily, brought to California in 1858. It is also grown in South Africa, Spain and Israel (Saunt, 1990). Tree morphology is slightly different from some other cultivars in having a less densely foliated, more spreading canopy. 'Eureka' is considered to be less frost-hardy in California than 'Lisbon'. Fruit are moderate- to small-sized and ovate with a moderately rounded apical mammilla. Fruit quality is excellent when grown in coastal, Mediterranean-type climates. The peel is smooth and thin and fruit have high juice and acid levels. Fruit usually have fewer than 9 seeds and are produced throughout most of the year. Yields of 'Eureka' are generally lower than those of 'Lisbon' in California or 'Fino' in Spain.

'Lisbon', like 'Eureka', is grown primarily outside the geographic Mediterranean region, namely in Australia, California and Argentina, where 'Lisbon' is better adapted to the climate variations of these locations. Trees are densely foliated and vary in degree of thorniness. The dense foliation delays radiation losses from the interior canopy more effectively than the less densely foliated 'Eureka'. Thus trees withstand some frosts better than 'Eureka' trees. However, there is probably no inherent physiological difference in freeze-hardiness between 'Lisbon' and other lemon cultivars. Fruit are excellent in quality, differing slightly from those of 'Eureka' under some growing conditions. The apical mammilla and areolar furrow are more pronounced on 'Lisbon', and under most conditions fruit of 'Lisbon' are very similar to those of 'Eureka'. Major harvest periods occur in winter and spring. Fruit usually have fewer than 9 seeds. 'Lisbon' has replaced 'Eureka' as a major cultivar in California because of superior yields. Moreover, the denser foliage provides some protection from wind and frost damage to the fruit. 'Lisbon' and 'Limoneira' are popular in Argentina because of their higher oil content than 'Genova' or 'Eureka'.

Pummelo (shaddock)
Pummelo (*Citrus grandis* [L.] Osb.) is probably native to southern China or the Malay and Indian archipelagoes. It is sometimes referred to as a shaddock and is grown to a limited extent in a few of the major citrus-producing regions. It is a popular fruit in much of China and southeastern Asia. Although the tree and fruit look similar to grapefruit, there are some important differences. Leaves, flowers and fruit are usually the largest of any citrus type, despite some variability. The lamina is ovate with a moderately serrated margin; the petiole is cordate and distinctly winged and separate from the lamina. The young twigs and leaf midribs are often pubescent. Flowers have the same basic structure as other true citrus, but they are the largest of the true citrus species. The fruit shape ranges from obovoid to pyriform with a pronounced flattening of the stylar end in most cultivars. Fruit diameter ranges from 10 to 30 cm, but may be significantly larger than this for some cultivars. The internal quality

of pummelos differs from that of grapefruit. The peel is extremely thick but easy to peel and the juice sacs are very pronounced and rubbery in appearance and texture. Generally, the rind and the segment walls are peeled before eating. Pummelo juice is not bitter nor as high in acid as that of grapefruit and thus it has a sweet, mild flavour. Seeds are large, plump and produce only zygotic progeny for nearly all cultivars.

There are many cultivars of pummelo but they are generally divided into the Thai, Chinese and Indonesian groups. Fruit in the Thai group are generally smaller than those in the Chinese group (Saunt, 1990). Major cultivars in the Thai group include 'Chander' (pink-fleshed), 'Kao Panne' and 'Kao Phuang' (white-fleshed); those in the Chinese group include 'Goliath', 'Mato' and 'Shatinyu' (all white-fleshed); those in the Indonesian group include 'Banpeiyu' (white-fleshed) and 'Djeroek Deleema Kopjar' (pink-fleshed). There are also two low acidity hybrids between pummelo and grapefruit, 'Melogold' and 'Oroblanco' (see grapefruit section above), which have intermediate characteristics between the two and reportedly have been grown to a limited extent in California and Israel.

Kumquat

Kumquat (*Fortunella* spp.) originated in northern China and due to several differences in morphology have been placed into the genus *Fortunella* rather than *Citrus* by several taxonomists. Kumquats are produced primarily in China and the Philippines with limited production in Spain and other areas of the world. Kumquats are eaten fresh or candied and are unique in that the entire fruit including the peel is generally eaten whole. The tree is moderately vigorous and upright to spreading in its growth habit. Leaves are small with lanceolate to elliptical laminae and greatly reduced petioles. The underside of the lamina has a characteristic silvery colour. Kumquats tend to bloom much later than other commercial cultivars and the leaves and wood are quite freeze-hardy when fully acclimated.

Fruit shape varies with species and cultivar from elliptical ('Nagami', '*Fortunella margarita*') to spherical ('Meiwa', '*Fortunella crassifolia*', 'Marum', '*Fortunella japonica*') and fruit size ranges from 2 to 3 cm in diameter. Fruit are composed primarily of peel, albedo and flavedo tissue and have a very limited amount of juice. Kumquats differ from other true citrus in having only 3 to 5 carpels, each containing 1 to 2 seeds. Seeds are small and have green cotyledons.

Hybrids of Citrus and Related Genera

The fact that citrus and related genera readily hybridize has created an interesting group of unusual plant forms and names. The existence of intergeneric hybrids is common in citrus and related genera but unusual in the plant kingdom, further suggesting the lack of clearly defined species as discussed previously. Swingle developed an elaborate system for naming hybrids, some of the most common of which are given in Fig. 2.4 (Davies and Jackson, 2009). Of this large and diverse group, the citranges (*C. sinensis* × *P. trifoliata*) and citrumelos (*C. paradisi* × *P. trifoliata*) are of greatest commercial importance primarily as rootstocks (see Chapter 3). Most of the original crosses were made by W.T. Swingle following the devastating 1894–1895 freezes in Florida.

Swingle's intention was to incorporate the freeze-hardiness of *P. trifoliata* into the sweet orange without introducing ponciridin, a bitter principle present in *P. trifoliata* that makes its fruit unpalatable. Unfortunately, while citranges and citrumelos have varying degrees of freeze-hardiness intermediate between the parents, the fruit remain unpalatable. Backcrossing F_1 hybrids improves the edibility of the fruit but also reduces freeze-hardiness. Scions propagated on citrumelos generally have greater freeze-hardiness than those on citranges but this factor may be more related to the vigour induced by the rootstock than to genetic differences in hardiness (see Chapter 4 for further information on freeze-hardiness).

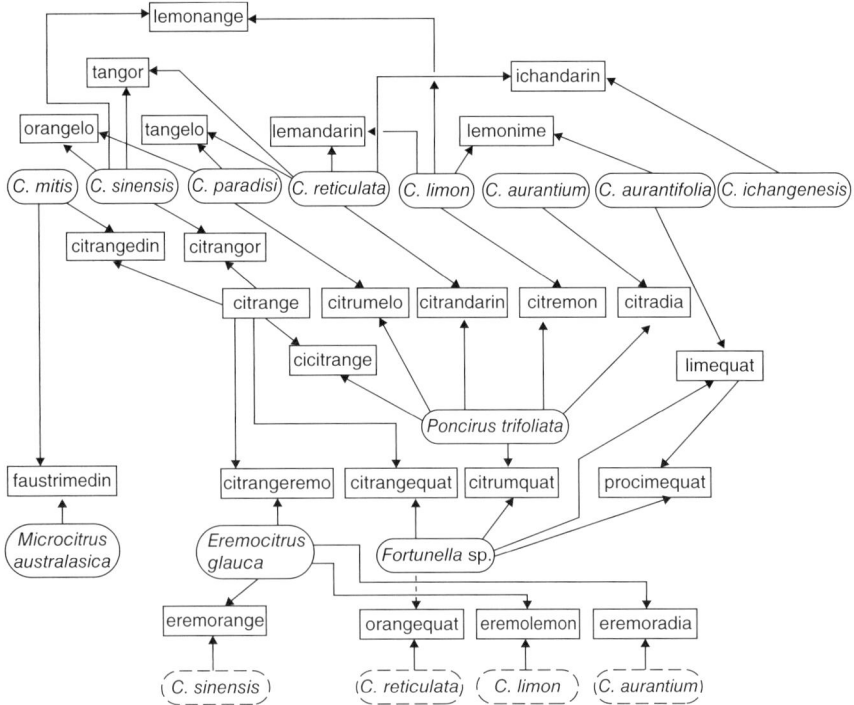

Fig. 2.4. Interspecific and intergeneric hybrids of citrus from breeding programmes (Davies and Jackson (2009). Reprinted with permission of the University Press of Florida).

Several other man-made and natural hybrids have achieved limited commercial importance. 'Orlando' and 'Minneola' tangelos (*C. reticulata* × *C. paradisi*) are grown in many areas of the world. In addition, several man-made hybrids have been developed using 'Clementine' mandarin, 'Temple' orange or pummelo as the seed parent because they produce only a single zygotic embryo, thus eliminating problems with nucellar progeny. Of these, 'Robinson' and 'Nova' ('Clementine' × 'Orlando'), 'Sunburst' ('Robinson' × 'Osceola') and 'Fallglo' ('Temple' × 'Bower') attained limited commercial importance worldwide, but they seem to be in more disfavour as time goes by and seedless fruit with good shipping characteristics are more in demand.

CITRUS GENETICS AND BREEDING

Citrus and related genera within the true *Citrus* group have 2 sets of 9 chromosomes ($2x = 18$). Chromosomes are small and variable in size and aberrations are moderately common (Raghuvanshi, 1968). Some naturally occurring triploids, tetraploids and hexaploids exist but generally occur at low percentages in the population. Some research suggests that percentage of polyploids is greater in large than in small seeds and may vary with fruit position on the tree.

Selection of new citrus and related cultivars has been occurring for thousands of years, probably beginning in ancient China where superior phenotypes were selected in the wild and eventually cultivated. However, systematic, mission-oriented breeding programmes first began in Florida in 1893 with Swingle and Webber. Since then, numerous programmes have been developed worldwide with a variety of objectives. Unfortunately, citrus breeding is difficult and time consuming. Most citrus and related species are very heterozygous and few important traits show single gene inheritance patterns; therefore F_1 hybrids tend to exhibit variability. Furthermore, the common occurrence of nucellar embryony and absence of characteristic morphological marker genes make selection of hybrids difficult, although this situation is improving through the increasing availability of biotechnology methodology and genetic markers. Also, the protracted juvenility period of seedlings in the field (5–15 years) makes citrus breeding a very long-term, costly and land-intensive proposition. Consequently, most cultivars of worldwide importance until now have resulted from mutation or natural hybridization in the wild. For example, pink and red grapefruit were selected from limb sports of white or other coloured grapefruit and navel oranges are probably limb sports of round sweet oranges. Several earlier maturing satsuma and 'Clementine' selections have also resulted from mutation. Grapefruit and some mandarins like 'Temple' orange or 'Murcott' mandarin are most probably naturally occurring hybrids of oranges and mandarins. Natural hybridization in the wild is common. Many hybrids are perpetuated by nucellar embryony, and this is a contributing factor to the confusing species situation in citrus.

Nucellar Embryony

The percentage of nucellar seedlings ranges from 0 for zygotic species such as *C. maxima* and of clementine cultivars and 'Temple' orange to virtually 100% for 'Dancy' and 'Kara' mandarins. Most citrus and related species are highly heterozygous and thus produce widely variant sexual (zygotic) progeny. The continued production of weak, noncompetitive zygotic offspring due to inbreeding depression may have led over time to selection for higher levels of nucellar embryony. The tendency to produce nucellar compared with zygotic embryos is simple inheritance in a 1:1 ratio in some instances, but deviates from this ratio in other crosses depending on the species involved (Cameron and Soost, 1979).

In the past, degree of nucellar embryony was determined largely by visual ratings. In some instances, nucellar and zygotic embryos are difficult to separate in this way due to the relative lack of distinctive morphological genetic markers, especially when hybridizing closely related cultivars or species such as 'Pineapple' and 'Valencia' sweet oranges or when zygotic seedlings result from self-pollination of a relatively homozygous parent. During the 1970s, methods were developed to distinguish between embryo types using isozyme analysis (Torres *et al.*, 1978). Although this method is not foolproof, it is usually more accurate than visual methods. Xiang and Roose (1988) compared the relative degree of nucellar embryony of some selected citrus rootstocks based on isozyme analysis. Among 10 rootstock accessions the percentage nucellar embryony ranged from 49.4 for 'Yuma' citrange to 94.5 for 'Indio' rough lemon. Moreover, percentage nucellar embryony varied for different seed lots from within the same species, e.g. 84.8 vs 94.0% for CPB4475 citrumelo. Selections with high percentages of nucellar embryos are important for the production of uniform, true-to-type citrus rootstocks (see Chapter 3). Moreover, nucellar trees are usually virus-free because most viruses and viroids are not translocated to the developing embryo (Timmer and Benatena,1977; D'Onghia *et al.*, 2000).

Breeding Objectives for Citrus

Despite the difficulties in citrus breeding, several breeding programmes continue to endeavour to solve important problems limiting citrus production worldwide. Many programmes have separate personnel doing the scion and rootstock breeding because the objectives of each are usually different, although overall breeding objectives may combine aspects of both programmes to solve more universal problems, e.g. improved freeze-hardiness or virus resistance. Therefore the researchers usually work closely together. Further, most programmes have broad goals as well as specific local priorities. For example, all breeding programmes in Florida and Brazil now emphasize tolerance or resistance to HLB. In addition, scion breeders emphasize such programmes as improvement of fruit quality (higher colour, TSS or seedlessness in particular), or selection of early- or late-maturing cultivars to expand the marketing season. Improvement of fruit structure and storability are also of importance to expand distribution potential, but usually have a secondary priority. Certainly, freeze-hardiness is of major concern in some subtropical regions such as the southern USA with efforts being made to either increase inherent hardiness, or to change the basic structure of the tree to avoid freeze-damage, i.e. develop a deciduous citrus tree with edible fruit. It is also essential that the scion and rootstock be graft compatible.

Rootstock breeding naturally emphasizes soil-related objectives. However, improved freeze-hardiness or tree size control are also important objectives in many programmes. Emphasis is being placed on salinity and drought tolerance, and resistance to soil-borne organisms such as *Phytophthora* root and foot rot, and citrus and burrowing nematodes. Tolerance or resistance to viruses like CTV, which is usually a bud-union related problem, is also important. For example, sour orange, which once was the most widely planted rootstock worldwide, is sensitive to CTV and the development of a rootstock with all of the favourable characteristics of sour orange (high quality fruit, freeze-hardiness, adaptability to saline soils) without CTV problems would be highly desirable. It is now known to be a pummelo × mandarin hybrid (Nicolosi *et al.*, 2000; Nicolosi, 2007) and several breeding programmes are trying to duplicate the hybrid but with citrus tristeza virus tolerance. Of course, improvements in disease or soil-related characteristics cannot be made at the expense of yield, fruit quality or other desirable characteristics. Moreover, rootstocks should have seedy fruit with a high degree of nucellar embryony to facilitate standard clonal propagation by seed on a commercial basis (Castle, 1987). Rootstocks may also be propagated clonally by tissue culture when seed is unavailable, although this method is more costly than propagation by seed. In Florida, the tissue culture method is now being employed for propagation of potentially important rootstocks that appear to have tolerance to HLB (Grosser and Gmitter, 2014).

Regional breeding objectives encompass the above, but also consider local climatic and edaphic conditions. In many areas, the most widely grown cultivars are local selections chosen for their adaptability over a long period of time. An excellent example of localized breeding and selection has occurred with the satsuma mandarin (*C. unshiu*) in Japan and China. Over hundreds of years, cultivars have been developed via breeding and selection that differ primarily in their time of maturity, often by as little as a few weeks. This procedure has provided a long season of continuous supply of similar fruit for local and export markets.

Breeding Techniques

Traditional techniques

Most citrus breeding programmes have relied on traditional methods for developing new cultivars or rootstocks based on making controlled crosses by hand and selecting superior types from literally thousands of seedlings in the field (Hearn, 1985). Scion breeding involves selection of parental types with favourable heritable characteristics. Often seed parents are selected that produce only zygotic progeny. Parents are usually selected for high combining ability and thus are more likely to transfer favourable characteristics to their progeny. For example, cultivars such as 'Murcott' mandarin and 'Page' orange, while having several favourable traits, have poor combining ability and are thus of low value in a breeding programme (C.J. Hearn, unpublished). After hybrid seeds have matured they are extracted from the fruit and planted in the field. Seedlings are maintained until they begin to fruit, which often requires 10 years or more. Programmes are now using a method of accelerating breaking of juvenility by growing seedlings as single shoots to up to 3 m in height as rapidly as possible and then turning the shoot downward. This stimulates flowering and fruiting in as little as 2–3 years (Snowball *et al.*, 1994; Chaires, 2013). During this juvenile period, trees are evaluated for disease and pest resistance, freeze-hardiness (in locations where needed), and for overall growth characteristics. Many rootstock characteristics are now screened for as the seedlings are germinated or immediately after germination. Examples of this screening include for extreme soil conditions (i.e. salinity and pH) and *Phytophthora*. In general, the population of seedlings is extremely variable in vegetative and fruiting characteristics due to the great heterozygosity of citrus. The next step is to bud seedlings with the desired fruit and tree characteristics onto several rootstocks and transfer them to a field location for further evaluation. These trees usually fruit within 4 to 5 years, after which further fruit and tree evaluations are made. Finally, budwood of the new cultivars is released to nurserymen for further field testing. This process obviously is extremely time consuming. For example, 'Robinson' tangerine was released by the USDA in 1959. The original hybrid between 'Clementine' mandarin and 'Orlando' tangelo was made in 1942, and 17 years is a relatively short time in traditional citrus breeding for a new selection. This selection still showed production problems once it was released. A new procedure for field testing (Fast Track) is being tried in Florida and perhaps other areas (Chaires, 2013). Growers can apply for access to new scion and rootstock releases and evaluate them on their own to enhance the testing process. These growers then have priority access to larger quantities of the selection when it is completely released.

Traditional rootstock breeding follows the same general procedure as scion breeding with some exceptions (Barrett, 1985; Hutchison, 1985). Some targeted disease resistance, salinity tolerance or other useful characteristics are evaluated in lab-greenhouse tests to eliminate unsuitable selections. The promising initial hybrids are increased (often by tissue culture) and planted out for a seed source and further increased (again often by tissue culture). Then, promising candidates are budded with suitable scion(s) material and observed until past fruiting by 5 to 6 years. Trees to produce seed must produce sufficient seed numbers/fruit to be useable in the nursery and equally importantly must have a high degree of nucellar embryony and seedling uniformity. Potential rootstocks with these favourable qualities are then planted out in a wider number of locations and field-screened for resistance to soil-borne problems such as *Phytophthora* and nematodes. In many cases, screening tests are used at an early stage of development to determine salinity, pH, nematode and *Phytophthora* resistance (Grosser and Gmitter, 2013). Seedlings that pass these requirements are then budded with a common scion and plants are grown out for an additional 5–10 years to determine if they are compatible with the scion to monitor rootstock effects on fruit quality, freeze-hardiness (where necessary), yields and disease and pest resistance.

HLB tolerance may be screened by budding infected HLB shoots (3 to 4 buds in length) to young plants, usually budded. This entire process may take over 20 years. Therefore, it is not surprising that citrus breeders have been searching for new methods to shorten this costly and time-consuming process.

Biotechnology
Technological advances are solving or eliminating some of the limitations involved with classical citrus breeding, starting with isozyme analysis to distinguish hybrids from nucellar seedlings. This method is quite accurate for identifying hybrids, including most intra-specific hybrids, but cannot distinguish between closely related cultivars. Hybrids can be further separated into groups using RFLP analysis (see Taxonomy, p. 20). The RFLP method was used to develop maps of the citrus genome that were useful in locating genes that have a specific function(s). Marker genes were identified on specific chromosomes, some of which are tightly linked to other genes responsible for factors like disease resistance or freeze-hardiness. By mapping these markers, suppositions were made as to the inheritance of a particular linked gene. Therefore, the ultimate goal is to identify gene locations using molecular biology techniques and then use this information to identify hybrids with the desirable genes and to discard those that do not possess them. Such a system partially eliminates the need for making numerous crosses and screening many seedlings in order to identify favourable characteristics.

There are at least three methods to transfer potentially useful genes from one organism to another. These include bacterial mediated transfer, osmotic pressure cell fusion or projectile (gun) injection of gene particles (Anon, 1997). Some of the details and complications of transformation in *Citrus* are reported by Febres *et al.* (2011).

This procedure, for example, is being used as a means of introducing CTV resistance into sour orange rootstock (Febres *et al.*, 2007). Although the techniques, termed 'genetic transformation', are complex, the basic features are described in Fig. 2.5. The gene to be introduced is transferred to a bacterial plasmid which serves as a carrier. A plasmid is a circular piece of DNA that can be cleaved at particular sites after which the 'new' gene is spliced into the structure. A specific bacterial strain is then used to introduce the gene and a small amount of bacteria DNA into a tissue culture of the plant, in this instance 'Carrizo' citrange. This DNA is incorporated into the 'Carrizo' DNA. The 'Carrizo' tissue then becomes 'transformed' and begins producing specific proteins coded by the introduced DNA, hopefully without producing any unwanted, undesirable changes in the plant. The transformed plant tissue is cultured on a selective media until shoots are regenerated. Shoots are then rooted in media, after which the transformed plant may be transferred to the greenhouse or field. Such transformed plants have been produced in tobacco, petunia and many other crops including citrus. Plants may also be transformed by introducing DNA directly into the target cell using a high velocity 'gene gun'. In this instance, pieces of DNA are literally fired into the target cell, thus eliminating the use of a bacterial strain for introduction. This technique has had limited use with citrus trees. An apparent successful HLB tolerant sweet orange has been produced by introducing a spinach bacterial resistance gene using the bacterial insertion method (E. Mirkov as quoted by R. Santa Ana, 2012). These plants have been evaluated for several years under controlled field conditions and still have HLB resistance.

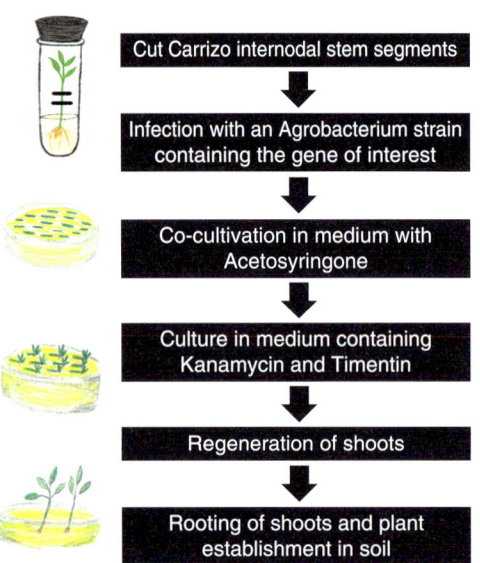

Fig. 2.5. An example of the basic processes involved with genetic transformation of citrus trees. *Agrobacterium* mediated genetic transformation of citrus epicotyl segments, M. Dutt and J. Grosser, University of Florida, CREC, Lake Alfred. FL, USA (as described by Dutt and Grosser, 2009).

Although many useful citrus genetically modified organisms (GMOs) have been produced, it is not clear if they can be used commercially since low consumer acceptance in many markets is still a major limitation to their use. Transforming rootstocks rather than scion varieties has some possible benefit for easier consumer acceptance since the fruiting part of the tree is not transformed.

Other biotechnology methods have also been developed which are currently being used to circumvent problems associated with conventional breeding. For example, some genera closely related to true citrus may have characteristics that would be desirable if introduced into citrus. An example is the *Phytophthora* and citrus nematode resistance found in *Severinia buxifolia*, a genus within the *Citrus* subtribe, but not in the true citrus group. Since *Severinia* and sweet oranges are sexually incompatible, another system had to be devised to produce hybrids. In the 1980s, techniques were developed to circumvent the sexual process by using protoplast fusion (Grosser and Gmitter, 1990). Protoplasts (individual cells lacking the cell wall) are usually extracted from ovule tissue from one and leaf tissue from the other of the two parental species. The protoplasts are cultured on artificial media in the laboratory under carefully controlled conditions. Protoplasts of the two species are placed in a mixture containing polyethylene glycol, which, when removed from the media, causes aggregation of the protoplasts. Protoplast fusion occurs, allowing for the unrestricted movement of genetic material between sexually incompatible species. The newly fused protoplasts are grown on selective media until embryos develop. The embryos are nurtured into plantlets and can then be transferred to the greenhouse where the plant is allowed to grow before transfer to the field for further testing. Protoplast fusion, like classical field breeding, is also a long-term proposition but offers a viable alternative to the problem of incompatibility within *Rutacae* genera of citrus. Hybrids developed in this manner are tetraploids (4×), however, and may not develop in the same manner as their diploid counterparts, but several useable rootstocks have been produced (Grosser *et al.*, 2003).

Mutation and selection
Citrus and related genera are, in general, extremely prone to mutate, although the extent of mutability varies among species. Mutations may occur in single buds or limbs, as a portion of an entire tree, or as the entire tree. Most mutations produce unfavourable or undesirable traits such as abnormally thick peels or dried sections within the fruit. Mutations may also arise, however, that produce earlier- or later-maturing fruit, or fruit with more acceptable juice colour or fruit characteristics.

There are also man-made mutations. Budwood or seeds are treated for short periods with gamma irradiation to induce mutations without killing the cells (Hearn, 1984, 1986). In general, seeds (LD_{50} = 0.1–0.15 Gy) are more tolerant of gamma irradiation than buds (LD_{50} = 0.05–0.09 Gy). The resulting seeds or buds are planted out or propagated by tissue culture techniques. A well-known example of the success of seed irradiation resulted from the 1959 irradiation of 'Hudson' red grapefruit seed by R. Hensz in Texas. Although many undesirable plants resulted, 'Star Ruby' grapefruit was selected in 1970 for its extremely deep red fleshed fruit colour. Other examples include the development of seedless oranges and grapefruit from seedy cultivars. The disadvantage, naturally, to this approach is that random chromosome damage can lead to undesirable characteristics and the need for large populations of plants and long-term field selection of the resultant mutations. Additionally, such mutations also may be unstable.

Chimeras
A chimera is a mixture of tissues of genetically different constitutions on the same part of the plant. This difference usually arises due to mutation, but also may occur due to irregular mitosis, somatic crossing-over of chromosomes, or artificial fusion of unlike tissue resulting from grafting (man-made). Several types of chimera exist in plants, which vary by virtue of the relative positions of the dissimilar tissues within the plant organ. The most commonly found forms in citrus are sectorial, periclinal and mericlinal chimeras. Sectorial chimeras involve dissimilarities among one or more sections within the tissue. They are most apparent in fruit where a specific region, not always corresponding to a segment, is of dissimilar tissue. Commonly, these sections will have different colouration or degrees of resistance to disease or mite damage. A sectorial chimera usually extends from the peel into the seed; consequently, seeds selected from these sectors are genetically similar to other tissues of the chimera. Chimeral selections from citrus with this characteristic have been made and are being further evaluated (Bowman *et al.*, 1991) but, with HLB priorities, low priority seems to be allocated to this work. Periclinal chimeras consist of an outer sector surrounding an inner sector of dissimilar tissue. Again, using fruit as an example, the peel tissue (exocarp) is of one type and the other internal tissues of other dissimilar types. In this instance, a plant grown from seed is genotypically like the inner, not the outer, tissue. Researchers are exploring tissue culture methods of removing cells from the outer tissues and culturing them *in vitro* with the idea of producing embryos and eventually plants having the genetic characteristics of the outer tissue (unpublished). Such a technique is potentially valuable in selecting and isolating tissues with superior colour or resistance to diseases, mites or insects. Mericlinal chimeras consist of incomplete sectoring not affecting the internal sectors of the tissue. Thus, although external tissues show differing characteristics, e.g. darker peel colouration, the seeds will produce plants similar to the inner tissues. For example, seeds of 'Pink Marsh' grapefruit will produce white 'Marsh' trees (mericlinal); whereas, seeds of 'Redblush' grapefruit produce 'Redblush' trees (sectorial). Chimeras are common in grapefruit and are important sources of pink and red peel and flesh colouration (see Grapefruit, p. 35).

A variation of forced mutations is to grow embryos from tissue culture and evaluate them for changes in characteristics of interest. An example is altered maturity dates and fruit characteristics in 'Valencia' oranges produced by somatic embryogenesis (Grosser et al., 1997). This process has produced selections of 'Valencia' orange that mature as early as 'Hamlin' with superior colour and soluble solids production (http://www.ftsp.net/varieties/citrus/valquarious-st14w-62).

With new methods and the progress that has been made, cultivars to meet the needs of the citrus industry are currently being produced and new selections by further expansion of methodology will be forthcoming. The recent breakthroughs in understanding the citrus genome and ability to create citrus types from this information may lead to new designer cultivars that meet very specific requirements.

3

ROOTSTOCKS

It is likely that, for thousands of years, citrus trees were grown as seedlings. Seedling trees, however, usually have a protracted juvenile period. Juvenility is undesirable because trees are nonproductive and excessively vigorous, resulting in upright growth with thorny branches. A protracted juvenility period is particularly undesirable in current times where the emphasis has been on producing a crop and returning an investment as soon as possible after planting. Most seedling trees are also susceptible to several soil-related problems, in particular *Phytophthora parasitica* and burrowing and citrus nematodes, and rootstocks may not be true-to-type for some cultivars with low rates of nucellar embryony (see Chapter 2).

Consequently, most citrus orchards worldwide consist of two-part (usually budded) trees that combine favourable attributes of the scion and rootstock. Rootstock selection is a major consideration in every citrus-growing operation. It is fundamental to the success of the orchard because the rootstock chosen will become the root system of the budded tree. Besides supporting the tree, the root system is responsible for absorption of water and nutrients, providing storage of carbohydrates produced in the leaves and synthesis of certain growth regulators, adapting the scion to particular soil conditions, and potentially providing tolerance to some diseases. More than 20 horticultural characteristics are influenced by the rootstock, including tree vigour and size, depth of rooting, freeze tolerance, adaptation to certain soil conditions, such as high salinity or pH, or excess water, resistance or tolerance to nematodes and diseases like *Phytophthora* foot rot and citrus blight, and fruit yield, size, texture, internal quality and to some extent maturity date (Castle, 2010). Some problems like sensitivity to excess copper or damage due to armillaria root rot cannot currently be remedied by choice of rootstock.

There is no perfect rootstock, even for a particular situation. Choice of rootstock should be based on the most important limiting factor(s) to production in a particular region, local climate and soil conditions, cultivar and intended use (fresh or processed) of the crop. For example, sour orange should not be used as a rootstock in areas such as Spain, Brazil, California, South

Africa or the Caribbean Basin where citrus tristeza virus (CTV) and its efficient vector, *Toxoptera citricida*, are prevalent and will kill or seriously debilitate trees. Rootstocks which import excessive vigour to the scion, like rough lemon or 'Palestine' sweet lime, should be avoided in areas like northern Florida and Texas that are susceptible to regular freeze-damage, and rough lemon rootstock should be avoided in areas known to have blight. Scions budded on *Poncirus trifoliata* and most citranges and citrumelos perform poorly in high pH soils. There are many other examples which will be discussed in the following sections, further demonstrating that identification of the most limiting factors in a given area related to rootstock is the first and probably most important step in choosing a rootstock. In almost all cases this selection requires weighing good and adverse characteristics of the rootstocks.

Local climatic and soil conditions are a first consideration in rootstock selection. Sour orange is used almost exclusively in Texas due to its adaptability to alkaline and saline soils. In contrast rough lemon, with its excellent drought tolerance, is widely planted in the sandy soils of South Africa and Australia. *Poncirus trifoliata* is well-adapted to the cool growing conditions and acid soils found in Japan and central China and 'Carrizo' citrange is an important rootstock where burrowing nematodes are a problem. Consequently, many citrus regions have limited their rootstocks to a few selections based on local conditions and traditions, and this is potentially very dangerous if a new problem arises that the current rootstock cannot tolerate.

Cultivar and intended use (fresh or processing) are also important for rootstock selection. 'Cleopatra' mandarin ('Cleo') is well-suited for use with mandarins and mandarin-hybrids ('Temple', 'Robinson', 'Nova', 'Sunburst'), but generally sweet orange and grapefruit cultivars on 'Cleo' produce small fruit and are not precocious. 'Cleo', however, provides CTV tolerance and moderate blight tolerance. Rootstocks that impart high vigour to the scion (e.g. rough lemon, '*Citrus volkameriana*', 'Palestine' sweet lime, and 'Rangpur' lime) generally induce high yields, but produce relatively poor quality fruit, lower in total soluble solids (TSS) and acids, with coarse peels, and are more acceptable for processing than for the fresh market (Castle, 1987). Conversely, grapefruit and sweet orange trees on sour orange, 'Troyer' and 'Carrizo' citranges and 'Swingle' citrumelo rootstocks typically produce high quality fruit suitable for the fresh and processing markets.

Important horticultural characteristics of the major rootstocks worldwide are discussed in the following section in more detail. Their characteristics as observed in Florida are summarized in a 2016 chart with horticultural, disease and pest ratings (Castle *et al.*, 2016; https://edis.ifas.ufl.edu/pdffiles/HS/HS126000.pdf). Ratings given in the table are based on years of observations and controlled comparisons of rootstock characteristics for 24 established rootstocks and for an additional 21 newer rootstocks from the USDA and University of Florida citrus rootstock breeding programmes. Most of the later additions have some HLB tolerance but limited field evaluation. These ratings under Florida conditions are not representative of every growing condition worldwide. Climatic conditions, cultivar and soil characteristics, in particular, may alter some of the horticultural characteristics of each rootstock. For additional information on citrus rootstocks refer to reviews by Wutscher (1979) and Castle (2010).

DESCRIPTION OF TRADITIONAL CITRUS ROOTSTOCKS

Rough Lemon

Rough lemon (*Citrus jambhiri* Lush.), which is native to northeastern India, is probably a natural hybrid because of its high degree of polyembryony compared with other lemon species (Barrett and Rhodes, 1976). The fruit has, as the name implies, a very coarse exterior, and rough lemon is unsuitable as a scion cultivar; however, it has been widely used in many countries as a rootstock. The adaptability of this selection as a rootstock for deep infertile sands prompted its use in the ridge (central) area of Florida, Australia and South Africa. However, the high susceptibility of rough lemon to blight and its lack of freeze-tolerance (Florida) has virtually eliminated its use in Florida and Brazil.

Sweet orange, grapefruit, mandarin and lemon trees on rough lemon rootstock are large, extremely vigorous and very productive in most rootstock trials worldwide, particularly in deep, sandy soils (Castle, 1987, 2010). Yields of mature grapefruit trees in areas like Florida have attained 80–100 tonnes ha^{-1} in some instances. The root system is extensive, sometimes reaching a depth of 4.6 m in deep sands (Castle and Krezdorn, 1973). Consequently, mature trees on this rootstock are also very drought tolerant. Rough lemon rootstock is moderately tolerant of high salinity. In addition, it is well-adapted to a wide range of soil pH.

Scions on rough lemon are very susceptible to freeze-damage due to their vigorous growth over a wide temperature range (Yelenosky and Young, 1977). Scions on rough lemon-type rootstocks have been consistently more damaged by freezes than those on other rootstocks if soil temperatures remain high, >15°C. Trees are poorly freeze-acclimated possibly because of the higher root conductivity of rough lemon at lower root temperatures compared to 'Carrizo' citrange (Wilcox and Davies, 1981). In Mediterranean-type climates where winter soil temperatures are low, citrus trees on rough lemon became moderately freeze-hardy. Regrowth of scions on rough lemon following freeze-damage is usually very vigorous and thus the tree recovers more rapidly than on other rootstocks. Rough lemon is tolerant to CTV, citrus exocortis viroid (CEV) and citrus xyloporosis viroid, although some dwarfing may occur in scions infected with the last two diseases (Wutscher, 1979). There is some evidence from South Africa, however, that severe strains of CTV may cause stem pitting even on rough lemon. It is highly susceptible to foot rot (*Phytophthora*), a factor which limits its use in many regions, and it is susceptible to damage by both the citrus and burrowing nematodes (O'Bannon and Ford, 1977). 'Milam', a probable rough lemon hybrid, produces similar characteristics in the scion as rough lemon but is resistant to the burrowing nematode. Most importantly, rough lemon is highly susceptible to blight, probably a virus disease responsible for extensive tree losses, particularly in Florida (Young *et al.*, 1982; Schneider *et al.*, 2015). This is the primary reason for rough lemon losing favour with Florida growers.

The excessive vigour imparted to the scion by rough lemon generally produces poor quality fruit. Both TSS and titratable acidity (TA) tend to be low in fruit of trees on rough lemon (Castle, 1987). For example, TSS for 'Valencia' on sour orange averaged 13.2% while on rough lemon TSS averaged 11.4% (Gardner and Horanic, 1961). Moreover, the peel tends to be thick and puffy, especially for fruit on vigorously growing young trees. Mandarin fruit in particular from trees grown on rough lemon rootstock tend to be puffy and hold poorly on the tree. Re-greening of 'Valencia' oranges is more severe on rough lemon than on other rootstocks. Fruit size is generally quite large for all cultivars grown on rough lemon. Use of rough lemon for inherently small-fruited cultivars like 'Dancy' mandarin was advantageous in some situations, but granulation tends to be more severe on rough lemon. In many citrus regions fruit produced on rough lemon does not attain sufficient quality for the export fresh fruit market but total kg-solids production per tree and hectare are higher for scions budded on rough lemon than for most other rootstocks (Castle, 1987). The high production of kg-solids (termed pounds- solids in Florida) per hectare also produces high returns for the processing market. Total kg-solids per hectare is a function of yield, juice content and TSS.

Citrus volkameriana

Citrus volkameriana Ten. and Pasq. ('Volkamer' lemon) is also a lemon hybrid which as a rootstock produces large, vigorous trees yielding large quantities of moderate to poor quality fruit like rough lemon (Castle, 1987). Scions on 'Volkamer' lemon are slightly more freeze-hardy than those on other lemon types. They are not susceptible to CTV, xyloporosis or CEV, but are susceptible to blight and the citrus and burrowing nematodes (Roistacker, 1991; Castle, 2015). However, cultivars on this rootstock are tolerant to malsecco and *Phytophthora* under most circumstances (Carpenter *et al.*, 1981).

'Volkameriana' is not widely used as a rootstock due to its similarities to rough lemon, particularly susceptibility to citrus blight. Nevertheless, some studies from Florida suggest that yields and net profits over the long-term are higher for trees on 'Volkamer' than for other less vigorous rootstocks such as 'Swingle' citrumelo even when tree losses due to blight or other factors are moderate (Castle, 2010).

Citrus macrophylla (Alemow)

Citrus macrophylla Wester is a hybrid species, possibly of *Citrus celebica* and *Citrus grandis*, native to the Philippines (Barrett and Rhodes, 1976). It is very similar morphologically and genetically to lemons and limes. Cultivars budded on *C. macrophylla* produce large, vigorous and high-yielding trees with growth characteristics similar to those on other lemon-type rootstocks under most growing conditions (Castle, 1987). In contrast, Carpenter *et al.* (1981) found 'Eureka' lemon trees on *C. macrophylla* were smaller than those on 'Swingle' or *C. volkameriana*. Fallahi and Rodney (1992) observed that 'Fairchild' mandarin trees were more precocious on *C. macrophylla* than on *C. volkameriana* or 'Carrizo' citrange. Moreover, yields of 'Valencia' orange on *C. macrophylla* were similar to, or better than, those on other lemon types and superior to those produced on trifoliate and trifoliate hybrids (Castle, 1987; Fallahi and Rodney, 1992). Lemons and limes also yield very well on this rootstock under most situations (Castle, 1987). Scions budded on *C. macrophylla* grow well on both sandy and high pH, calcareous soils. Trees on *C. macrophylla* have a deep, dense root system that imparts drought tolerance to the scion.

C. macrophylla itself and scions budded to it are freeze-sensitive. Scions budded on *C. macrophylla* are more freeze-tender than those on rough lemon and are far less freeze-hardy than scions on sour orange, 'Cleopatra' mandarin, trifoliate orange or 'Swingle' citrumelo (Castle, 1987). As with rough lemon, regrowth is very rapid following moderate freeze-damage.

Scions on *C. macrophylla* (except lemons) are susceptible to CTV and xyloporosis, although to a lesser extent than sour orange or 'Palestine' sweet lime, respectively (Castle, 1987). *C. macrophylla* is more tolerant to foot rot than true lemons (Carpenter *et al.*, 1981), but is not tolerant of citrus or burrowing nematodes, and is moderately susceptible to blight (Castle, 1987, 2015).

Fruits of sweet oranges and grapefruit budded on *C. macrophylla* are generally large. They may become over-large and puffy especially on young trees. Lemon and lime fruits grown on this rootstock are also large, which is a favourable characteristic. Fruit quality is moderate to poor, similar or slightly better than that on other lemon-type rootstocks. Generally, TSS, TA and TSS:TA ratios are lower on this rootstock than on sour orange, 'Carrizo' citrange or 'Swingle' citrumelo (Castle, 2015).

C. macrophylla is an excellent rootstock for lemons and limes but is not widely used for sweet oranges, grapefruit and mandarins. However, 'Eureka' lemon trees budded on *C. macrophylla* may develop a disorder called *C. macrophylla* rootstock necrosis. Blockage of the sieve elements leads to necrosis and eventually tree decline and death (Schneider *et al.*, 1978). There is evidence that *C. macrophylla* is better adapted to cool, dry climates. When used as a rootstock in such climates, e.g. Spain, many cultivars on *C. macrophylla* out-yield those on citranges.

'Rangpur'

'Rangpur' (*Citrus reticulata* hybrid) is a mandarin-type hybrid (Wutscher, 1979). 'Rangpur' is not widely planted in most citrus areas with the exception of Brazil where it is the most important rootstock primarily because of its tolerance to CTV, drought and acid soils. Cultivars budded on 'Rangpur' are moderately vigorous and yields are similar or slightly lower than those produced on rough lemon (Castle, 1987) but greater than yields found on most citranges, trifoliata and 'Cleopatra' mandarin rootstocks. 'Rangpur' is moderately tolerant of high saline and calcareous soil conditions.

Fruit quality and freeze-tolerance are intermediate to those of lemon types and sour orange. 'Rangpur' is resistant to CTV, but susceptible to CEV, the citrus and burrowing nematodes and moderately susceptible to foot rot. Observations in Brazil show that 'Rangpur' is very susceptible to blight (Lima, 1982). Although sweet orange trees on 'Rangpur' have a number of favourable characteristics it has not become an important rootstock outside of Brazil.

'Palestine' Sweet Lime

'Palestine' sweet lime (*Citrus limettoides* Tan.), which is probably a hybrid rather than a true lime, is of minor importance worldwide as a rootstock (Castle, 1987). Cultivars budded on sweet lime are quite similar in vigour, yields and fruit quality characteristics to rough lemon and marginally superior to lemon types in freeze-tolerance. Trees on 'Palestine' are moderately susceptible to CTV and foot rot, and may be stunted by CEV and particularly xyloporosis (Wutscher, 1979). In fact, 'Palestine' sweet lime, formerly a major rootstock in Israel, was largely replaced due to its susceptibility to xyloporosis. Trees on 'Palestine' sweet lime are also susceptible to citrus blight.

Sour Orange

Sour orange (*Citrus aurantium* L.) was the most widely planted rootstock in the world. However, susceptibility of sweet orange on sour orange to CTV has greatly decreased its use for new plantings in Australia, Argentina, Brazil, California, Spain, South Africa, Florida and southern Caribbean countries. Sour orange is an excellent rootstock for areas free of CTV and particularly for fresh fruit production (Castle, 1987).

Cultivars grown on sour orange produce trees of moderate vigour and moderate to large size (Hutchison, 1977). Trees on sour orange grow somewhat more slowly than those on rough lemon types but are certainly not dwarfed. Yields on a mass or kg-solids basis are also adequate, but less than those on rough lemon (Hutchison, 1977), *C. macrophylla* or *C. volkameriana* on sandy soils (Castle, 1987).

Sour orange produces a deep and moderately branched root system and scions budded on this rootstock are moderately drought-tolerant. Scion vigour on deep sandy soils is certainly not as good as on more fertile soils where water and mineral elements are not limiting. Sour orange rootstock is commonly used in heavy or poorly drained soils due to its moderate *Phytophthora* tolerance, although it is not as physiologically tolerant of flooding as rough lemon rootstock (Syvertsen *et al.*, 1983). Trees on sour orange rootstock are particularly well-adapted to high pH and high salinity soils (Wutscher, 1979).

Cultivars budded on sour orange are about as freeze-hardy as those growing on 'Cleopatra' mandarin, *P. trifoliata*, or 'Swingle' citrumelo when fully acclimated, but are appreciably superior to lemon types and citranges (Yelenosky and Young, 1977). Regrowth of freeze-damaged scions, however, is slower for trees on sour orange compared with those on rough lemon or other vigorous rootstocks (A.H. Krezdorn, unpublished).

Tree stunting, bark sloughing or stem pitting commonly associated with CEV and xyloporosis do not occur when sour orange is used as a rootstock. *Phytophthora* foot rot is usually a minor problem for trees on sour orange rootstock. Nevertheless, in some countries such as Mexico and Spain trees are budded high to avoid foot rot problems, and sour orange trees are moderately susceptible to root rot (Timmer *et al.*, 1991). Sour orange is susceptible to damage by the burrowing and citrus nematodes (O'Bannon and Ford, 1977) but is one of the rootstocks least susceptible to citrus blight (Young *et al.*, 1982).

Fruit size of cultivars on sour orange is somewhat smaller than on rough lemon, but larger than on 'Cleopatra' mandarin. Both the TSS and TA of the juice are high, thus sour orange has been the preferred rootstock in fresh fruit-producing regions such as Spain, the Indian River area of Florida, and Texas (Castle, 1987). High TA frequently results in fruit meeting maturity standards later than fruit from trees on more vigorous rootstocks. In contrast, in certain cultivars like 'Hamlin' orange for which the limiting maturity factor is low TSS, fruit will attain earlier maturity on sour orange than on a rootstock like rough lemon. Total soluble solids for fruit from scions on sour orange often average 0.5 to 1.5% higher than for those on rough lemon. This characteristic along with moderate *Phytophthora* tolerance has made sour orange the preferred rootstock in tropical countries where CTV is not limiting. After losing most sour orange rooted trees in Florida to CTV, some growers are replanting on sour in the hope that these trees will survive. They are assuming that brown citrus aphid control will be equal to that of the citrus psyllid, the vector for HLB, due to the heavy spray programme for the psyllid. This is probably not a wise decision.

The peel of fruit from cultivars grown on sour orange is generally smooth and thin. For this reason, excessive splitting can occur, although the problem is less severe than for the same cultivars on 'Cleopatra' rootstock (A.H. Krezdorn, unpublished). While splitting can be a serious problem with sweet oranges and some mandarins or mandarin-hybrids, it is not important in either grapefruit or 'Temple' oranges and generally is not severe enough to limit its use as a rootstock.

There has been interest in using selections of sour orange such as 'Bittersweet' and *Citrus taiwanica*. 'Bittersweet' has been available as a rootstock for many years. Studies from the 1960s in Florida suggest that most sour orange selections have similar effects on scion fruit quality and performance, and therefore there appears to be no advantage to using one type of sour orange over another. 'Bittersweet' is not tolerant of CTV but may be more tolerant of *Phytophthora* than sour orange and *C. taiwanica* was thought to be CTV-tolerant, although this has not been substantiated (Castle, 1987). Wutscher and Dube (1977), in summarizing 30 years of rootstock studies with red grapefruit in Texas, observed slightly higher yields for 'Bittersweet' over *C. taiwanica* and sour orange.

Recent genetic findings that sour orange is a mandarin × pummelo hybrid have led to attempts to recreate sour orange with its good characteristics but with CTV tolerance. An example of such an approach is somatic hybridization of mandarin + pummelo (Grosser *et al.*, 2004).

'Smooth Flat Seville' (Australian Sour)

'Smooth Flat Seville' (pronounced se'vil), which originated in Australia, is probably a hybrid of pummelo, sweet orange and sour orange (Barrett and Rhodes, 1976). Its characteristics have not been widely compared to those of other rootstocks, and 'Seville' has been tested as a rootstock in only a few experimental plantings. Yields and vigour of trees budded on this rootstock are moderate for sweet oranges and moderate to good for grapefruit. Fruit quality is slightly poorer than on sour or trifoliate orange but superior to that on lemon types. Trees on 'Seville' are not susceptible to xyloporosis and CEV but are moderately susceptible to CTV and *Phytophthora*, although there is some question about its tolerance to these diseases (Castle, 2015). 'Smooth Flat Seville' is also susceptible to citrus and burrowing nematode damage. Scions budded on Seville appear to have good blight tolerance.

'Cleopatra' Mandarin

'Cleopatra' mandarin (*C. reticulata* Blanco) is of minor importance as a rootstock on a worldwide basis; however, it has several favourable attributes which have increased its use. Scions budded on 'Cleopatra' mandarin are large and moderately vigorous (Hearn and Hutchison, 1977), producing a deep, densely branched root system that imparts moderate drought tolerance. 'Valencia' and 'Parson Brown' sweet oranges on 'Cleopatra' rootstock had moderate yields during a 17-year study in Florida (Gardner and Horanic, 1961), being less than those for trees on rough lemon but greater than yields on sour orange rootstock. Similarly, yields of 'Ruby Red' grapefruit were comparable with those on sour orange but considerably lower than those on rough lemon. Moderate yields resulted from poor fruit set and size and splitting of mature fruit (A.H. Krezdorn, unpublished). Scions on 'Cleo' are not precocious, which is a major factor limiting the selection of 'Cleo' as a rootstock, but they attain moderately large size and yields 10–15 years after planting (Gardner and Horanic, 1961). In south Florida, improvements in irrigation and fertilization practices resulted in faster growth and attainment of moderate yields at an earlier age for trees on 'Cleopatra' mandarin rootstock (see Chapter 5).

Trees on 'Cleopatra' mandarin are as freeze-hardy as those on sour orange, *P. trifoliata*, or 'Swingle' citrumelo when fully acclimated, and are appreciably superior to trees on rough lemon or 'Carrizo' citrange (Yelenosky and Young, 1977). Regrowth following freeze-damage is less than that of rough lemon but comparable to that observed for sour orange.

A major advantage of 'Cleo' over many other rootstocks is its tolerance of the major citrus virus (viroid) diseases (Wutscher, 1979). 'Cleopatra' mandarin is tolerant of CTV, CEV and xyloporosis, displaying none of the typical symptoms associated with these problems. There are observations that 'Cleopatra' is not completely tolerant of xyloporosis, but such cases seem to be exceptions. 'Cleopatra' is susceptible to both burrowing and citrus nematode damage and has moderate *Phytophthora* foot rot and poor root rot tolerance because damaged roots regrow very slowly. 'Cleopatra' usually reaches an age of 12–15 years before losses to blight occur, but after this age blight losses can become fairly high (Young *et al.*, 1982).

'Cleopatra' mandarin rootstock is also adapted to a wide variety of soils ranging from light sands to heavy clays, although scions budded on it are most productive on heavier soils. It is resistant to high salinity and is more tolerant of high pH, calcareous soils.

Fruit size of cultivars on this rootstock is consistently smaller than that of trees on other commercially important rootstocks (A.H. Krezdorn, unpublished). Juice of fruit produced on 'Cleopatra' is of moderately high quality. Total soluble solids are usually intermediate between those of sour orange and rough lemon (Castle, 1987), although Economides (1976) found that TSS was higher for 'Marsh' grapefruit on 'Cleo' than on sour orange and rough lemon in Cyprus. The peel is smooth and thin, which is apparently related to the excessive splitting commonly found in fruit produced on 'Cleopatra'. Splitting can be a serious problem with sweet oranges in some years, but is unimportant for grapefruit.

'Cleopatra' is not widely used as a rootstock for sweet oranges and grapefruit, or for some small-fruited mandarins like 'Dancy'. In contrast, 'Cleopatra' is an excellent rootstock for 'Temple' and is widely used for mandarin hybrids such as 'Orlando', 'Nova', 'Murcott' (Honey), 'Robinson', 'Sunburst' and 'Minneola' (Hearn and Hutchison, 1977). The size and quality of the fruit of these cultivars on 'Cleopatra' are excellent; however, with the exception of the self-fruitful 'Murcott' and 'Temple', the yields are low unless adequate cross-pollination with a compatible cultivar is provided.

Although sweet oranges and grapefruit are not precocious when budded on 'Cleo', tree survival and longevity are usually very good, particularly in areas where citrus blight and CTV are prevalent. Brazilian and Venezuelan growers have planted 'Cleo' as an alternative to more citrus blight susceptible rootstocks since sour orange cannot be used because of its sensitivity to CTV. In these cases, 'Cleo' rooted trees have usually been exceptionally vigorous and not good bearers.

Sweet Orange

Use of sweet orange (*Citrus sinensis* [L.] Osb.) as a rootstock is based on previous favourable performance in California, its moderate to high tolerance to citrus blight and CTV (Young *et al.*, 1982), and the possibility of re-growing trees from their own roots after freezes. Sweet orange-rooted trees are not precocious but eventually become vigorous and moderately productive. Gardner and Horanic (1961) observed that yields of 'Valencia' sweet oranges on sweet orange were similar to those on 'Cleopatra' mandarin, sour orange and grapefruit, but considerably less than those on rough lemon rootstock. Fruit quality is intermediate between lemon types and sour orange rootstocks. Sweet orange has not been commonly used as a rootstock in most of the world except in Brazil and Australia because of limited drought tolerance and extreme foot rot susceptibility (Castle, 1987). Nevertheless, use of improved micro-irrigation systems and systemic fungicides may permit expanded use of sweet orange, but most interest is on newer rootstocks. In limited tests in Florida and Texas, own-rooted cuttings of sweet orange have performed fairly well, although long-term data are lacking. Scions budded on sweet orange are not susceptible to CEV or xyloporosis, but most selections are adversely affected by burrowing and citrus nematodes (O'Bannon and Ford, 1977).

Many old groves in Florida were seedling trees over 70 years of age in the 1970s. No data were collected on their productivity but some groves were still productive until the 1980 freezes.

Trifoliate Orange

The trifoliate orange (*P. trifoliata* [L.] Raf.) is widely used as a rootstock for satsuma mandarins and sweet oranges in citrus areas in Japan, China, Argentina and Australia. A large collection of trifoliate selections was assembled in Australia, but apparently not highly evaluated. Most orchards on trifoliate selections in Central China have relatively small trees, although partly from limited available soil on typical hillsides used for planting. Trifoliate orange differs from other rootstocks because many selections are available that have variable effects on scion characteristics. This fact probably accounts for variations in yields and disease responses reported for scions budded on trifoliate orange. Trifoliate orange can be a dwarfing rootstock as in the case of 'Flying Dragon'. A few selections of trifoliate produce very large trees, observed in Florida and Brazil. Dwarfing of scions on this rootstock, however, is often due to infection with CEV (Cohen, 1968), although certainly most scion cultivars budded on trifoliate orange are less vigorous than those on lemon-type rootstocks. Moreover, some strains of CEV do not cause bark scaling but still dwarf the tree. In fact, inoculation with CEV has been used commercially to dwarf trees on trifoliate orange in countries such as Israel and Australia (Bevington and Bacon, 1977). Yields of most scion cultivars budded on trifoliata are less than those on rough lemon, 'Rangpur' and sour orange in tropical and subtropical areas, primarily due to differences in tree vigour (Castle, 1987). Nevertheless, yields of satsuma mandarin on trifoliata are moderate to good in cooler growing regions such as Japan and central China where yields approach 80 tonnes ha^{-1}.

Trifoliate orange itself is deciduous, becoming very dormant and extremely freeze-hardy. Trifoliate orange trees survive as far north as Long Island, New York (USA) (42°N latitude). This has led to the common misconception that trees budded on trifoliate orange are always much more freeze-hardy than those on other rootstocks. In subtropical regions such as Florida, trees become quiescent slightly later when on this rootstock than when budded on 'Cleopatra' mandarin or sour orange (Young, 1977). Thus, early in the winter scions budded on trifoliate orange may be slightly more sensitive to freeze-damage than when on the other two rootstocks (Young, 1977). Moreover, Yelenosky and Hearn (1967) observed that a series of freezes in a given year had a more detrimental effect on scions on trifoliate than on rough lemon-type rootstocks. In contrast, during periods of cool night-time temperatures (<10°C), trees on trifoliate orange cease root growth and become fully quiescent and equal or greater in hardiness to trees on sour orange, 'Cleopatra' mandarin or 'Swingle' citrumelo.

Trees budded on *P. trifoliata* are not affected by CTV or xyloporosis, but are susceptible to blight. Young *et al.* (1982) found less spread of citrus blight for citrus trees on trifoliata than those on rough lemon, but orchard-to-orchard variability was very high in the limited numbers of orchards sampled. However, plantings of trifoliate in Florida and Brazil are not widespread enough to make conclusive statements about its citrus blight susceptibility. Trifoliate orange in general is resistant to the citrus nematode but not to the burrowing nematode (O'Bannon and Ford, 1977). Not all selections of trifoliate orange, however, are resistant to the citrus nematode. Trifoliate orange is highly resistant to foot rot and other problems associated with poorly drained soils.

Scion cultivars budded on trifoliate orange grow poorly on infertile, sandy soils and are not drought-tolerant but grow quite well on moderately fertile sands; trifoliate orange is better adapted than most rootstocks to heavy, poorly-drained soils. Trifoliate orange, however, is not well adapted to high salinity and high pH, calcareous soils (Von Staden and Oberholzer, 1977). Leaves of scions budded on to trifoliate become quite chlorotic in high pH soils and growth and yields are decreased.

Fruit size of cultivars budded on trifoliate orange varies with soil type. Some studies indicate that fruit of trees on trifoliate orange is exceptionally large, but in general trees on this rootstock bear relatively small fruit. This may result from the fact that trifoliate sets large crops of fruit even under drought conditions with a concomitant reduction in fruit size. The juice of fruit produced on trifoliate orange is of excellent quality, rating as good or better than that on any other rootstock (Von Staden and Oberholzer, 1977). Cohen and Reitz (1963) observed that TSS and TA of 'Valencia' orange and 'Ruby Red' grapefruit were similar to those for fruit grown on sour orange and superior to those on rough lemon or 'Rangpur' rootstocks. In some instances, TA is higher for fruit grown on trifoliata, a factor which may delay maturity. The fruit peel is smooth and thin, resulting in somewhat more splitting than with fruit on rough lemon-type rootstocks.

Citranges

Citranges are intergeneric hybrids of sweet orange and trifoliate orange. The original crosses were made in Florida by W.T. Swingle, beginning in 1897 following the severe freezes of 1894–1895 to incorporate the freeze-hardiness of trifoliate orange into sweet orange. Several citranges have been tested as rootstocks, including 'Rusk', 'Morton', 'Savage', 'Benton', 'C-35', 'Carrizo' and 'Troyer'. The last two actually arose from the same cross between 'Washington' navel orange (seed parent) and *P. trifoliata* (pollen parent) made in 1909 (strictly speaking, these are citruvels, not citranges) (Savage and Gardner, 1965). Although seedlings of 'Troyer' and 'Carrizo' appear identical, some horticultural characteristics, like tolerance to burrowing nematode, differ, with 'Carrizo' being more tolerant. 'Troyer' is widely used in California and Spain, and 'Carrizo' has been a commonly used rootstock in Florida. Several other citranges have proved promising in rootstock trials worldwide but currently are not widely planted.

Scion cultivars budded on citranges produce moderately vigorous to vigorous trees, somewhat similar or larger than trees on sour orange, but generally smaller than those on vigorous rootstocks like rough lemon. 'Carrizo' and 'Troyer' citranges have become widely planted rootstocks for oranges and grapefruit for several reasons. Fruit are seedy and have a high incidence of nucellar embryony and thus are easily propagated as rootstocks (Table 3.1).

Table 3.1. Seed per fruit and percentage nucellar embryos for ten citrange rootstock cultivars (Hutchison 1977).

Cultivar	Scientific name	Seed/fruit	Nucellar embryos (%)
Citranges	*C. sinensis* (L.) Osb.		
Carrizo	× *P. trifoliata*	23	100
Cunningham		4	94
Morton[1]		1	100
Rusk		5	96
Savage[1]		14	100
Troyer		20	98
Uvalde		9	100
Willits		3	90
L-44-4		2	100
L-44-7		2	94

[1]The actual parentage is the reciprocal of the cross indicated.

In contrast, 'Morton' and 'Rusk' citranges are also highly nucellar but produce few seeds, a factor that limits the economic practicality of using these citranges as rootstocks (Hutchison, 1977). Trees on 'Carrizo' will grow moderately well on sandy and sandy-loam soils; however, they grow poorly on high pH soils (Wutscher, 1979) and apparently do better at pH of 6 to 6.5 particularly in the presence of HLB disease (Graham and Morgan, 2018). Citranges, like trifoliate orange and other trifoliate hybrids, produce a 'bench' (an overgrowth of the rootstock) at the bud union with most scion cultivars.

Wutscher and Dube (1977) found that yields of red grapefruit on 'Morton' and 'Troyer' citranges were similar to those of sour orange and 'Swingle' but superior to those of 'Milam' lemon when grown on soils with a pH ranging from 6.7 to 7.6. Similarly, sweet oranges grown on 'Troyer' citrange in South Africa (Von Staden and Oberholzer, 1977) or 'Carrizo' in Florida (Hutchison, 1977) also produced moderate to high yields. C-35 citrange from California has been evaluated in some trials.

The original reason for developing citranges was to produce a hybrid with edible fruit which was more freeze-hardy than sweet orange. However, citranges are far from commercial edible quality and scions budded on 'Carrizo' are generally of intermediate freeze-hardiness, depending on the time of a freeze (Yelenosky and Young, 1977). Freeze-hardiness is less than that of trees on sour orange, 'Cleopatra' mandarin or 'Swingle' citrumelo but generally better than on lemon-type rootstocks (Young, 1977). Trees on 'Carrizo' are not tolerant of early winter or late spring freezes in subtropical climates because they are slow to acclimate in the winter, and readily de-acclimate early in the spring.

Scion cultivars budded on most citranges are stunted by CEV; however, trees are not affected by CTV or xyloporosis. The extent of stunting varies with the strain of CEV present. 'Carrizo' citrange is tolerant to burrowing nematode (O'Bannon and Ford, 1977), i.e. nematode populations decline around the roots with time. Nevertheless, some 'Carrizo' seedlings, which are probably gametic, are burrowing nematode susceptible (D. Kaplan, unpublished) and therefore the source of rootstock material is important. 'Troyer' citrange is not tolerant of burrowing nematode. Neither citrange is tolerant of citrus nematode (O'Bannon *et al.*, 1977). 'Carrizo' is moderately susceptible to *Phytophthora* foot rot, particularly in young plantings. Citrus blight susceptibility is intermediate between rough lemon and sour orange (Young *et al.*, 1982), although tree losses due to blight have been extreme in some coastal flatwood areas of Florida (M. Cohen, unpublished), and as a replant for a tree lost to blight, trees on 'Carrizo' have been affected as early as 4 years of age (L.G. Albrigo, unpublished). Nevertheless, sweet oranges budded on 'Carrizo' are among the more profitable combinations over the long term in Florida (Castle *et al.*, 2010a,b).

Citrumelos

Citrumelos are intergeneric hybrids of grapefruit and trifoliate orange. The original crosses were made in Florida by Swingle in 1907, but since then several citrumelos have been produced. Currently, 'Swingle' citrumelo is still the most widely propagated rootstock in Florida and has gained in popularity worldwide since its release in 1974 (Hutchison, 1974). There is conflicting information about the vigour of trees on 'Swingle'. 'Swingle' has a semi-dwarfing effect on sweet orange trees, while grapefruit trees on 'Swingle' are quite vigorous (Wutscher and Shull, 1975). Orange trees on 'Swingle' tend to be larger than those on sour orange or mandarin rootstocks, however. These differences may be due to the grapefruit parentage resulting in better compatibility with grapefruit cultivars or due to the presence or absence of viruses, although 'Swingle' apparently is not adversely affected by CTV, CEV or xyloporosis. Lemon and lime cultivars do not yield well on 'Swingle' rootstock.

Scion cultivars budded on 'Swingle' grow well on sandy and loamy soils, but grow poorly on clays, high pH soils or in poorly-drained areas (Wutscher, 1979). Recent observations in Florida, with HLB presence suggest that 'Swingle' might grow even better at lower pH, 6 to 6.5 (Graham and Kelly, 2018). 'Swingle' has moderate salinity tolerance and is moderately drought tolerant (Hutchison, 1974). Wutscher and Shull (1975) conducted a 9-year study in Texas that showed yields of red grapefruit on 'Swingle' to be better than those on sour orange and rough lemon. In Florida, yields of 'Valencia' oranges are lower on 'Swingle' than rough lemon or sour orange rootstocks, but yields on a per unit area are quite high, indicating that trees on 'Swingle' are quite productive at high densities on suitable soils.

Scion cultivars on 'Swingle' are quite freeze-hardy, similar in tolerance to trees on sour orange and superior to those on rough lemon types or 'Carrizo' citrange rootstocks (Castle, 1987). This enhanced hardiness appears to be a major advantage of 'Swingle' over 'Carrizo' as a rootstock in chronically cold growing regions.

Although trees on 'Swingle' are tolerant to CTV, CEV and xyloporosis (Hutchison, 1974), trees infected with tatter leaf virus are stunted and may produce a bud union crease. Susceptibility to other viruses is unknown. 'Swingle' is not tolerant to burrowing nematode but is immune to citrus nematode (O'Bannon *et al.*, 1977). 'Swingle' is very *Phytophthora*-tolerant and citrus blight tolerance is moderate to good (Castle, 2015). As a nursery seedling, 'Swingle' leaves are very susceptible to citrus bacterial spot, which prompted the burning of millions of nursery trees in Florida in the 1980s. This susceptibility, of course, does not carry over to scions budded on 'Swingle' and is not related to citrus canker.

Fruit size of sweet oranges and grapefruit are comparable with those produced on sour orange and 'Carrizo' citrange rootstocks. In addition, Wutscher and Shull (1975) and Wutscher and Dube (1977) found TSS and TA of red grapefruit on 'Swingle' to be similar to that on sour orange and to be higher than that on rough lemon-type rootstocks.

'Swingle' is potentially one of the better all-purpose rootstocks for grapefruit and sweet oranges. It is highly nucellar, producing only about 10–15% zygotic seedlings, is moderately seedy and very vigorous in the nursery. Its disease- and freeze-tolerance are very advantageous qualities in the field. 'Swingle' is less suitable than sour orange on clay or saline soils and it does not have the burrowing nematode tolerance of the 'Carrizo' citrange. The desirable characteristics of 'Swingle' have made it a widely planted rootstock in Florida on all but calcareous or high pH clay soils. However, some studies suggest that net income for sweet orange trees on 'Swingle' is less than for more vigorous rootstocks over a 14-year period (Castle *et al.*, 2010a,b).

Several other citrumelos have been evaluated in rootstock trials in Texas (Wutscher, 1977) and Florida (Hutchison, 1977). Some of these rootstocks produce scion characteristics different from those of trees on 'Swingle'. Further field testing, however, of these selections is necessary and 'Swingle' remains the major citrumelo in use as a rootstock (Kesinger, 2012). A comprehensive comparison of 12 rootstocks for 'Valencia' orange trees in Florida reported that 'Volkamer' had highest tree losses but was more profitable than 'Carrizo' due to higher yields (Castle *et al.*, 2010b).

Other Citrus Rootstocks

Established crosses
Several hundred other rootstocks have been evaluated worldwide but in general none has consistently surpassed the overall performance of the rootstocks

discussed previously in this chapter. In particular, a number of mandarins and mandarin-hybrids have been evaluated including 'Orlando' tangelo, 'Changsha', *Citrus depressa*, 'Sunki', 'Sun Chu Sha', etc. (Wutscher and Dube, 1977). Most do not have the consistent yields and fruit quality of currently used rootstocks. The 'Hongju' red tangerine, however, is widely used in central China as a rootstock for sweet oranges. It is well adapted to the rocky, calcareous soils and cool growing conditions of this region. Several rootstocks with characteristics similar to those of sour orange but with CTV tolerance, including *C. obovoidea*, were tested in Florida with favourable CTV tolerance results (Castle *et al.*, 1992). *Citrus taiwanica* and *Citrus myrtifolia* have also been suggested for rootstock use.

'Sun Chu Sha' mandarin may have promise in some citrus regions. A 14-year study in Florida suggested a high survival rate for 'Valencia' oranges on this rootstock on the east coast, where citrus blight is a major concern (D.J. Hutchison, unpublished). 'Sun Chu Sha' also appears to be tolerant of CTV and *Phytophthora* spp. Tree vigour of 'Valencia' oranges on 'Sun Chu Sha' after 14 years was similar to that of sour orange and 'Cleopatra', but greater than that of 'Carrizo' citrange. Yields were similar to those on 'Carrizo', sour orange and rough lemon, but superior to those on 'Cleopatra' mandarin. In contrast, Wutscher and Dube (1977) found that grapefruit yields were lower on 'Sun Chu Sha' than on 'Swingle' citrumelo, sour orange or 'Troyer' citrange. Tests of 'Sun Chu Sha' under a variety of growing conditions suggest the rootstock may have promise for 'Murcott' in calcium soils (Castle and Baldwin, 2006) and sweet orange ('Hamlin') where blight is a major concern (Castle *et al.*, 2010a).

'Rangpur' × 'Troyer' was in the lower half of rootstocks tested for 'Hamlin' sweet orange in Florida and California (Castle *et al.*, 2010a). Trees budded on 'Rangpur' × 'Troyer' are semi-dwarfing and may be suitable for high-density plantings. Sweet orange trees budded on 'Rangpur' × 'Troyer' are 30–50% smaller than those on standard rootstocks, and yields are comparable with those on trifoliate orange (Castle, 1987). Scions budded on 'Rangpur' × 'Troyer' are precocious and begin yielding fruit 2 years after planting. This hybrid has poor *Phytophthora* tolerance and is sensitive to high salinity and susceptible to citrus blight. It is not affected by CTV, but it is susceptible to xyloporosis and CEV. Scions budded on it are fairly freeze-tolerant for oranges. Fruit quality is superior to that produced on lemon types and slightly inferior to that produced on sour orange, 'Carrizo' citrange, or 'Swingle' citrumelo rootstocks.

Grapefruit was a rootstock of interest in the 1950s and 1960s; although tree vigour was excellent, yields were lower than on rough lemon but similar to on sweet orange and 'Cleopatra' mandarin (Gardner and Horanic, 1961). In addition, fruit produced on grapefruit rootstock were small and had poor internal quality. Yields of rooted cuttings of red grapefruit were lowest of any rootstock studied (Wutscher and Dube, 1977). Therefore, grapefruit is not widely used as a rootstock worldwide.

Several citrus hybrids and related species have also been evaluated for use as rootstocks or interstocks for dwarfing of commercial citrus including species of *Clymenia*, *Hesperthusa*, *Citropsis*, *Microcitrus*, *Eremocitrus*, *Severinia*, *Fortunella* and *Swinglea* (Bitters *et al.*, 1977). However, none have proven commercially acceptable. The Chinese box orange (*Severinia buxifolia*) is well adapted to the alkaline soils of Texas, but yields were lower than on other commonly used rootstocks (Wutscher and Dube, 1977).

Newer techniques like protoplast fusion have provided several tetraploid hybrids including some of these relatives with *Citrus* as rootstocks in Florida (Grosser and Gmitter, 2014). This effort has resulted in 16 rootstocks being released to the Florida industry. Characteristics of this group include sour orange like cultivars, citrange, citandarins and citrumelos with tree dwarfing and HLB tolerance. Another more conventional breeding programme at the USDA has produced at least seven rootstocks released since 2001 with tree size control, disease and insect tolerance, including HLB tolerance, that are often more adapted to shallow soils (Bowman, 2001, 2014a, 2014b). All of these selections have been released to growers but have not been evaluated extensively. However they are expected to be thoroughly evaluated by growers in the 'Florida Fast Track' system.

Dwarfing rootstocks
There has been limited success in developing rootstocks that reduce citrus tree size, permitting the use of ultra-high density plantings like those used for apples or peaches. In regions such as the USA, Brazil and Mexico, sufficient arable land has been available for large, low density plantings, thus there was little incentive to use dwarfing rootstocks. In areas where land is at a premium, such as Japan, Spain and Italy, tree size is mainly controlled by pruning. Similarly, hedging and topping is used to control tree size in other citrus-growing regions, but this has limited success unless the scion rootstock combination has moderate to low vigour. Interest in dwarfing rootstocks increased as land and water became limiting.

Several methods have been developed to reduce citrus tree size, including dwarfing rootstocks, inoculation of rootstocks with CEV and use of interstocks. Dwarfing rootstocks include 'Rangpur' 'Troyer', which was discussed in the previous section, 'Rubidoux' trifoliate, 'Rusk' citrange, 'Koethen' sweet orange × 'Rubidoux', and procimequat ([*C. aurantifolia* × *Fortunella japonica*] × *F. hindsii*). The tree size of 'Valencia' orange and 'Marsh' grapefruit on these rootstocks was smaller than that of trees on rough lemon, but the size reduction was usually less than 50%. However, in Florida and California, sweet orange trees have been grown commercially on 'Rangpur' × 'Troyer' that are more than 50% smaller than a standard size tree. Fruit must be removed in the first 2 or 3 years to establish a tree canopy rather than a bush. 'Flying Dragon' rootstock (a trifoliate orange) causes extreme dwarfing and may be useful in ultra-high density plantings, particularly if early fruit removal is used to allow year 1 and 2 canopy growth.

Recent breeding work in Florida (Bowman, 2001; Grosser and Gmitter, 2014) has produced several rootstock selections that impart tree size control to scion cultivars budded on these selections.

Inoculation of trifoliate and trifoliate hybrid rootstocks with budlines containing strains of CEV has also proved effective for dwarfing and has been used commercially in Australia, California and Israel (Ashkenazi and Oren, 1988; Hardy *et al.*, 2007). Extent of dwarfing was most pronounced for navel orange trees on trifoliata, intermediate for 'Troyer' and 'Carrizo' citranges, and least for trees on 'Rangpur' (Bevington and Bacon, 1977). However, even for the trifoliate rootstock, canopy surface area was reduced by only 51%. There was a positive correlation between time of inoculation and final tree size, with those trees inoculated early being more dwarfed than those receiving inoculation as larger trees. Studies indicate that the graft transmissible dwarfing agent used in Israel consists of at least five viroids rather than CEV alone (Bar-Joseph, 1993).

Bitters *et al.* (1977) tested several citrus relatives as interstocks for lemons with varying degrees of dwarfing ranging from 25 to 75%. While some of these relatives have promise as interstocks, this practice is not widely used to promote dwarfing of commercial citrus because of the cost and time involved to establish an interstock tree.

MORPHOLOGY AND ANATOMY OF CITRUS ROOTSTOCKS

The citrus primary root, the radicle, is the first plant part to emerge from the seed and produces a taproot which supports the tree. Lateral roots then develop, most of which are located in the upper soil layers. Fibrous, or feeder roots, develop from the laterals, producing a dense mass of fine roots with a high surface area which increases the capacity of the root for water and nutrient uptake. Fibrous roots often produce root hairs, although in some cases hairs are not clearly visible (Castle, 1978). Root hairs may improve nutrient uptake, particularly of immobile ions such as phosphorus.

Depth and density of the citrus root system varies with rootstock, environmental conditions (see Chapter 4), soil type, and irrigation and drainage practices. Fibrous root density of citrus growing in Florida varied from 0.5 to 1.3 g dm^{-3} in sandy loam soils to 1.9 g dm^{-3} in deep sands to 9.3 g dm^{-3} in shallow, poorly drained soils (Castle, 1978). The surface area of fibrous roots of 'Carrizo' citrange 13 months after planting was 3137 cm^2 with an overall length of 152 m (Bevington and Castle, 1982). Differences among rootstocks become apparent even in the nursery. Seedlings of *C. volkameriana*, *C. macrophylla* and 'Palestine' sweet lime had pronounced taproots and were most vigorous after 2 years in a field nursery in Florida. Rough lemon, sour orange and 'Cleopatra' mandarin seedlings were intermediate in vigour and *P. trifoliata* and 'Carrizo' seedlings had poorly developed, compact root systems (Castle and Youtsey, 1977).

Under current nursery operations where trees are grown in pots in protecting greenhouses or screenhouses, trees almost always lose their tap root. Development of the final root system in the field may therefore be altered, but this effect has not been well studied.

The rootstock influence on rooting depth and distribution also occurs for mature budded trees in the field. 'Orlando' tangelo trees on rough lemon or 'Palestine' sweet lime rootstocks growing in well-drained, sandy soil had 50% of their fibrous roots at depths below 76 cm (Castle and Krezdorn, 1973). In contrast, 'Rusk' citrange and trifoliate rootstocks produced 60% of their fibrous roots shallower than 76 cm. Maximum rooting depth of trees on rough lemon rootstock approached 5 m, while 'Orlando' trees on trifoliate orange and 'Rusk' citrange rootstocks were the shallowest-rooted of the group. As expected, trees on the most vigorous rootstocks such as rough lemon produced the largest plants, whereas those on less vigorous rootstocks produced the smallest plants. Deeper-rooted species such as rough lemon are typically more drought-tolerant than shallower-rooted species like *P. trifoliata* because they access a greater soil volume. Irrigation with microsprinkler or drip systems often alters rooting depth and distribution, which may alter previous concepts of rootstock performance based on potential rooting depth using infrequent overhead sprinkler irrigation systems. Generally, on well drained sandy soils, rooting depth is much shallower today than when infrequent deep irrigation was practised. In Florida bedded groves, roots are seldom deeper than 0.4 m and on deep sands with microsprinklers rooting depth is usually not more than 0.75 m.

Soil conditions, in particular soil structure, also affect rooting depth and distribution. In soils with high clay content, compacted soils, or those with impervious hardpans or high water tables, 75% of the roots may be located in the upper 15–45 cm of the soil surface (Castle, 1978). Studies from Cyprus suggest that feeder root growth is severely restricted in soils with clay content greater than 80%, and in South Africa citrus planting is not recommended in soils having greater than 50% clay content. Therefore, rootstock influence on rooting depth and distribution becomes less apparent as the potential volume of rooting becomes restricted by these factors.

PHYSIOLOGY OF CITRUS ROOTSTOCKS

Rootstocks differ not only morphologically but also physiologically with respect to water and nutrient uptake and salinity tolerance. Leaf water potentials (Ψ) of 'Orlando' tangelo trees on rough lemon and 'Palestine' sweet lime were less negative, indicating less water stress than those on sour orange or 'Cleopatra' mandarin (Crocker *et al.*, 1974). The enhanced water status of fruit from trees on rough lemon probably accounts for the dilution of TSS and TA commonly observed for this rootstock (Albrigo, 1977). These differences in leaf Ψ occurred not only due to the greater rooting depth of the former two rootstocks but also due to greater root hydraulic conductivities which increased water uptake per unit of root mass. Syvertsen (1981) found that root hydraulic conductivity was greatest for 'Carrizo' citrange, intermediate for rough lemon and lowest for 'Cleopatra' mandarin and sour orange (Fig. 3.1). The greater root conductivity of rough lemon over sour orange may be partly responsible for the more vigorous growth of scions on this rootstock as well as for reduced freeze-hardiness (Wilcox and Davies, 1981). In general, there is an inverse relationship between vigour and freeze-hardiness of citrus (Young, 1977).

The general effects of citrus rootstocks on leaf mineral element content are summarized in Table 3.2 and represent years of research on this subject (Wutscher, 1989). It is difficult to make overall generalizations for the various rootstocks and elements, because in some instances a particular rootstock induces high levels of one element and low levels of another. However, it appears that rootstocks inducing vigorous growth in the scion such as rough lemon or 'Palestine' sweet lime (Castle and Krezdorn, 1973) generally have higher leaf levels of N and K than those low vigour-inducing rootstocks like trifoliate orange. Syvertsen and Graham (1985) also observed that leaf N and P contents were positively correlated with root hydraulic conductivity of five citrus rootstocks. Leaf N and P levels and root conductivities were lowest for sour orange and highest for trifoliate orange seedlings, but this trend is not always

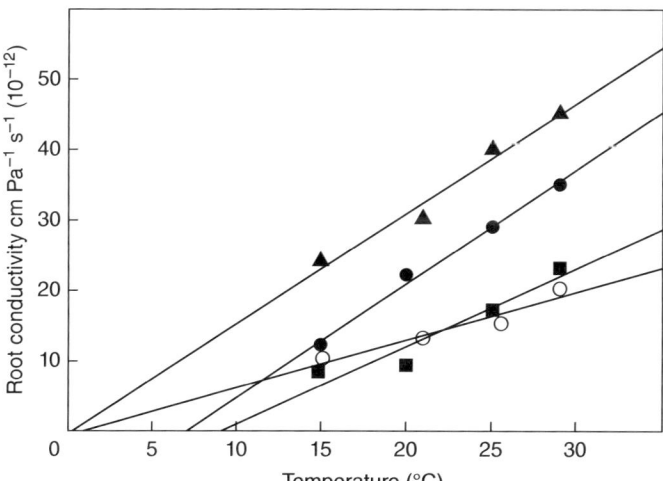

Fig. 3.1. The effect of temperature on the hydraulic conductivity of intact 12-month-old seedling root systems. Each point is the mean of four replicate plants of each rootstock. Linear regression lines have been fitted to data from each rootstock (R = 0.42 to 0.83). ● Rough lemon; ■ Sour orange; ▲ 'Carrizo'; ○ 'Cleopatra' (Syvertsen, 1981).

Table 3.2. Rootstock effects on mineral element levels in citrus leaves (Wutscher 1989).

Element	Rootstocks inducing high levels	Rootstocks inducing low levels
N	Rough lemon Sweet orange 'Rusk' citrange Alemow 'Rangpur' lime	Sour orange Trifoliate orange 'Cleopatra' mandarin Grapefruit
P	Sweet orange Trifoliate orange Rough lemon Grapefruit *S. buxifolia* 'Swingle' citrumelo	Sour orange 'Cleopatra' mandarin 'Troyer' citrange 'Morton' citrange 'Milam' rough lemon
K	Grapefruit 'Sampson' tangelo *S. buxifolia* 'Milam' rough lemon	'Cleopatra' mandarin Rough lemon 'Rusk' citrange 'Morton' citrange 'Troyer' citrange
Ca	'Cleopatra' mandarin Rough lemon 'Troyer' citrange Sour orange	Sweet orange Grapefruit Alemow *S. buxifolia*
Mg	Trifoliate orange 'Carrizo' citrange 'Cleopatra' mandarin 'Sun Chu Sha'	Grapefruit Sour orange *S. buxifolia* 'Rangpur' lime
Na	'Rusk' citrange 'Yuzu' Rough lemon	Sour orange Sweet orange 'Morton' citrange
S	Rough lemon Grapefruit	'Cleopatra' mandarin Trifoliate orange
Fe	Rough lemon 'Yuzu' 'Rusk' citrange Sour orange	Grapefruit Trifoliate orange 'Taiwanica' orange 'Swingle' citrumelo
Mn	Rough lemon Alemow 'Yuzu' *S. buxifolia* 'Sunki' mandarin	Grapefruit Sour orange Sweet orange 'Swingle' citrumelo
Zn	Grapefruit Rough lemon *S. buxifolia* 'Cleopatra' mandarin	Sour orange Sweet orange 'Carrizo' citrange Trifoliate orange

Continued

Table 3.2. Continued.

Element	Rootstocks inducing high levels	Rootstocks inducing low levels
Cu	'Rusk' citrange	Sour orange
	Sweet orange	Rough lemon
	S. buxifolia	'Troyer' citrange
	'Swingle' citrumelo	'Cleopatra' mandarin
Cl	'Troyer' citrange	'Sunki' mandarin
	'Carrizo' citrange	'Cleopatra' mandarin
	Trifoliate orange	'Milam' rough lemon
	Sweet lime	Sour orange
B	'Cleopatra' mandarin	S. buxifolia
	Sweet lime	Alemow
	Trifoliate orange	Sour orange
	Grapefruit	'Carrizo' citrange

consistent. Leaf N levels were not necessarily correlated with tree growth, as levels were among the highest for trees on 'English Small' trifoliate, but these trees were among the smallest (Castle and Krezdorn, 1973). Similarly, trifoliate seedlings accumulated higher levels of labelled ^{15}N than 'Swingle' citrumelo, yet growth was less (J.A. Lea-Cox, unpublished). Consequently, rootstocks that produce a small tree may have higher N on a percentage basis but lower total N than rootstocks producing large trees with more leaf canopy. Slower growing scion/rootstock combinations are more likely to have excess N remaining in the soil that is potentially vulnerable to leaching (Lea-Cox et al., 2001).

A rootstock's ability to exclude nutrients may be of more importance than its ability to accumulate them. Sour orange, 'Cleopatra' mandarin and 'Rangpur' (Kirkpatrick and Bitters, 1968), for example, are sodium and chloride excluders; qualities which make them more tolerant of high salinity. *Severinia buxifolia*, which has limited potential as a rootstock, is an effective Cl and B excluder. In contrast, trifoliate orange, which is salinity sensitive, is a chloride accumulator. Leaf nutrient content, however, is not always a function of uptake rates as nutrients are also stored in the roots and may be translocated at different rates to the canopy. The physiological basis for differences in nutrient uptake by citrus rootstocks has not been thoroughly studied at the cellular level and the mechanisms for the observed differences in nutrient uptake are still largely unknown. A useful characteristic is better uptake and reduction of Fe^{+3} (Castle et al., 2009). Grouped in descending order of Fe^{+3} reduction rates were 'Volkamer' lemon / 'Rangpur' / sour orange selections / *Citrus macrophylla* / mandarins and mandarin hybrids / citranges / citrumelos / trifoliate orange. Of the citrus relatives tested in solution culture, only those in the genera *Glycosmis, Citropsis, Clausena*, and *Murraya* had high Fe reduction rates with good seedling growth and new leaves developing a light yellow colour or showing no loss of greenness.

MYCORRHIZAE

Some species of citrus rootstocks commonly became stunted in the nursery following methyl bromide fumigation. The cause of stunting was initially unknown but was later found to be due to an absence of vesicular arbuscular mycorrhizae (VAM), which increase uptake of immobile nutrients such as Zn, Cu and particularly P (Kleinschmidt and Gerdeman, 1972). Several mycorrhizal fungi are associated with citrus roots but most are *Glomus* species (Nemec, 1978). Endomycorrhizae are also commonly associated with citrus roots in the field, although the degree of colonization varies with availability of mycorrhizae, tree age, P content of the soil and root, general soil nutrition and rootstock. Young trees generally have less extensive colonization than older ones due simply to differences in inoculation time. Mycorrhizae use metabolites which diffuse from the roots as an energy source. When root P content is high there is less leakage of metabolites and less germination of the fungus, with the inverse relationship occurring at low P content (Graham and Syvertsen, 1985). The energy expended by the plant in maintaining mycorrhizae is considerable. As much as 6–10% of labelled ^{14}C was translocated from sour orange seedlings to mycorrhizal fungi within 2 h after incubation (Koch and Johnson, 1984). Mycorrhizal 'Carrizo' citrange and sour orange seedlings have higher root conductivity than nonmycorrhizal seedlings (Graham and Syvertsen, 1984).

Rootstocks differ significantly in their dependence on mycorrhizae. Sour orange and 'Cleopatra' mandarin are most dependent, with sweet orange, rough lemon and 'Rangpur' intermediate, and 'Carrizo' citrange least dependent (Nemec, 1978). The degree of dependence is related to root density, hydraulic conductivity and ability to take up P. 'Carrizo' and *P. trifoliata* have denser fibrous root systems and greater hydraulic conductivities than sour orange or 'Cleopatra' mandarin. Therefore, these rootstocks are less dependent on the additional surface area provided by mycorrhizal fungi (Graham and Syvertsen, 1985), and thus the symbiotic relationship between the tree and fungus does not develop.

Citrus tree stunting in the nursery may be reduced by adding P or reintroducing mycorrhizae. At low levels of P, nonmycorrhizal seedlings of sour orange and 'Carrizo' citrange were stunted compared with mycorrhizal seedlings. However, as P levels were increased from 0 to 560 kg ha^{-1}, plants became of similar size (Menge *et al.*, 1977). Mycorrhizae may be introduced by seed inoculation, banding in the planting row, or by layering in the soil (Ferguson and Menge, 1986). These practices are fairly effective under carefully controlled conditions. In practice, however, it is difficult and costly to culture the fungi because no *in vitro* cultural system exists. Generally, fungi must be cultured on an alternative host such as Sudan grass. In addition, colonies must be handled carefully to prevent desiccation in the field. Therefore, mycorrhizal fungi are not widely used in commercial nurseries, particularly as field nurseries have decreased (see Chapter 5).

RECENT ADVANCES IN ROOTSTOCK DEVELOPMENT

Rootstock breeding and field testing are time-consuming processes, often requiring 20 years or more before new rootstocks can be released. The mechanics of traditional breeding of citrus rootstocks are described in Chapter 2. In the 1980s and 1990s new methods of producing rootstocks were being developed based on biotechnology. These included protoplast fusion and genetic engineering (see Chapter 2 for details). Protoplast fusion allows for hybridization of species that might otherwise be sexually incompatible or which produce a high percentage of nucellar seedlings, making the development of new hybrids difficult. For example, Grosser and Gmitter (1990) developed several hybrids which were field tested in the 1990s and led to the recently released hybrid rootstocks (Grosser and Gmitter, 2014). They have successfully developed hybrids of rough lemon and sour orange with the potential of developing a hybrid which has the CTV tolerance of rough lemon and the blight tolerance of sour orange, but have also produced sour orange-like rootstocks from the same species that sour orange originally came (*C. maxima* × *C. reticulata*). Traditional hybridization methods for these two species have proved largely unsuccessful due to the high percentage of nucellar embryos produced in the progeny. Other groups are also trying to produce sour orange-like progeny from various breeding methods. Several researchers are also developing methods of improving CTV tolerance in sour orange by inserting the gene for the coat protein of the virus into the sour orange genome (see Chapter 2). Although none of the new biotechnology techniques have yielded an improved rootstock to date, the potential for solving some of the major rootstock-related problems of citrus remains high.

ENVIRONMENTAL CONSTRAINTS ON GROWTH, DEVELOPMENT AND PHYSIOLOGY OF CITRUS

Citrus may be grown successfully over a moderately wide range of environmental and edaphic conditions. Most commercial production, however, is limited to regions between 40° north–south latitudes where minimum temperatures are generally greater than –7°C. Several microclimatic regions exist within these latitudes and all citrus production occurs only in tropical or subtropical climatic regions. For example, many areas in the eastern USA south of 40°N latitude are too cold for citrus production. Environmental factors associated with these climates have a pronounced influence on growth, development and yields of citrus trees, and are largely responsible for the range of yields in mature orchards that have been as high as 100 tonnes ha^{-1} in the subtropics to as low as 10–15 tonnes ha^{-1} in the tropics for healthy trees.

TROPICAL REGIONS

Tropical regions lie between 23.5° north and south of the equator where average annual temperature is above 18°C. Minimum temperatures never fall below 0°C except at the highest elevations. Tropical regions, especially those within 10° north or south of the equator, experience small fluctuations in day length at all elevations and diurnal temperatures in low and mid-elevations. Local climatic conditions vary considerably within the tropical regions based on elevation or proximity to water or mountain ranges, which affect wind patterns and rainfall. Therefore, tropical regions may be further subdivided into lowland, midland or highland tropics and into wet (humid) or dry (arid or semiarid) regions.

Lowland tropical regions are located between sea level and about 500 m elevation and have the highest average temperatures and thus the greatest annual heat unit (hu) accumulation. The elevation separating the three regions is not clearly defined and may vary related to average temperature in each zone. An average temperature of less than 24°C, for example, is necessary to induce flowering (hours below 20°C required, Valiente and Albrigo, 2004) and may also be used to separate lowland from mid- and highland regions. Heat units are calculated as the amount of time (h) multiplied by the average temperature difference from the minimum for citrus vegetative growth of 12.5°C (Mendel, 1969). For example, a growing region with an average temperature of 18.5°C would accumulate 180 hu for a 30-d month (18.5 − 12.5 = 6.0, 6.0 × 30 = 180 hu). Heat unit accumulation is strongly correlated with vegetative growth rate and indirectly with poorer fruit quality at high hu levels, provided that water and nutrients are not limiting (see fruit development section in this chapter, p. 109). Annual heat unit accumulation in lowland tropical regions like Mombasa, Kenya, Giradot, Colombia or Mannar, Sri Lanka, is greater than 5000 (Table 4.1). Such high hu accumulation increases respiration which decreases fruit solids and acid levels. In moving to mid-elevation tropical areas for citrus (500–1500 m) average temperatures and annual hu accumulation are less. For example, annual hu accumulation at Palmira, Colombia (elevation 1000 m), is 3500 hu, whereas in highland areas (1500–2500 m) such as Conocoto, Ecuador, annual hu accumulation is only 1000 hu. Average annual temperatures at elevations >1500 m are often <12–13°C, which is at or below minimum levels for tree growth. Citrus trees are seldom grown at elevations above 2600 m and fruit are used for local consumption only (Camacho-B., 1981). Average annual temperatures may be <10°C under these conditions and are not conducive to tree growth or development of high quality fruit.

Low, mid- and high tropical regions vary not only in temperature (hu accumulation) but also in rainfall and interception of sunlight. Many lowland areas such as in parts of Costa Rica, Ecuador and Malaysia are characterized by high relative humidity (RH) and rainfall with excessive cloud cover. However, it is also common to find distinct wet–dry cycles in equatorial Africa, Central America and particularly on islands such as Jamaica. In many tropical regions two distinct wet–dry cycles occur. With the onset of each rainy season the greatest number of flowers are produced, although some flowering occurs throughout the year (Cassin et al., 1969). Mid-elevation tropical regions also vary in rainfall and humidity with distinct wet–dry seasons as in areas of Central America, Venezuela and Colombia. Highland and some midland areas are often distinctly different from the other regions due to the regular presence of ground fog. Fog reduces light intensity during the day, thus reducing temperature and net CO_2 assimilation. At night, fog also may reduce longwave radiation losses from the soil, thus increasing temperatures. Moreover, although light intensity may be reduced due to the fog, ultraviolet light may be very intense due to reduced particulate matter in the atmosphere at these high elevations. Ultraviolet light at high levels may cause leaf distortion and reduce growth of citrus trees at these high elevations, particularly for shade leaves (Basiouny and Biggs, 1974; Mirecki and Teramura, 1984). A few lowland areas may also have heavy rainfall and fog-like conditions leading to many disease problems.

Table 4.1. Data on temperature and annual heat units (above 12.5°C) for various citrus-growing areas (Mendel, 1969).

Area and location	Latitude	Elevation above sea level (m)	Annual heat units (°C)	No. of months with average temp.	
				<12.5°C	<17.5°c
Tropical regions					
Trinidad (Piarco Airport)	10°40'N	10	5000	0	0
Colombia (Arcataca)	10°30'N	30	5500	0	0
Colombia (Giradot)	4°20'N	400	5700	0	0
Colombia (Palmira)	3°30'N	1000	3500	0	0
Colombia (LaFlorida)	4°40'N	1800	1700	0	10
Ecuador (Santa Ross)	3°30'S	10	4400	0	0
Ecuador (Conocoto)	0°15'S	2200	1000	0	11
Kenya (Mombasa)	4°00'S	20	5200	0	0
Kenya (Nairobi)	1°20'S	1600	2500	0	1
Uganda (Jinja)	0°30'N	1100	3330	0	0
Sri Lanka (Mannar)	9°00'N	30	5700	0	0
Sri Lanka (Nuwara Elyia)	7°00'N	1900	1000	0	0
Subtropical regions					
Spain (Valencia)	39°30'N	30	1600	3	6
California (Riverside)	34°00'N	260	1700	3	6
California (Indio)	33°40'N	–10	3900	1	4
Israel (Degania)	32°40'N	–200	3600	0	4
Israel (Rehovot)	31°50'N	50	2600	1	4
Florida (Orlando)	28°40'N	30	3700	0	2
Texas (Weslaco)	26°05'N	40	3900	0	2
Brazil (Limeira)	22°30'S	700	3000	0	1

SUBTROPICAL REGIONS

The hu concept also applies when comparing tropical regions with subtropical ones. Subtropical regions are located between 23.5 and 40° north and south latitudes. Although not all areas within these latitudes are suitable for growing citrus, the area encompasses the major citrus-growing regions in the world, namely, portions of Brazil, the USA, Spain, Italy, Japan, Israel, Argentina and much of Mexico and China. Cassin *et al.* (1969) further divided the subtropics into those areas between 30 and 40° latitude and those between 23.5 and 30° latitude (semitropics). Annual hu accumulation also varies considerably within the subtropics from 1600 hu in Valencia, Spain, to 3900 hu in Indio, California and Weslaco, Texas, USA (Table 4.1). Hu accumulation in intermediate subtropical and mid-elevation tropical regions such as Orlando, Florida (3700) and Palmira, Colombia (3500), is intermediate to low tropical and Mediterranean subtropical areas. The distribution is more uniform at the intermediate locations allowing for greater net CO_2 assimilation during the winter. Moreover, subtropical regions tend to have more months of average temperatures less than 17.5°C than tropical regions, except at high elevations (Table 4.1).

Subtropical regions are characterized by mean annual temperatures between 15 and 18°C but have greater diurnal temperature fluctuations than tropical regions. Moreover, many subtropical areas are exposed to temperatures below 0°C on a regular basis with temperatures as low as –7°C. Minimum air temperatures of –10°C have occurred in Florida and central China on rare occasions. These low temperatures have had a devastating effect on citrus production in the USA over the years but particularly during the 1980s and also have periodically damaged fruit and trees in parts of Europe, northeastern Mexico, Australia, Argentina and central China. In contrast, freeze-damage rarely occurs in the most productive growing regions of Brazil (São Paulo State). Although freezing temperatures occur in Japan and central areas of China, rarely do extensive crop or tree losses occur due to regional microclimatic effects such as proximity to large bodies of water that moderate temperatures. Also, extended periods of cool temperatures occur in these locations prior to freezes allowing maximum freeze-hardiness development. Japan's citrus industry is located between 35 and 40°N latitude, yet the citrus-growing region of Texas at latitude 25°N has experienced far more devastating freezes over the years. High-pressure arctic air masses develop in Canada and move unimpeded across the central plains of the USA into Texas and Florida causing the advective freezes that have resulted in severe damage to much of the citrus industry. Furthermore, most of Japan's citrus industry consists of satsuma mandarin on trifoliate rootstocks, which when fully freeze-acclimated, is an extremely freeze-hardy combination for citrus.

Subtropical regions may be further subdivided into humid, semiarid and arid regions. The humid regions such as Brazil and Florida have relatively high humidity and rainfall and smaller fluctuations in diurnal temperatures than semiarid and arid regions such as southern California, Spain, Italy, Greece, Israel, South Africa and Australia (New South Wales, Murray River). These latter areas, among others, have what is termed a 'Mediterranean-type' climate, which is characterized by hot days, cool nights and low RH in the summer and colder winters with most of the rainfall. Although humid subtropical regions have greater annual rainfall than Mediterranean-type climates, rainfall distribution may be seasonal, thus producing drought conditions at certain times. Citrus trees in semiarid or arid regions such as those in southern California or southeastern South Africa are exposed to intense, dry winds which may reduce tree vigour and, more importantly, cause wind scarring of the fruit. In South Africa, wind scarring (see Chapter 8) is the major cause of peel blemishes which prevents fruit from being marketed fresh. Wind scarring is also the major fruit blemish in humid subtropical areas like Florida (Miller and Burns, 1992).

It is logical to expect that such pronounced differences in hu accumulation, humidity, rainfall and wind will have a marked effect on citrus tree growth, development, productivity and fruit quality. Even between subtropical regions there are sufficient differences in climate to have a large effect on production of fruit and its quality. These environmental factors also influence incidence of pest and disease problems and thus indirectly affect cultural practices within a region.

ENVIRONMENTAL CONSTRAINTS ON VEGETATIVE GROWTH AND DEVELOPMENT

Seed Growth and Development

The citrus seed (Fig. 4.1A) consists of a seed coat surrounding a much reduced nucellus and endosperm (exalbuminous). The seed contains two cotyledons usually below ground and from one to as many as seven embryos. Only one embryo is derived from sexual fusion of the sperm and egg cells with additional embryos originating from nucellar tissue which is genetically the same as the diploid maternal tissue. Nucellar embryony is very rare in plants and is of importance in production of true-to-type, virus-free rootstocks (see Chapter 3).

Seed germination commences with the emergence of the radicle (primary root) through the micropylar (pointed) end of the seed and is dependent on moisture and temperature. Seed germination is hypogeal, i.e. the cotyledons remain underground (Fig. 4.2b) with the tap root extending downward and branching occurring above. The temperature for first emergence of the radicle ranges from 9 to 38°C and varies with cultivar (Fig. 4.2) (Wiltbank *et al.*, 1995). For example, seeds of *Poncirus trifoliata* will emerge at 9°C, while those of rough lemon require a minimum of 15°C. This difference is probably related to the tropical origin of rough lemon (India) as compared with the subtropical origin of *P. trifoliata* (central China). Days to first emergence range from nearly 80 at 15–20°C to fewer than 14–30 d range at 30–35°C for most cultivars. Removal of the seed coat prior to emergence shifts the curve so that emergence occurs at lower temperatures and more rapidly than when seed coats are present. Seed coat removal also improves percentage emergence. This practice was used by some citrus nurserymen to improve percentage emergence and hasten emergence time, but is only practical where labour costs are low.

Light intensity does not affect emergence or germination, but seedlings developing in the dark will be etiolated and spindly. Further, seedling growth of trifoliate related rootstocks is enhanced by light interruption at night and warmer night-time temperatures (Brar and Spann, 2014a, 2014b).

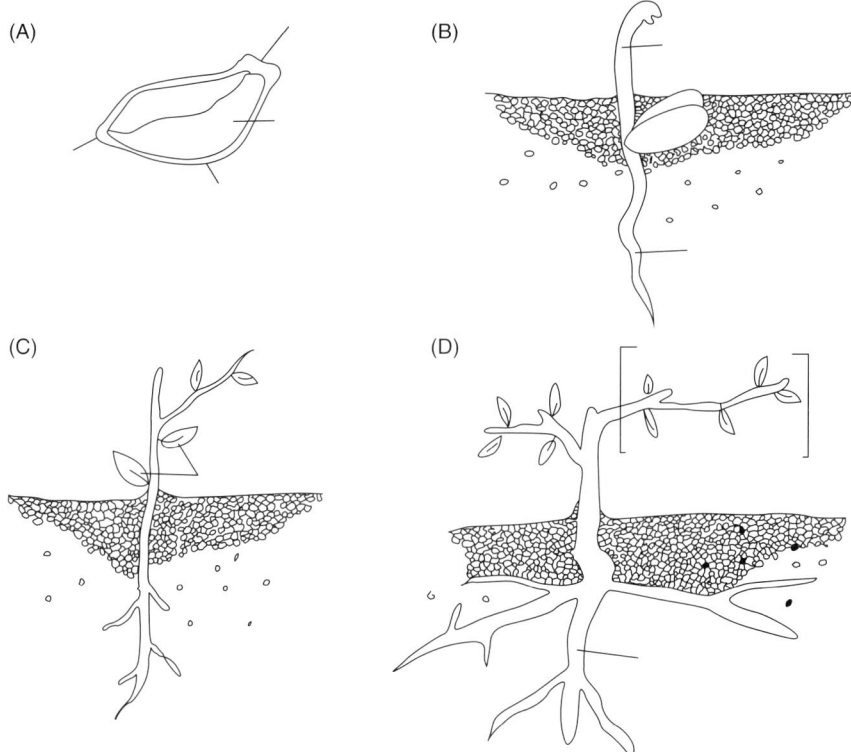

Fig. 4.1. Life cycle of a citrus seedling. (A) Seed morphology before emergence; (B) Seedling growth about 2 weeks after radicle and epicotyl emergence; (C) Development of lateral branches; (D) Development of a sympodial growth habit (Wiltbank et al., 1995).

The Juvenile Plant

Citrus seedlings undergo a relatively long juvenility period (time until first flowering) following germination, depending on species and growing conditions. The seedling typically grows from a single apical meristem during the first few weeks after germination (Fig. 4.1B) but then begins to produce lateral meristems in the leaf axils from which subsequent branching may occur. Juvenile citrus trees are thorny (modified stems from auxiliary buds) and generally have a very upright, unbranched growth habit. As the tree grows, the apical meristem abscises and lateral buds break along the central axis. New shoots emerge from vegetative lateral buds (Fig. 4.1C), and in turn the apical meristem of these shoots will abscise and other lateral buds emerge. This zigzag growth pattern, termed sympodial growth, is characteristic of most citrus and related species (Fig. 4.1D). Since the terminal bud abscises, citrus trees have a determinate growth habit.

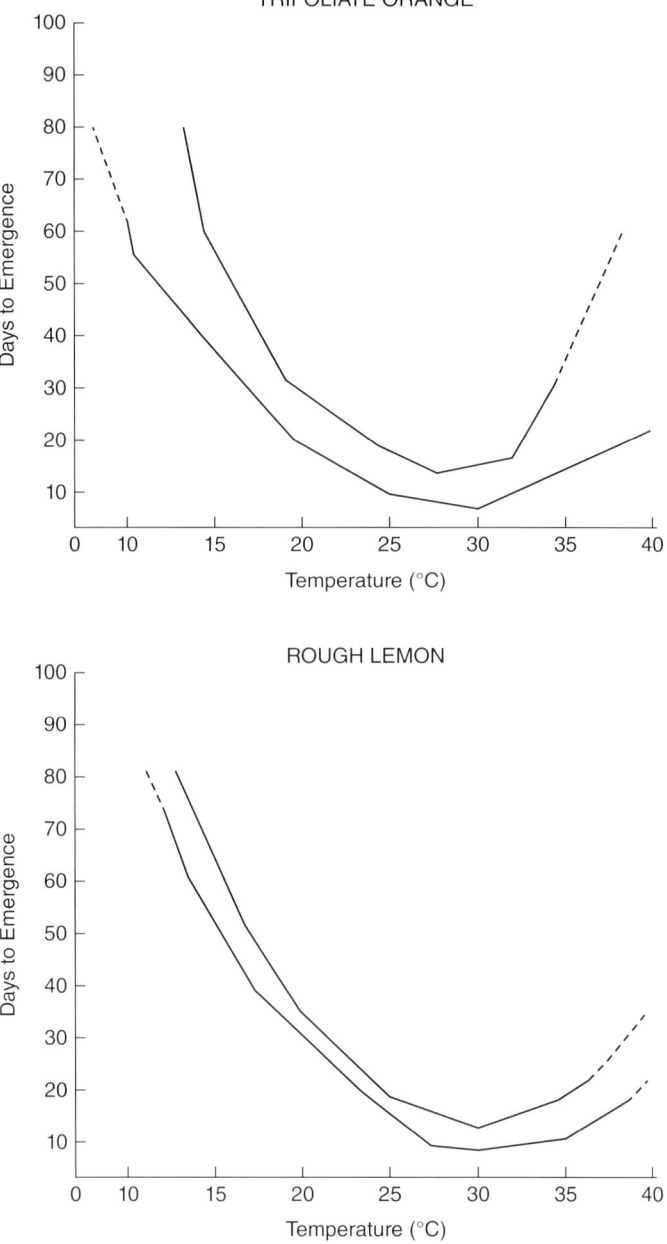

Fig. 4.2. Effect of temperature on seedling emergence of two citrus species. For each graph upper line = + seed coat and lower line = – seed coat (Wiltbank et al., 1995).

As the new shoots emerge, the apical meristem produces leaf primordia at regular intervals around the stem as it elongates. Citrus species have characteristic patterns of phyllotaxis. For example, sweet orange has a 3/8 phyllotaxis and grapefruit a 2/5 phyllotaxis. In 3/8 phyllotaxis, eight leaves are produced around the stem in a spiral pattern with the eighth leaf emerging directly above the first leaf of the spiral. This occurs within three revolutions around the stem. For grapefruit, five leaves are produced within two revolutions around the stem. These patterns are then repeated as the shoot develops. Although phyllotaxis occurs in a somewhat regular manner within a citrus species, environmental factors may alter the pattern, e.g. where shoots have developed under water stress or excessively vigorous growing conditions.

The duration of juvenility varies within species and with environmental factors. Generally, the juvenility period is inversely related to tree vigour and hu accumulation, again provided other factors are not limiting. Vigorous species such as limes and lemons have juvenility periods of less than 2 years under subtropical growing conditions, while juvenility periods of 5 to as long as 13 years may occur for mandarins, sweet oranges and grapefruit when grown from seed. Marcotts (air-layers) of 'Tahiti' lime often produce fruit within 1 year of planting. Duration of juvenility is dramatically affected by temperature, moisture, and in some cases edaphic and cultural conditions. For example, in lowland tropical areas with high rainfall the juvenility period is considerably shorter than in arid subtropical regions with suboptimal irrigation.

Citrus breeders in particular are interested in shortening the juvenility period to decrease the time required to breed and select new cultivars. Calamondin and 'Key' lime flowered in fewer than 18 months from seed by growing them in controlled climate chambers at 30°C days, 25°C nights and 16 h day lengths (Snowball et al., 1988). Under these conditions plants grew continuously (not in distinct growth flushes) attaining a height of 2 m. Growth was then arrested using paclobutrazol (a growth retardant) and extensive lateral branching and flowering occurred. Unfortunately, this technique does not produce flowers for all citrus species but appears to be useful for shortening the juvenility period of some species. An alternative method of shortening the juvenility period is to grow seedlings in a greenhouse or covered structure as single, long stems. The top of the stem is bent toward the base above 2 or 3 m and this area usually flowers.

The Budded Plant

The juvenility period is rarely a problem in most citrus-growing regions since most commercial citrus trees in the world are grown as plants consisting of two parts, the scion and the rootstock. The scion originates from budwood of preferably disease-free mature bearing trees and is budded onto a rootstock (see Chapter 5). This practice promotes earlier fruit production than found in seedlings as well as taking advantage of scions and rootstocks with desirable and uniform fruit and tree characteristics (see Chapters 2 and 3).

Growth and development of budded nursery trees is also dependent on environmental factors. For example, nursery trees grown outdoors in Riverside, California, developed to a marketable size in about 30 months at 1700 hu, while the same rootstock/scion combination developed in only 15 months at Mannar, Sri Lanka, at 5700 hu (Mendel, 1969). Similarly, a 3-year-old 'Washington' navel orange tree attained a height of 4 m under lowland tropical conditions (5700 hu), while a 4-year-old tree attained only 3 m under highland conditions (2000 hu). The same type of disparity in growth rates occurs when comparing citrus trees grown under arid subtropical and humid subtropical conditions. Citrus tree growth rates are much slower in Valencia, Spain (semiarid subtropical), than in Orlando, Florida (humid subtropical). This difference occurs not only due to differences in heat unit accumulation (1600 vs 3700) but also to the greater rainfall and RH in Orlando (Mendel, 1969).

Light intensity and quality also affect vegetative growth and development of citrus trees in several ways. Light intensity has a direct effect on net CO_2 assimilation and an indirect effect on leaf temperature. Vegetative growth of citrus trees is closely related to net CO_2 assimilation provided that other factors such as temperature, nutrition and water are not limiting. Net CO_2 assimilation increases linearly as photosynthetic photon flux (PPF) increases from 0 to about 700 µmol m^{-2} s^{-1} and plateaus above this level known as the light saturation level (Syvertsen, 1984) (Fig. 4.3). Maximum net CO_2 assimilation for most citrus species is achieved at 30–35% of full sunlight. A PPF of 2000–2200 µmol m^{-2} s^{-1} represents full sunlight near sea level; however, cloud cover can significantly reduce PPF to below light saturation in many tropical locations, particularly in the inner canopy. The longer a citrus tree is at or above light saturation, potentially the greater the net CO_2 assimilation, again provided that water, temperature, nutrition or other factors are not limiting photosynthesis. Therefore, vegetative growth of citrus trees is usually greatest where day lengths are moderate to long (>12 h), maximum hus are attained and water is not limiting, e.g. in high rainfall, lowland tropical regions. In contrast, it is not uncommon for PPF to be 100 µmol m^{-2} s^{-1} or less within the canopy of a mature citrus tree in arid regions of South Africa. Consequently, net CO_2 assimilation levels are often less than 2 µmol m^{-2} s^{-1} with a concomitant reduction in flower number and yield in this region.

In many arid regions such as Israel or Indio, California, extremely high light intensity may decrease net CO_2 assimilation due to increased radiation load on the leaf. Under extreme conditions leaf temperature may be 7–10°C greater than air temperature and may approach 55°C for sun-exposed leaves (Ketchie, 1969). Optimum temperatures for net CO_2 assimilation for citrus are species-dependent and range from 28 to 30°C in humid air (Kriedemann and Barrs, 1981), with temperatures greater than 35°C severely limiting the activity of ribulose 1,5-bisphosphate carboxylase/oxygenase (Rubisco), and probably causing midday stomatal closure. At low humidity, temperature

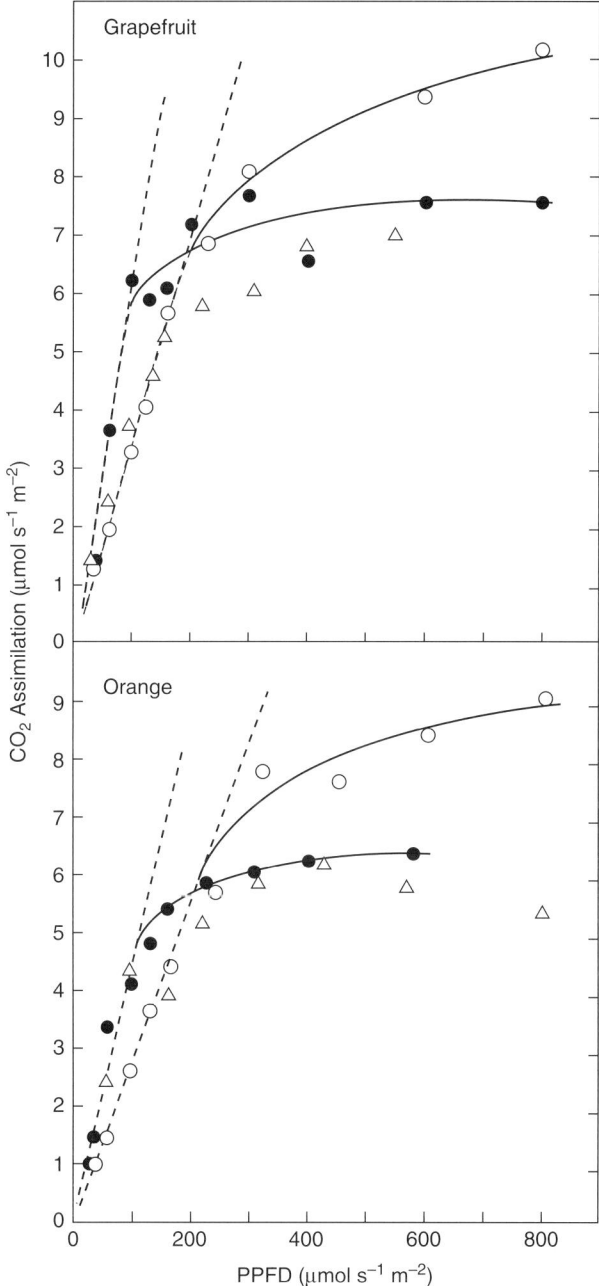

Fig. 4.3. Light responses of net CO_2 assimilation rate and apparent quantum yield (μmol CO_2 assimilated per incident μmol PPF, dashed lines) of grapefruit and orange leaves grown for 5 months under low (●) or (○) PPF and after being moved from low into high (△) PPF for 14 d. Curved lines were fitted by eye (Syvertsen 1984).

optima range from 15 to 22°C. As temperatures increase leaf-to-air vapour pressure deficit (VPD) also increases, lowering stomatal conductance. Citrus trees have moderate to low net CO_2 assimilation compared with other fruit tree species, such as cherry or apple; values range from 9 to 12 µmol m^{-2} s^{-1} of CO_2 assimilated under optimum environmental conditions (Syvertsen, 1984) compared with 25–30 µmol m^{-2} s^{-1} for cherry. Lower than optimum temperatures also decrease net CO_2 assimilation probably due to effects on VPD and enzyme activity. Net CO_2 assimilation, however, did not exceed 6 µmol m^{-2} s^{-1} under the high VPD conditions found in the arid climate of northeastern South Africa. Nevertheless, annually citrus fixes a moderately high amount of CO_2 because it is an evergreen, unlike deciduous cherry or apple. Freezing, but not lethal, low temperatures also significantly decrease net CO_2 assimilation for several days following a freeze by causing stomatal closure and decreasing activity of enzymes involved in photosynthesis (Young, 1969).

There is also some suggestion that light quality affects vegetative growth of citrus trees. Studies in Israel (Mendel, 1969) suggest that shoot elongation is enhanced by far-red light and inhibited by red light. Generally, far-red light levels are greatest inside the tree canopy. High ultraviolet-B light has an inhibitory effect on shoot growth of some citrus species (R.H. Biggs, unpublished). In high light, high temperature environments, shading may be beneficial in reducing leaf temperatures, specifically mesophyll temperatures, thus increasing photosynthetic CO_2 accumulation (Syvertsen *et al.*, 2003). Finely powdered clay can also reduce leaf temperature to improve leaf function (see Chapter 6 for reducing insects on leaf surfaces).

Shoot and leaf growth
Shoot elongation of citrus trees usually occurs in two to five distinct growth flushes annually in subtropical regions but may occur on an almost continuous basis in lowland tropical areas and in some coastal subtropical areas as long as soil moisture is available, particularly for lemons and limes (Mendel, 1969). Commencement of shoot growth is regulated by temperature (>12.5°C) in subtropical regions and by availability of water in tropical regions (Cassin *et al.*, 1969). Seasonal cumulative shoot elongation or dry matter accumulation is generally greatest under moderate, consistently long days and relatively high mean day and night temperatures typical of low tropical areas.

The distribution and extent of shoot growth is also affected by temperature. In subtropical regions, the spring growth flush, which generally occurs in March–April in the northern hemisphere and September–October in the southern hemisphere, usually occurs from many growing points producing many shoots with short internodes. Mean temperatures generally range from 12 to 20°C during this period. In contrast, the summer growth flush (which occurs in late May–July in the northern hemisphere and January–February in the southern hemisphere) occurs from fewer growing points but produces shoots with longer internodes, larger leaves and greater stem diameter. Temperatures during this period range from 25 to 30°C. Late summer or autumn growth flushes also tend to occur from fewer growing points, and are minimal on well yielding trees.

Potentially, shoots are produced throughout the season in tropical regions due to high mean temperatures year-round if water is not limiting. But water is often limiting in tropical regions with distinct wet–dry climatic cycles and seldom is irrigation provided. Shoots are generally produced from many growing points with internode length dictated by plant water status, temperature, nutrients and competition between shoots and from other organs. Therefore, a citrus tree appears to have a genetically predetermined level of growth that can occur in a few major flushes or almost continuously as minor growth flushes.

On each new shoot, new growth does not occur until the shoot is fully developed and the leaf axillary buds have had time to mature, usually about 2 months (Albrigo and Chica, 2011). Young non-bearing trees flush more often (commonly five flushes in a humid subtropical climate with no water limitation) while bearing trees normally flush three times. Under these conditions, more leaf area is produced in the first summer flush on bearing trees (Albrigo, 1996). In a Mediterranean climate, most of the leaf area is produced in the spring flush with a minimal summer flush and a late summer, early autumn flush (Spain, Garcia-Marí *et al.*, 2002 and Albrigo, personal observations in China). The limited summer flush and a later flush in Mediterranean climates is presumably due to the relatively short growing seasons and possibly competition from other organs.

Citrus trees, being evergreen, are densely foliated within a few years in the field, which causes extensive shading of interior portions of the canopy. For example, leaf numbers and surface areas of a citrus tree growing in California increased from 16,000 leaves of 34 m^2 area at age 3 years, to 37,000 of 59 m^2 at 6 years, 93,000 of 146 m^2 at 9 years and 173,000 of 203 m^2 at 29 years (Turrell, 1961). The rate of canopy development may differ from these values depending on climate. Many mature trees produce over 350 m^2 of leaf surface area with a leaf area index (LAI) of 12. LAI is a measure of the total leaf surface area per unit of land covered by the plant. In large trees, most of the fruiting occurs in the outer metre of the canopy because radiant energy is reduced to nearly zero at depths into the canopy greater than this and PPF is typically less than 100 ($\mu mol\ m^{-2}\ s^{-1}$). LAI is curvilinearly or log-linearly related to PPF in the canopy (Jahn, 1979). Obviously, at high LAI, light is limiting for shoot and flower production (PPF < 50 $\mu mol\ m^{-2}\ s^{-1}$) in the interior of the canopy, and most new growth occurs in the outer canopy. The problem of shading becomes even more acute for high-density plantings. Therefore, a judicious yearly pruning or hedging and topping programme is important to maintain light penetration and promote fruit production in mature trees (see Chapter 5).

Newly developing leaves are generally carbon importers until two-thirds to fully expanded after about 3–4 weeks in humid subtropics (Albrigo, 1996) and

4–6 weeks after full bloom (Erner and Bravdo, 1983). Stomata, which occur on the abaxial surface, are not fully developed and stomatal control over transpiration is poor. Net CO_2 assimilation continues to increase until about 6 months later at which time net CO_2 assimilation becomes stable until the latter stage of leaf ageing. Healthy citrus leaves can remain on the tree for as long as 3 years depending on tree vigour, but disease, pest pressures and low light levels significantly reduce leaf longevity (see Chapters 6 and 7).

Root growth

Temperature also regulates root growth and development, and water and nutrient uptake. Root and shoot growth have different temperature thresholds with root growth occurring at temperatures higher than about 7°C. Root growth, like shoot growth, occurs in flushes which often but not always alternate with shoot growth flushes (Bevington and Castle, 1985). Studies from Florida suggest that major root growth periods occur from May to June and particularly from August to September in the northern hemisphere (Bevington and Castle, 1985) (Fig. 4.4). Moreover, mean elongation rate for citrus roots is strongly temperature dependent for both pioneer and fibrous roots, showing a linear positive increase in growth from 17 to 30°C. Most of the root growth cycle data

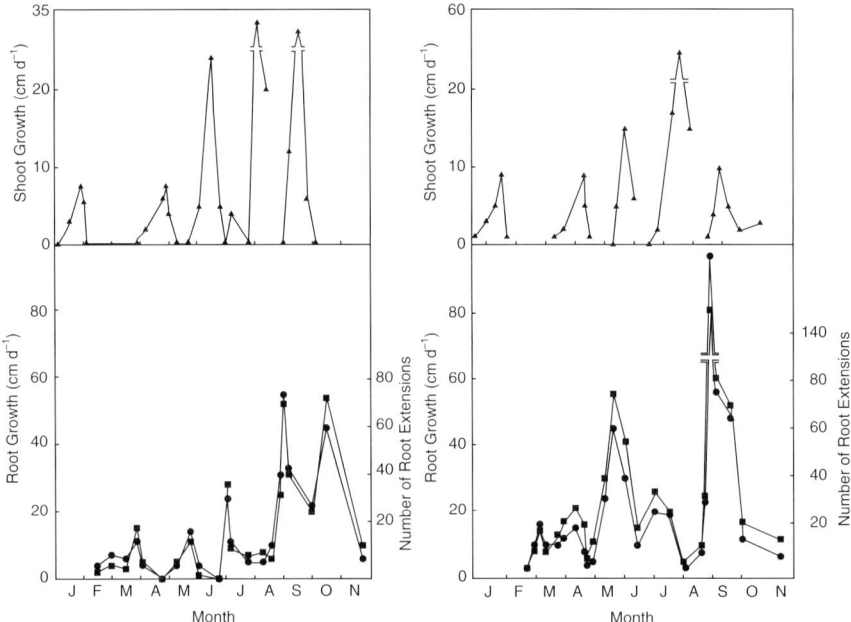

Fig. 4.4. Pattern of root and shoot growth during 1982 for 'Valencia' orange trees on rough lemon (left) and 'Carrizo' citrange (right) rootstocks. Data are shown for replicates of each rootstock. ▲ Shoot growth; ■ Root growth; ● Number of roots (Bevington and Castle, 1985).

has been obtained on young, non-bearing trees and the timing and extent of flushing may differ for mature bearing trees, likely having fewer root flushes than young trees as flushing does. Monselise (1947) did report on mature tree root flushes under various nutritional conditions in a dry Mediterranean climate.

Water and nutrient uptake rates are also positively correlated with root temperatures. Root hydraulic conductivity, a measure of water uptake, increases considerably from 10 to 30°C (Wilcox and Davies, 1981). Since uptake of nutrients is related to water uptake and respiration, it is also temperature dependent. Winter chlorosis of citrus leaves occurs in many subtropical regions classified as Mediterranean, probably due to cold winter soil temperature-induced reduction in nutrient uptake.

Plant and soil moisture status also affect root growth. Soil matrix potentials less than –0.05 MPa inhibited root elongation and the production of new roots (Bevington and Castle, 1985). Upon re-watering, root elongation rate increased but the primary increase in root area was due to production of new roots. Root volume of young citrus trees and production of new roots decreased significantly as soil moisture percentage decreased to less than 45% of available water for a sandy soil in Florida (Marler and Davies, 1990). Sufficient soil moisture is also necessary to support functions of non-growing roots. Nevertheless, excessive water for even as little as a few days may cause root death at high soil temperatures due to production of sulphur dioxide by soil-residing bacteria (see Drainage and Flooding, Chapter 5, p. 144).

Stress Responses of Citrus

An important aspect of citrus growth is its response to stress. These stresses may be biotic or abiotic, but the plant presumably responds through physiological processes. Many of these stresses are manifested through water stress. Reductions in root system, plugging of the xylem or simply exposure to drought may exhibit similar canopy response of more negative leaf water potentials and more extensive stomatal closure. Eventually, this can lead to reduced photosynthesis and fewer carbohydrates for fruit, canopy and root development. Stresses that impact the canopy or phloem may first reduce photosynthesis and/or carbohydrates, which will impact fruit production and carbohydrates for root growth and function. Root systems are the furthest from the leaves and are usually affected most if carbohydrates are limited. Excessive crop load is a stress that can lead to alternate bearing. The root system is usually the part of the plant that shows the greatest reduction in growth (Smith, 1976; Goldschmidt and Golomb, 1982).

A physiological action that plays a role in citrus response to water stress is the production of 1-aminocyclopropane-1-carboxylic acid (ACC) by root systems under stress. ACC is translocated to the canopy where it is converted to ethylene which can cause mature leaf and fruit abscission (Iglesias *et al.*, 2007). Gibberellin can inhibit this action.

ENVIRONMENTAL CONSTRAINTS ON FLOWERING AND FRUITING

Environmental factors, particularly water and temperature, regulate the time and extent of flowering in healthy citrus trees. Therefore, intensity and duration of flower production also varies with climatic region. Moreover, environmental factors largely regulate the type of flower inflorescences produced, their distribution on the tree, the percentage of fruit set and ultimately the resulting yield. Citrus will flower in most climates under a variety of photoperiods and several studies have reported that photoperiod does not regulate citrus flowering (Moss, 1969). Studies that suggested photoperiod played a role in citrus flowering were confounded by different durations of night cool temperature (Lenz, 1969). A recent study has substantiated general knowledge (Roistacker, personal communication) that light interruption of night duration will enhance vegetative growth of nursery trees (Brar and Spann, 2014b). Further evaluation of the possible effects of this on older citrus trees that are able to flower would be interesting. Flowering of citrus consists of induction, growth initiation and differentiation (evocation) periods which precede anthesis.

Flower Induction, Initiation of Growth and Differentiation

Flower bud induction commences with a cessation of vegetative growth during the winter 'rest' (non-apparent growth) period in subtropics or dry periods in tropical regions. Generally, on mature trees, shoot growth ceases and root growth rate decreases as temperatures decrease into the winter even though temperatures are not below 12.5°C. During this period vegetative buds develop the capacity to flower. Therefore, induction involves the events directing the transition from vegetative growth to production of inflorescences (Davenport, 1990). Davenport (1990) and Garcia-Luis *et al.* (1992) proposed that bud initiation may precede induction, but experimental evidence is contrary to this for *Citrus* species. Abbott (1935) showed that initiation of flower bud growth was not detected until after inductive temperatures had occurred. Increasing cool temperatures during induction increases the amount of flowering (Moss, 1969; Valiente and Albrigo, 2004) Cold and water stress are the primary inductive factors, with cold being the primary factor in subtropical climates and water stress in tropical climates (Cassin *et al.*, 1969). Temperatures below 20°C for several weeks appear to be required for induction of flower buds in significant quantities (Moss, 1969; Inoue, 1990).

In Florida, cold induction may start as early as October but more usually in November. Accumulated hours below 20°C may be as low as 600 to over 1200 depending on the year (Valiente and Albrigo, 2002). The induction process may be interrupted by warm periods – if warm enough and long enough after some cool inductive temperatures these periods stimulate initiation of flower bud differentiation of waves of buds that have different potential for flowering, with terminal buds being the easiest to induce (Valiente and Albrigo, 2004). Typically, under Florida conditions, two to three waves of flower buds will be developing with these waves flowering from early February to mid-April depending on these relevant climatic conditions in a given year (Chica and Albrigo, 2011). In the processing orange production area of São Paulo and Minas Geris States of Brazil a mixture of cool temperature and drought induction occurs with more cool temperature induction in the South and mostly drought induction in the North (Albrigo and Carrera 2012). The duration and intensity of these factors varies from year to year, but the goal is to achieve 60 to 70 d of drought and hope for over 600 h of cool temperatures. In 2009 and 2010, the southern areas of São Paulo's citrus area received over 1000 h below 20°C and an excellent bloom resulted. However, in 2000 drought was excessive, cool temperatures were insufficient and yields were the second lowest of an 8-year period. The lowest year, 2002, had neither sufficient drought nor cold induction (Albrigo et al., 2006b). Drought periods lasting more than 80 d are debilitating to the trees and reduce productivity (Albrigo and Carrera, 2015). Cassin et al. (1969) estimated from their observations that drought periods in excess of 70 d were detrimental. In the field, drought periods longer than 60 d are usually required to induce a significant number of flower buds (Cassin et al., 1969). The degree of induction is proportional to the severity and duration of stress (Southwick and Davenport, 1986). Under tropical climate field conditions 45 (Borroto and Rodríguez, 1979) to 60 d (Cassin, et al., 1969) are required to produce an economic crop. Water stress has been used as a practical means of inducing flowering in citrus for many years. In Italy, water is withheld from lemon trees during the summer until trees become severely water stressed. During this time flower buds are induced but rarely develop. The trees are then irrigated, 'forcing' (*forzatura*) them to flower in the autumn, thus producing a crop the next summer ('Verdelli' lemons). The Verdelli crop might be produced with less stress based on the Brazilian experience of using minimal water stress to induce flower buds and prevent growth without severe wilting.

Generally, trees following drought stress will flower 3–4 weeks after irrigating in a lowland tropical climate. This method has also been used to force off-season flowering in Israel and Spain and for 'Tahiti' limes in Florida. Application of gibberellic acid (GA_3) during the flower induction period will reduce or prevent induction and inhibit subsequent flowering (Monselise and Halevy, 1964; Davenport, 1990). GA_3 applied up to initiation of bud growth will revert induced flower buds to vegetative buds and buds near fruit usually do not flower (Chica, 2011), presumably due to high GA levels produced by the fruit.

Recent work in *Arabidopsis* (Blazquez, 2005) and then in *Citrus* (Nishikawa *et al.*, 2007) has shown that environmental triggers promote signal production (FT) that travels from leaves to the lateral meristems and acts with other genes produced in the meristems to up-regulate the flowering genes. *Citrus* and probably all plants have a similar array of flowering genes (Pillitterri *et al.*, 2004), but different environmental triggers for flowering. While photoperiod is operative in some of the earliest findings in *Arabidopsis*, it appears that other plants, including *Citrus*, have adapted a similar mechanism to other environmental triggers – specifically, cool temperatures and drought stimulate production of the flowering signal FT in leaves of *Citrus* (Chica and Albrigo, 2013a, 2013b). Although *Citrus* does not respond to photoperiod as *Arobidopsis*, it will not produce FT in constant light or constant dark (Chica and Albrigo, 2013b). This behaviour may be caused through the Constans gene which wasn't measured. Buds near fruit have lower FT than those further away (Chica, 2011). Goldberg-Moeller *et al.* (2013) found lower bud levels of FT after GA treatment, probably explaining the GA inhibitory effect on citrus flowering.

Initiation of growth and differentiation of flower buds

After sufficient inductive conditions have occurred, initiation of flower bud growth occurs with continuous warm temperatures and adequate soil moisture (Cassin *et al.*, 1969; Valiente and Albrigo, 2002). This can typically occur after 400–600 h of inductive temperatures with a week to 10 d of temperatures above 26°C (Albrigo *et al.*, 2002). Differentiation involves the histological and morphological changes in the vegetative meristem to become a floral meristem (Davenport, 1990). The dome-shaped apical meristem broadens and flattens and organogenesis begins with formation of the sepal primordia followed by carpel development. Once sepal primordia are developing, the flower bud will not revert to the vegetative condition even with application of GA_3 (Lord and Eckard, 1987). The anatomical status of the terminal apex determines the sequence and disposition of the lateral buds (Lord and Eckard, 1987). If the terminal apex forms sepals, the lateral buds will also form flowers; if the apex forms leaves the lateral buds may form thorns, which are modified stems, not leaves or flowers.

The rate of flower development from bud break to anthesis is independent of flower position or inflorescence type and is positively correlated with degree days (Lovatt *et al.*, 1984; Valiente and Albrigo, 2002). The typical progression of flower induction through anthesis in subtropical California (northern hemisphere, Mediterranean climate) is presented in Fig. 4.5 (Lord and Eckard, 1985). The resting bud undergoes microscopic bud break (bud scale loosening) as early as November and December; macroscopic bud differentiation occurs during December and January, with flower buds becoming visible as early as February. Anthesis then occurs in April. However, time of flower bud induction

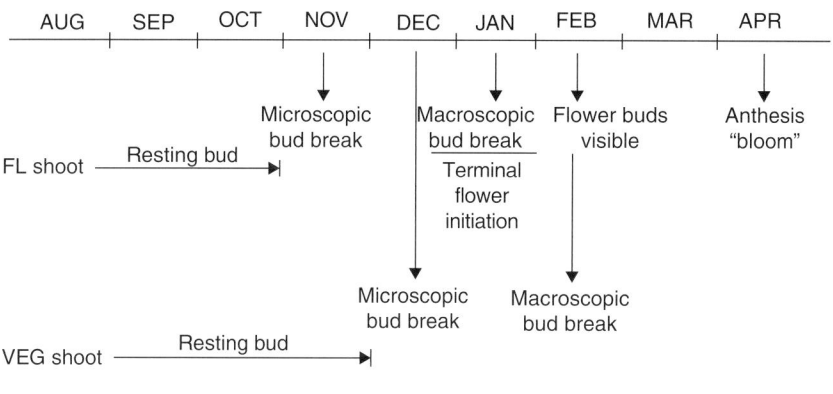

Fig. 4.5. Diagrammatic representation of the progression of events leading to anthesis in citrus in a subtropical climate in the northern hemisphere (Lord and Eckard, 1985).

and anthesis varies considerably from season to season depending on temperature and water availability, particularly in humid subtropical regions such as Florida and Brazil. Flower bud development from initiation to anthesis occurs in as little as 24 d in lowland tropics (Cassin *et al.*, 1969) to as much as 120 d in a Mediterranean climate (Moss, 1973) and from 55 to 85 d over a 20-year period in subtropical Florida (Valiente and Albrigo, 2002).

Anthesis

Anthesis (flowering) occurs after induction and differentiation when favourable temperature and soil moisture conditions exist. The best-fit minimum threshold temperature for flowering was 9.4°C or considerably lower than the minimum for vegetative growth, in a study by Lovatt *et al.* (1984). Citrus flowers are borne on cymes with the terminal flower bud breaking first followed by the most basal flower bud on the shoot. The second flower bud to the apical (terminal) position is the last to open, probably due to apical dominance (Jahn, 1973). Lord and Eckert (1985) also observed that the terminal flower was first to open for navel oranges. Lateral flower buds in positions 6 and 7 opened next in this study. Lack of additional bud break was probably due to insufficient induction for more flower buds. Under Florida conditions of moderate induction only 3–4 lateral buds consistently flower in most years (Valiente and Albrigo, 2004). Flower size generally decreases from the terminal to the last flower to open. Thus, the second flower position below the apex usually produces the smallest flower but also has the highest percentage fruit set on the shoot (Jahn, 1973) (Table 4.2). Late-opening flowers grow faster and persist longer than early-opening flowers (Lovatt *et al.*, 1984). Numbers

Table 4.2. Number of fruit set by position on the inflorescence and by the number of flowers in the inflorescence of 'Hamlin' and 'Valencia' oranges, 1970 (Jahn, 1973).

Fruit position	No. of flowers/inflorescence									No. of flowers	Per cent fruit set
	1	2	3	4	5	6	7	8	9		
'Hamlin'											
1[1]	40	1	1	1						731	5.9
2		14	18	4	10	5	3	1	1	327	17.1
3			3	3	3	2		2		257	4.7
4				6	3	3	2			185	7.6
5					2	1				127	2.4
6						1				56	1.8
7										21	0.0
8										8	0.0
9										1	0.0
No. of flowers	404	140	216	232	355	210	91	56	9	1713	
Per cent fruit set	9.9	10.7	9.7	6.0	4.8	5.7	4.4	3.6	11.1		7.5
'Valencia'											
1	4	2		2		1				640	1.0
2		9	6	7	7	3				337	9.5
3			4	4	3	1	1			268	4.9
4				3	6	3		2		212	6.6
5						3				130	2.3
6						2				68	2.9
7							1			19	5.3
8										4	0.0
No. of flowers	303	138	168	328	310	294	105	32		1678	
Per cent fruit set	1.3	8.0	6.0	4.9	5.2	4.4	1.9	6.3			4.4

[1]Position nearest the apex of the inflorescence.

of flowers per inflorescence are usually lower in humid subtropical locations like Florida (Valiente and Albrigo, 2004) than reported in Mediterranean climates like California (Lord and Eckert, 1985). This is probably related to the greater number and consistency of inductive hours of cool temperatures in Mediterranean climates (Albrigo, unpublished).

Five basic types of growth arise during flowering: (i) generative shoots (leafless or bouquet bloom); (ii) mixed shoots having a few flowers and leaves; (iii) mixed shoots having several flowers and a few leaves; (iv) mixed shoots having a few flowers and many leaves; and (v) vegetative shoots having leaves only. All types are borne primarily only on the previous season's growth

(Moss, 1969; Guardiola, 1981). All of the mixed shoot types produce flowers and leaves in the new growth flush (leafy blooms) in the spring or beginning of new growth rainy cycles. Leafy blooms set a higher percentage of flowers than leafless blooms (Guardiola, 1981). Generally, shoots with a high leaf to flower ratio, such as category (iv), produce and hold the greatest percentage of fruit to maturity. Vegetative shoots, category (v), produce the greatest shoot growth during the season and the generative shoots the least, with other categories intermediate and related again to the amount of fruit set. It is unclear whether the greater fruit set of leafy blooms is due to increased net CO_2 assimilation and carbohydrate levels provided by the newly developed leaves, or to improved vascular connections to the developing fruit mediated by hormones from the newly developed leaves or a greater sink capacity of mixed buds (Bustan *et al.*, 1995; Garcia-Luis *et al.*, 2002). Improved vascular connections could improve carbohydrate flow to inflorescences and reduce water stress in leafy compared with leafless blooms. All these hypotheses have merit, although under some conditions net CO_2 assimilation is similar for both leafy and leafless blooms and carbohydrate levels have not always correlated with fruit set. Alternatively, newly developing leaves 4–6 weeks after anthesis may decrease fruit abscission rather than increase fruit set (Erner and Bravdo, 1983).

Generally under Mediterranean conditions, more leafless blooms are produced on a tree than leafy blooms and a higher percentage of the fruit produced arises from leafless blooms (Erner and Bravdo, 1983). Percentage of leafless blooms of 'Marsh' grapefruit and 'Valencia' orange trees varied from 55–60 to 70–75 in 2 consecutive years in Florida (Jahn, 1973). There has been interest in finding practical methods of producing a higher percentage of leafy blooms, thus increasing fruit set and potentially yields. GA_3 sprays near the end of the induction period appear to reduce the number of flowers per inflorescence resulting in leafier blooms (Guardiola *et al.*, 1982). The relative abundance of leafy and leafless blooms seems to be related to temperature. Seasons with lower winter temperatures of long duration lead to development of more leafless blooms and more leafy blooms occur in seasons with higher winter temperatures. This no doubt relates to higher levels of FT produced by the leaves in response to more cool temperatures and more FT accumulation in the flower buds that are being induced (Chica and Albrigo, 2013b). Low to moderate temperatures during bloom (<20°C) produce a protracted bloom, while temperatures of 25–30°C produce a shorter bloom and fruit set period (Moss, 1970; Valiente and Albrigo, 2002). Relatively high temperatures during flower bud development may lead to earlier flower/ fruitlet drop and eventually better set (Moss, 1970, 1973). Application of urea sprays near the end of flower induction can increase the number of flowers (Ayalon and Monselise, 1960; Lovatt *et al.*, 1992; Ali and Lovatt, 1994; Albrigo, 1999) and possibly more leafy blooms, but only if more buds are induced to flower.

At least in an intermediate humid subtropical climate, flower bud induction, initiation of bud growth and full bloom can be monitored (Albrigo et al., 2002, 2006a). During the flower bud induction to bloom period (October–April) a monitoring system for Florida citrus is available (http://disc.ifas.ufl.edu/bloom) and related advisories are also available at the Citrus Research and Education Center website from November to April (http://www.crec.ifas.ufl.edu/).

Tree Internal Factors Associated with Flowering

Numerous studies have been conducted over the years to determine which physiological factors control flowering in citrus. Earlier information was reviewed in detail by Davenport (1990). The most likely control factors are carbohydrates, hormones, temperature, water relations and nutrition. The carbohydrate theory has its basis on the fact that branch or trunk girdling increases flower induction, fruit set and starch levels in the branch. This probably occurs because girdling inhibits phloem transport of carbohydrates to the roots. In contrast, several studies have found no correlation between starch levels in leaves and twigs and flowering of citrus (Oslund and Davenport, 1987; Davenport, 1990). Carbohydrate levels in roots, however, are in some instances associated with flowering in alternate bearing mandarins. Extremely low levels of carbohydrates in the roots due to excessive crop loads have been associated with limited shoot and flower production the following year (Goldschmidt and Golomb, 1982). This condition is especially severe in 'Murcott' ('Honey') mandarin trees which in some instances produce so much fruit that root carbohydrates become so depleted that trees die ('Murcott' collapse, Smith, 1976). In normal reproductive cycles, roots are also a spring source of stored carbohydrates, therefore once again the correlation between carbohydrate levels and flowering may not be causal. A critical level for carbohydrates is at least implied even if higher levels of carbohydrates do not elicit greater flowering. This appears to be a major factor in alternate bearing of (at least) citrus (Goldschmidt and Golomb, 1982).

Hormonal control of citrus flowering has also been extensively studied for many years (Davenport, 1990). Some studies involve application of endogenous hormones to citrus shoots followed by an evaluation of the extent of flowering. For example, application of GA_3 to citrus shoots before differentiation inhibits flowering (Monselise and Halevy, 1964). Therefore, GA_3 appears to regulate some aspects of flowering. However, studies on changes in endogenous levels of GA_3 indicated no significant relationship between GA_3 and type of shoot (generative or vegetative) produced (Davenport, 1990). GA_3 does inhibit production of the flowering signalling gene (FT) in citrus (Chica, 2011; Goldberg-Moeller et al., 2013). It is presumed that GA_3 produced by fruit is the cause of reduced or no flowering near a current maturing fruit since FT is reduced (Chica, 2011).

Plant nutrition is directly and indirectly associated with flowering of citrus trees. High levels of leaf N in particular for young citrus trees may induce excess vigour and produce a vegetative rather than flowering tree. In contrast, low leaf N levels may promote extensive flowering, although fruit set and yields are poor. Severely N-deficient trees produce few flowers. Therefore, maintaining leaf N levels in the optimum range (2.5–2.7%) produces a moderate number of flowers but the greatest fruit set and yields. Nitrogen as ammonia may directly increase flowering via regulation of ammonia and polyamine levels in the bud (Lovatt *et al.*, 1988), but further work seemed to implicate N metabolism after initiation of growth, perhaps in fruit set (Lovatt *et al.*, 1992). Water or low temperature stress increases leaf ammonia levels and flowering. Moreover, a winter application of urea during the stress period to 'Washington' navel orange trees in California increased leaf and bud ammonia levels and number of flowers per tree; the number of flowers produced was positively correlated with the duration of low-temperature induction (Table 4.3). Also, spray applications of 1% urea 6–8 weeks before bloom increased flowering and yields of 9-year-old 'Shamouti' orange trees (Rabe and van der Walt, 1992) and a similar study in Florida found that urea as well as phosphorous acid sprayed at the end of the induction period increased flowering and yields of 'Valencia' oranges in south Florida (Albrigo, 1999).

It is obvious from the previous sections that the physiological basis of flowering in citrus is not clearly understood, although many pieces of the puzzle have been assembled. Flower buds are induced by cool temperatures or drought stress and flowering level is related to level of induction. Buds differentiate when warm temperatures or soil moisture release the stress. Time

Table 4.3. Effect of low temperature stress and foliar application of urea on the leaf NH_3–NH_4^+ content and on the flowering of 5-year-old rooted cuttings of the 'Washington' navel oranges. (Lovatt *et al.*, 1988).

Duration of low temperature stress (weeks)	Leaf NH_3–NH_4^+ content[1]		Average no. of flowers per tree	
	Without urea[2]	With urea[3]	Without urea	With urea
0	456a	–	6a	–
4	559b	928 (166%)	117b	227 (194%)
6	583b	1253 (215%)	131b	310 (230%)
8	672a	900 (134%)	347a	437 (126%)

[1]Ammonia (µg g^{-1} dry weight) determined during first week after transfer to warmer temperature.
[2]Mean separation by Duncan's multiple range test, 5% level.
[3]Low biuret urea applied at the rate of 1.5 g per tree at the end of the low temperature treatment. Figures in parentheses represent percentage of the value recorded in the trees where no urea was added.

to flowering is also temperature dependent. Although these processes may be under hormonal control, only GA_3 has clearly been shown to play a role (inhibitory), but cytokinins appear to be needed for final flower development (Garcia-Luis *et al.*, 1989; Albrigo, unpublished data from excised shoot flower forcing experiments). Part of the difficulty in understanding the control of flowering in citrus is because some buds appear very easy to induce to flower while other buds, on the same branch, appear to require much more stress to be induced. It appears that this happens because, as FT signal protein is produced by the leaves in response to inductive conditions, it moves toward the shoot apex and as more is produced sufficient amounts can then accumulate in buds below the apex to eventually trigger flowering gene upregulation further down the shoot.

Pollination and Fruit Set

Most commercially important citrus species do not require cross-pollination to set and produce a crop. The exceptions to this include some mandarin hybrids such as 'Orlando' tangelo and 'Sunburst' tangerine. Pollination is essential, however, for seed production, or in stimulating ovary growth in nearly parthenocarpic cultivars such as 'Hamlin' sweet oranges. Parthenocarpy is the capacity to produce fruit without the stimulus of sexual fertilization. Some strongly parthenocarpic cultivars like 'Marsh' grapefruit, however, produce fruit even if the stigma and style are detached prior to pollination. This may be true for clementine cultivars also, but unlike grapefruit clementine cultivars do not set well in warmer climates conducive to grapefruit set.

Temperature has a significant effect on pollination efficiency, either indirectly by affecting bee activity in the orchard (bees are the primary pollinator for citrus), or directly by affecting pollen tube growth rate. Bee activity in an orchard is minimal at temperatures below 12.5°C. Once pollen grains land on the stigma their germination and growth rates through the style are enhanced at high temperatures (25–30°C) and reduced or totally inhibited at low temperatures (<20°C). Pollen tube growth through the stylar channels may take from 2 d to as long as 4 weeks depending on cultivar and temperature (Frost and Soost, 1968).

Accomplishing pollination is also affected by several other factors that determine bee activity and availability of suitable pollen (Albrigo and Russ, 2002). Bees prefer cultivars with more nectar, while mandarin types that require cross-pollination usually have small flowers (Fig. 4.6) and low nectar volumes. The small-flowered cultivars have flowers of about 50% less weight, 35% less width and produce about a quarter of the nectar of large-flowered cultivars. These cultivars also flower later when better nectar and pollen sources have passed their peak, but bees will continue to visit these preferred cultivars if the colonies are not moved. More resourceful strategies are needed to get good pollination of the mandarin cultivars.

Fig. 4.6. Comparison of small flowers typical of many mandarin cultivars with *C. reticulata* parentage ('Fallglo') vs large flower of a *C. sinensis* cultivar ('Hamlin').

More and more mandarin hybrids are produced so that no seeds are produced (truly seedless, newer selection often irradiated or triploids) because of consumer preference (see Chapter 1). Therefore, much of this problem is much less important – in fact, avoiding bee pollination is a more serious commercial problem for most of the clementine cultivars than getting acceptable cross-pollination, but increased use of triploid, irradaited or other naturally seedless cultivars will make seedy mandarin production a non-problem.

Initial fruit set
Initial fruit set, subsequent fruit drop and ultimately fruit yields are affected by several environmental and physiological factors. Most commercially important citrus cultivars bloom prolifically in cool winter climates producing as many as 100,000–200,000 flowers on a mature tree; however, fewer than 1–2% of these flowers produced harvestable fruit on mature orange trees in a Mediterranean climate (Erickson and Brannaman, 1960) (Table 4.4). An initial drop period occurs from flowering until 3–4 weeks postbloom. A second drop period occurs from May to June in the northern hemisphere and November to December in the southern hemisphere. The initial drop period involves the abscission of 'weak' flowers and fruitlets with defective styles or ovaries, or flowers which did not receive sufficient pollination (where applicable). During initial phases of abscission (up to 6–8 weeks after bloom) most fruit abscise at the zone between the pedicel and the stem. Citrus fruit have two abscission zones, one at the base of the pedicel and the other at the base of the ovary.

Hormones are probably involved with the capacity of fruit to persist during initial fruit set based on circumstantial evidence. Spray application of GA_3 increases initial fruit set for weakly parthenocarpic species like 'Orlando' tangelo (Krezdorn, 1969), although GA_3 does not improve fruit set of many other citrus cultivars. Application of gibberellin did improve fruit set of 'Fortune'

Table 4.4. Abscission of reproductive structures per tree from 'Washington' navel and 'Valencia' orange, Riverside, 1958–1959 (Erickson and Brannaman, 1960).

	'Washington' navel						'Valencia' orange				
			No. of fruit						No. of fruit		
Ovary diam. (mm)	No. of buds	No. of flowers	With pedicel	Without pedicel	Total no.		No. of buds	No. of pedicel	With pedicel	Without pedicel	Total no.
<1	59,635	674	1,002	12	61,323		20,020	341	893	173	21,427
2	22,790	14,939	13,953	344	52,026		4,099	565	5,468	2,621	12,753
3	10,804	15,295	25,043	541	51,683		1,365	1,031	11,151	1,880	15,427
4	3,114	2,321	13,469	589	19,493		499	219	7,610	2,533	10,861
5	–	6	3,853	460	4,319		–	7	1,956	2,161	4,124
6	–	–	2,399	634	3,033		–	–	702	1,898	2,600
7	–	–	1,250	643	1,893		–	–	137	1,164	1,301
8	–	–	586	450	1,036		–	–	59	582	641
9	–	–	417	663	1,080		–	–	21	699	720
10	–	–	179	462	641		–	–	5	611	616
11	–	–	77	357	434		–	–	2	453	455
12	–	–	37	275	312		–	–	1	392	393
13	–	–	20	194	214		–	–	5	297	302
14	–	–	7	137	144		–	–	–	237	237
15	–	–	7	118	125		–	–	–	158	158
16	–	–	3	79	82		–	–	–	106	106
17	–	–	2	71	73		–	–	–	83	83
18	–	–	2	55	57		–	–	–	65	65

				2		40	42						34	34
19	—	—	—	—	—	40	42	—	—	—	—	—	34	34
20	—	—	—	—	—	32	32	—	—	—	—	—	23	23
>21	—	—	—	—	—	232	232	—	—	—	—	—	61	61
Total	96,343	33,235	62,308	6,388	198,274	25,983	2,163	28,010	16,231	72,387
Average no. of mature fruit/tree					419					708
Total flower buds/tree					198,693					73,095
Distribution	48.5%	16.7%	31.4%	3.2%	crop=0.2%	35.5%	3.0%	38.3%	22.2%	crop=1%

mandarin in Spain (Duarte and Guardiola, 1996), and often navels, but benzyladenine (BA), a cytokinin, usually did not.

Time of anthesis is also linked to percent initial fruit set. Studies from California suggest that flowers opening early in the bloom period have much lower set than those opening later. Low temperatures early in the season may limit bee activity or pollen tube growth as discussed previously. In addition, fruits developing early during bloom on leafless shoots grow more slowly than those developing on leafy shoots later during bloom due to higher temperatures at that time (Lovatt *et al.*, 1984). Extremely high temperatures (>40°C), however, cause excessive fruit drop, particularly for cultivars like navel oranges which are inherently more sensitive to stress than most other citrus species (Davies, 1986a). Similarly, very high temperatures during the early fruit development period also reduce set of round oranges in Brazil's São Paulo processing orange industry (Albrigo, personal observations).

Physiological drop
Although the initial drop period in citrus apparently is primarily for physiological reasons, the term 'physiological drop' generally has been reserved for the drop wave that occurs in May to June in subtropical regions in the northern hemisphere and in November to December in the southern hemisphere. Physiological drop is also called 'June' (or 'November') drop and describes the abscission of fruitlets as they approach 0.5–2.0 cm in diameter (mandarin to oranges or grapefruit result in the size range). Initial set is tied to pollination and fertilization of the ovule with gibberellin playing an important role in successful set (Talon *et al.*, 1998). Fruit drop occurs from the abscission zone at the base of the fruit leaving the pedicel attached to the tree temporarily. Physiological drop is most probably related to competition among fruitlets for carbohydrates, water, hormones and other metabolites and occurs in most perennial crops. The problem, however, is greatly accentuated by stress, especially high temperatures and/or water deficit. Consequently, physiological drop is usually most severe where leaf temperatures may reach 35–40°C and where water stress is a problem, as in arid regions of southern California or South Africa and when drought occurs at this time in São Paulo, Brazil. One hypothesis is that high temperatures and severe water stress cause stomatal closure with a concomitant reduction in net CO_2 assimilation. Fruit abscission then results because the fruit maintain a negative carbon balance. Direct heat damage may also be a factor. Overhead sprinkling reduces temperature and has been successfully tested in some regions to reduce physiological fruit drop (Brewer *et al.*, 1977), but little use of this practice has occurred. Differences in hormone and/or carbohydrate levels also are usually involved (Talon *et al.*, 1998; Ruiz *et al.*, 2001; Iglesias *et al.*, 2003, 2007). Nutrition can affect set since the new flush and fruitlets are in competition for a limited supply of nutrients (Sanz *et al.*, 1987). The relatively small amount of carbohydrates required

per fruitlet in mandarins compared to larger fruit like grapefruit contributes to the heavy set of mandarins that can occur, leading to significant alternate bearing (Bustan and Goldschmidt, 1998).

Alternate bearing
Many perennial crops including citrus have cultivars that tend to have heavy crops one year and very light crops the next. This may continue for several years (Fig. 4.7) (Wheaton, 1986). Initiation may occur due to an unfavourable environmental condition (freeze, drought, late harvest of the current crop, etc.) that greatly reduces the crop early in its development or extends the on-tree time of a current crop. Seedy mandarins are very susceptible because the early development of each very small fruit requires very little resources, unlike grapefruit or even round oranges (Bustan *et al.*, 1996). After the physiological drop the many mandarins that often set require substantial nutrient and carbohydrate resources. Other organs, particularly roots, become carbohydrate deficient and shoots that would flower the next year have few flowers, leading to a small crop and alternate bearing (Goldschmidt, 1999). An alternative scenario is for the harvest to be late, particularly on a late maturing cultivar like 'Valencia' orange, which overlaps with the new crops flowering and fruit set. Competition between the crops results in fewer flowers, but particularly poorer set, thus leading to a reduced crop the following year (Jones *et al.*, 1964). Fruit thinning of potentially heavy crops with naphthaleneacetic acid (NAA) at the beginning of the physiological drop period can alleviate this problem (Wheaton and Stewart, 1973), but results depend on tree vigour and climatic conditions making results erratic. Chemical thinning has been little used because of variable results. Earlier harvest of heavy crops can also minimize alternate bearing.

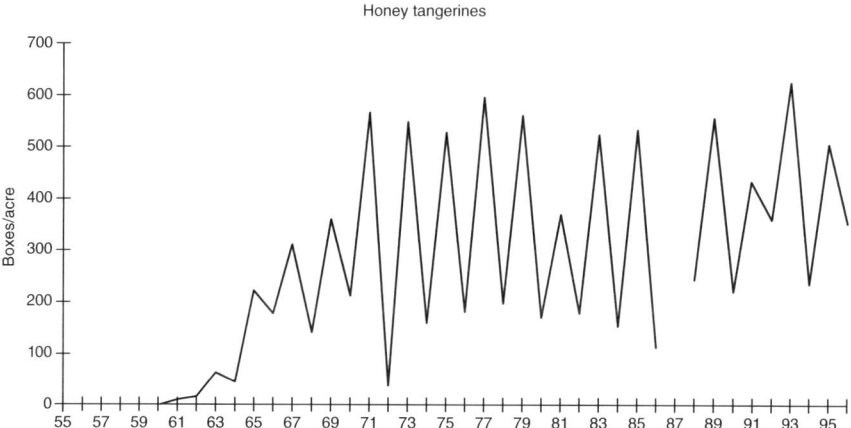

Fig. 4.7. Thirty years of alternate bearing in a central Florida 'Murcott' ('Honey' tangerine) block. Data was provided courtesy of Dr John Attaway (Wheaton, 1997).

ENVIRONMENTAL CONSTRAINTS ON FRUIT YIELDS

Because climate affects flowering, fruit set, fruit drop and cumulative fruit number per tree, it is logical to expect that fruit yields will also be affected. In addition, fruit size, another component of yield, is affected by climate, in particular moisture and temperature. Cultural practices, as well as the choice of cultivar and rootstock (Chapters 2 and 3), and nutrition (Chapter 5), also affect yields but are discussed in those chapters.

Generally, the years (tree age) to production of an economic crop are shortest for the low tropical and humid subtropical regions and longer for arid, cooler subtropical areas. There is then a period where tree size (canopy bearing volume) and yield increase asymptotically for all three regions. Due to shorter growing seasons, yields in low tropical regions become maximum 10–15 years after planting, while yields in the humid subtropical regions continue to increase, reaching a maximum at 15–20 years. Yields from the Mediterranean climate increase more gradually, with little winter growth, peaking around 20–25 years. Some production practices like modified open hydroponics (Kadyampakeni *et al.*, 2013a) and tree spacing can significantly shorten the time to maintenance tree size (see Chapter 5). Orchard longevity is usually greater in Mediterranean-type climates than humid subtropical or low tropical climates. Some orchards in Spain and Italy, for example, have been or were productive for over 200 years, while most orchards in humid subtropical and low tropical regions are now much less than 100 years old. In the face of huanglongbing (HLB), orchard life is often less than 10 to 15 years for a new planting. Disease and pest pressures account for much of these differences in orchard longevity.

Excluding freeze events, yields of commercially important citrus cultivars vary considerably from year to year, primarily due to climatic factors. Research from South Africa (DuPlessis, 1984) suggests that a large part of the yearly variation in yields of navel oranges is due to climatic factors, in particular temperature during the physiological drop period. Similarly, 38 years of data on navel oranges in California also clearly demonstrates the irregularly cyclic nature of yields related to climate (Jones and Cree, 1965) (Fig. 4.8). Major factors associated with this variation include temperature during bloom, physiological drop, and throughout the growing season. In general, excessively high temperatures during these times reduced yields. Soil moisture was not a major factor in this instance since most of the orchards were irrigated.

The maximum yield and year-to-year variation obtained from a mature citrus orchard is also a function of climate, although factors such as soil type, cultivar and rootstock selection, technological capacity and disease constraints influence yields within a climatic region. Tropical regions generally do not have adequate cool temperature induction and drought induction is erratic, therefore flowering is usually limited. Humid subtropical areas, notably Florida and Brazil, in many years have inadequate cool temperature or

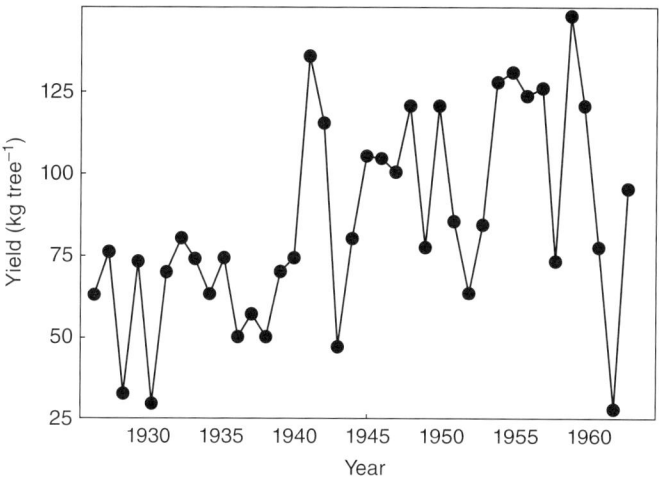

Fig. 4.8. Yields of 'Washington' navel in California over a 38-year period (Jones and Cree 1965).

drought induction of flower buds, or in the case of locations in Brazil, excessive drought debilitates the tree near flowering time. Induction hours in Florida can differ by as much as 200 h from one year to another, which has resulted in 60% more flowers and 40 million boxes between 2 years in the early 2000s (http://disc.ifas.ufl.edu/bloom and Albrigo, unpublished data). These are examples of major limits to yield in many years in these areas (Valiente and Albrigo 2002, 2004; Albrigo et al., 2006a; Albrigo and Carrera, 2015). With drip irrigation, a mild drought level can be maintained through the flower induction period and continued until the rainy season begins in most years (Albrigo and Carrera, 2015).

Soil moisture as well as temperature become major factors regulating yields in the humid subtropics (Koo, 1963) and tropical areas, and in arid and semi-arid regions (Hilgeman, 1977) later in fruit development, particularly during fruit set. Furthermore, maximum yields are related to cultivar and rootstock, with grapefruit generally being most productive, followed by sweet oranges, lemons and mandarins. The most productive citrus-growing regions generally occur in the humid subtropics when tree health is not limiting. For example, in Florida, average yields for grapefruit and sweet oranges were 41 and 31 tonnes ha^{-1}, respectively. However, maximum yields may exceed 120 tonnes ha^{-1} for grapefruit and 100 tonnes ha^{-1} for sweet oranges. In contrast, per ha yields in Brazil, also located in the humid subtropics and tropics, have been about half that of Florida primarily due to a lack of adequate irrigation in most major growing regions of São Paulo State. When irrigated, yields are similar to those in Florida and more orchards are now provided with drip irrigation. Similarly, per ha yields in humid subtropical regions of China were less than half those of

Florida in this instance also due to a lack of irrigation and severe tree losses due to viroids like citrus exocortis viroid (CEV) or in southern China to citrus yellow shoot disease (HLB, citrus greening) (see Chapter 7). Most citrus nursery trees in China are now disease free, are grown in protected cover, and many northern area orchards include irrigation. With production practices to overcome climatic factors like drought during fruit set and using disease-free plant material, yields may not be limited in these areas. Over time there has been an enormous increase in yields through technological advances, such as proper irrigation and fertilization and pest control. An example of recent changes has been associated with an 'Advanced Citrus Production System' that integrates open hydroponics with high density planting to get trees into high production at an early age in the face of HLB disease (Schumann *et al.*, 2016). Yields are still limited in regions with excessive cloud cover or under near-tropical conditions where flower bud induction is limiting.

By comparison, average and maximum yields in semiarid or arid subtropical regions such as California, Spain, Italy, South Africa, Israel and Australia are often lower than those in Florida or irrigated regions of Brazil. In most cases, yields are less because of smaller tree size due to less hu accumulation, less winter photosynthesis and in some instances due to more intense physiological drop in arid regions. Water quality (especially salinity) and quantity may also limit yields in areas such as Australia. This is not to say, however, that yields are always lower in semiarid or arid subtropical regions than in humid subtropical ones. Yields of 80 tonnes ha^{-1} or more for sweet oranges can occur in many of these regions such as Israel when water and diseases are not limiting, but maximum yields are more in the 40 to 60 MT ha^{-1} range. Therefore, blanket generalizations concerning yields in various regions are not possible.

Yield potential in tropical regions is also influenced by poorly drained or nutrient-deficient soils in some cases and severe pest and disease pressures in most regions (see Chapters 6 and 7). Yields in lowland tropical regions, in particular, are reduced significantly by the latter problems. Furthermore, trees tend to be excessively vegetative under the temperature and moisture conditions of many low tropical regions. The excessive vigour is not conducive to flower bud induction, thus decreasing yields. For example, in tropical areas such as southeastern Mexico, sweet orange yields average only 15–20 tonnes ha^{-1} due primarily to a lack of sufficient irrigation, fertilization and pest control. In contrast, mid-elevation tropical regions like those in Central and South America are potentially as productive as humid subtropical regions if factors like soil moisture and cloud cover are not limiting. Unfortunately, most of these areas are not irrigated to manage drought periods. Both flower induction and fruit development periods often suffer from excessive drought.

The yield potential of an orchard refers to the maximum amount of fruit mass that is produced per unit of land, but does not consider juice quantity and quality or percentage marketable fruit, a factor of considerable importance to fresh fruit producing countries. Percentage marketable fruit is generally greater in semiarid or arid, subtropical climates such as Spain, California, Italy, etc., than in humid subtropical or tropical locations due to less intense disease and pest pressures, fewer fruit blemishes and development of more intense and desirable peel colour (see Chapters 6–8).

ENVIRONMENTAL FACTORS AFFECTING FRUIT GROWTH, DEVELOPMENT AND QUALITY

Climate has a significant effect on fruit growth and quality, as clearly demonstrated by Reuther and Rios-Castano (1969) when they compared various fruit quality factors in tropical regions of Colombia with arid and coastal subtropical regions of California. Regions in Colombia were further subdivided into lowland, mid- and highland areas and the California regions into arid and coastal regions differing in average and extreme temperatures, rainfall and humidity.

Fruit Growth

Fruit growth of most citrus cultivars follows a sigmoid pattern which may be subdivided into three phases (Bain, 1958). Phase I is the cell division and differentiation phase in which nearly all the cells of the mature fruit will be produced. It is this initial cell number that primarily will ultimately determine final fruit size. The duration of this phase ranges from initiation of flower bud development until about 1 to 1.5 months following bloom depending on climatic conditions and cultivar. During phase I, cells also differentiate into the various tissue types such as sections (juice vesicles and segment walls), albedo, flavedo, etc. (Bain, 1958, Burns et al., 1992). Phase II, the cell enlargement phase, produces a rapid increase in fruit size and percentage total soluble solids (TSS). During this time cells may increase in volume by 1000 times. Phase III duration varies with cultivar from 2 to 3 months for lemons and limes to more than 6 months for late maturing sweet oranges and grapefruit. For a late maturing cultivar such as 'Valencia' orange, the duration of phase III may also vary from 3 to 4 months in lowland tropics (Cartagena, Colombia) to 10 months under cool, coastal subtropical conditions (Santa Paula, California). Phase III also includes a drop in acidity and peel colour changes.

Peel colour begins to change from green to yellow or orange (except for oranges in lowland tropical regions) towards the end of phase III, the maturation phase, which is typified by a levelling-off of growth and a slight, gradual increase in TSS along with a more rapid decrease in total acidity (TA). The maturation phase may continue for 9–12 months for 'Valencia' oranges under some Mediterranean subtropical conditions, but is 1–2 months shorter in humid subtropical and considerably shorter yet under lowland tropical conditions. Total time from bloom until attainment of an acceptable TSS:TA ratio ranges from 6–7 months in the low tropics to 14–16 months in Mediterranean-type climates for 'Valencia'. Cultivars that mature more quickly have a more compressed development time with too little time to accumulate soluble solids when grown in a warm tropical climate. After inherent differences in maturation time between cultivars, the time interval required for fruit development is a function of temperature.

Fruit growth rate within each climatic region is primarily a function of temperature during each developmental stage and soil moisture, particularly during phase III (Fig. 4.9). The highest mean temperatures provide the fastest fruit growth rates (Cartagena, Colombia, lowland tropical) and lowest mean temperatures the slowest (Santa Paula, California, semiarid subtropical, coastal) which is consistent with the heat unit concept discussed previously (Table 4.1). However, adequate soil moisture via rainfall or irrigation significantly improves fruit size during phase III but a corresponding dilution of soluble solids results (see the projected fruit growth rate [dotted line] for Cartagena). Fruit growth rates are intermediate and similar for high tropic (Medellin, Colombia) and arid subtropical (Indio, California) regions. Fruit

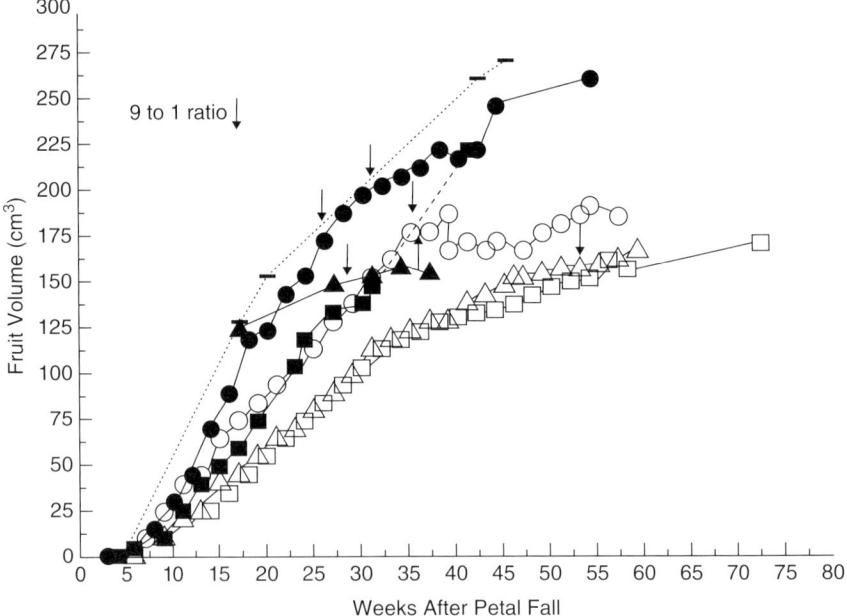

Fig. 4.9. Comparison of growth rates of 'Valencia' oranges in tropical and subtropical locations. The mid-bloom or 'zero' petal fall dates were estimated as follows: ∆ Riverside, 4 May 1961; ○, Indio, 6 April 1962; □, Santa Paula, 11 May 1964; ▲ Cartagena, about 25 June 1966; ■ Medellin, 20 January 1967. The Palmira data (●) are for the local cultivar 'Lerma' which appears to be a mid-season type (Reuther and Rios-Castano, 1969).

growth rates for humid subtropical regions like São Paulo, Brazil, or Orlando, Florida, are similar to those of Palmira, Colombia.

External Quality

External quality factors including peel colour, incidence of blemishes and fruit shape are significantly affected by climate. Peel colour of citrus fruits results from a combination of pigments including chlorophyll, carotenoids, anthocyanins and lycopene among others. Initially the peel cells contain high levels of chlorophyll allowing the fruit to produce some photosynthetic metabolites. Typically the peel provides no more than 10% of the total carbon requirements of the young fruit and much less for the mature fruit, with the remainder being imported from the leaves. In growing regions where the average temperatures remain high all year (e.g. lowland tropical regions), chlorophyll levels remain high for oranges and mandarins and the fruit peel remains green. However, as air and soil temperatures fall below 15°C during fruit maturation, chlorophyll is degraded and chloroplasts are converted to chromoplasts containing yellow, orange or red pigments (carotenoids, lycopene, etc.) (Thomson et al., 1967). Carotenoid synthesis is reduced above 35°C or below 15°C but still occurs at temperatures conducive to chlorophyll degradation (Meredith and Young, 1969; Young and Jahn, 1972). Peel colour in grapefruit results from yellow carotenoids and red lycopene synthesis in pink and red cultivars (Erickson, 1968; Meredith and Young, 1969). Lycopene synthesis proceeds even at moderate average temperatures, but synthesis is retarded at low temperatures. Grapefruit attain a yellowish to red-orange peel colour, depending on cultivar, even in low tropical regions after slow chlorophyll breakdown, but internal colour fades during cool winters in humid subtropical climates. Of interest, the peel of certain cultivars like 'Valencia' orange can undergo reversion from orange to green, a process termed 're-greening'. In this instance, chromoplasts revert to chloroplasts (Thomson et al., 1967). Re-greening generally occurs in late-harvested cultivars after fruit have remained on the tree late into their normal harvest period and spring high temperatures and adequate soil moisture result in stimulation of chlorophyll synthesis. The blood oranges, popular in Italy (see Chapter 1), contain anthocyanin pigment and require prolonged cool winter temperatures for good colour development.

Besides climate (temperature), tree vigour also has a pronounced effect on fruit colour. Generally, vigorously growing trees produce more poorly coloured fruit than slower growing trees. As a result, any factor that enhances vigour delays the development of peel colour. Fruit from young vigorous trees or from those on vigorous lemon-type rootstocks are generally more poorly coloured than those from slower growing trees within a particular growing region. Further, trees receiving excessive nitrogen tend to have poor peel colour. As light is necessary for carotenoid and anthocyanin synthesis, shaded fruit will be more poorly coloured than exposed fruit. Orchards that have trees grown into hedgerows produce few, but poorly coloured, fruit toward the interior of the tree.

Blemishes are major factors that prevent citrus fruits from being marketed fresh. Blemishes are produced due to abiotic and biotic factors (see Chapter 8). Incidence of blemishes caused by biotic factors is usually much greater in humid subtropical and tropical regions than in semiarid or arid subtropical regions due to increased pest and disease pressures. Insect, mite and fungus diseases may be extremely intense and difficult to control in humid lowland areas of the tropics (see Chapters 6 and 7).

Fruit shape is adversely affected in grapefruit by high temperatures during the cell division stage. Excessive cell division in the albedo near the stem end produces an elongated, 'sheepnosed' fruit. In addition to high spring temperatures, light crop load, higher N and lower K tend to increase sheepnosing (Syvertsen *et al.*, 2005).

Internal Quality

Water comprises a major portion of the fruit mass (85–90% by weight), with carbohydrates contributing 75–80% of the TSS (see Chapter 8). As a consequence, regulation of water and carbohydrate loading into the citrus fruit has a great impact on internal fruit quality. Fruit growth is a function of tree water status and carbohydrate partitioning as well as temperature. Fruit shrink and swell diurnally as the water relations of the tree change. Ultimate fruit size is increased by irrigation and/or rainfall (Taylor and Furr, 1937). The fruit in addition serve as a storage organ for water. Leaves on a detached fruitless twig wilt within hours, while those on a detached fruited twig remain turgid for many hours. A majority of the water translocated to the leaves is stored in the fruit peel. Changes in fruit volume have been used to schedule irrigation in some areas such as Australia, California and Arizona. This practice, while quite reliable, is not widely used due to the amount of labour involved (Hilgeman, 1977), although the process can be automated using electronic monitoring equipment it is usually done on tree trunks rather than fruit (Ortuño *et al.*, 2006).

TSS accumulate most rapidly in the fruit under lowland tropical conditions and most slowly under cool coastal Mediterranean conditions (Fig. 4.10). Maximum levels of TSS are usually attained in the midtropics (Palmira, Colombia) and in humid subtropical regions with warm winters such as São Paulo, Brazil, or Florida, USA. Levels of TSS are intermediate in semiarid or arid subtropical and highland tropical areas such as Riverside, California, or Medellin, Colombia.

Of equal or more importance to attainment of commercial edibility of citrus fruit than TSS is the rate of decline in TA. Under hot lowland tropical conditions such as those in Cartagena, Colombia, acidity decreases rapidly from 2.0% (for sweet oranges) to below 0.5% (Fig. 4.11). Orange juice from

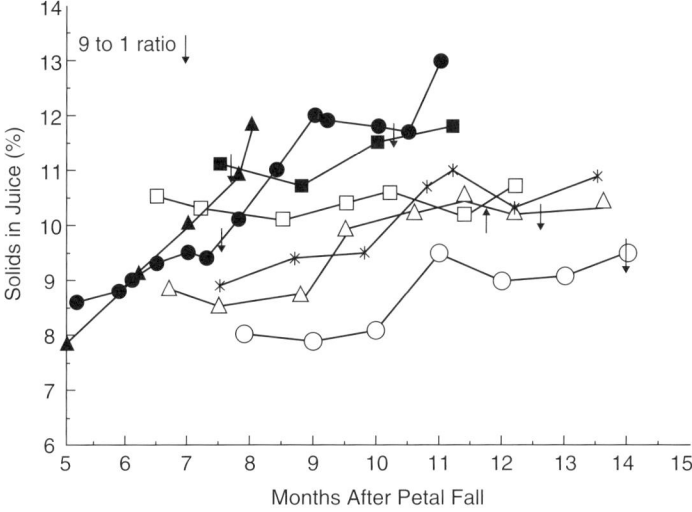

Fig. 4.10. Comparison of trends in per cent TSS in the juice of 'Valencia' oranges grown in Colombia and California. ▲ Cartagena; ● Palmira; ■ Medellin; ✶, Indio; □ Lindsay; △ Riverside; o Santa Paula (Reuther and Rios-Castano, 1969).

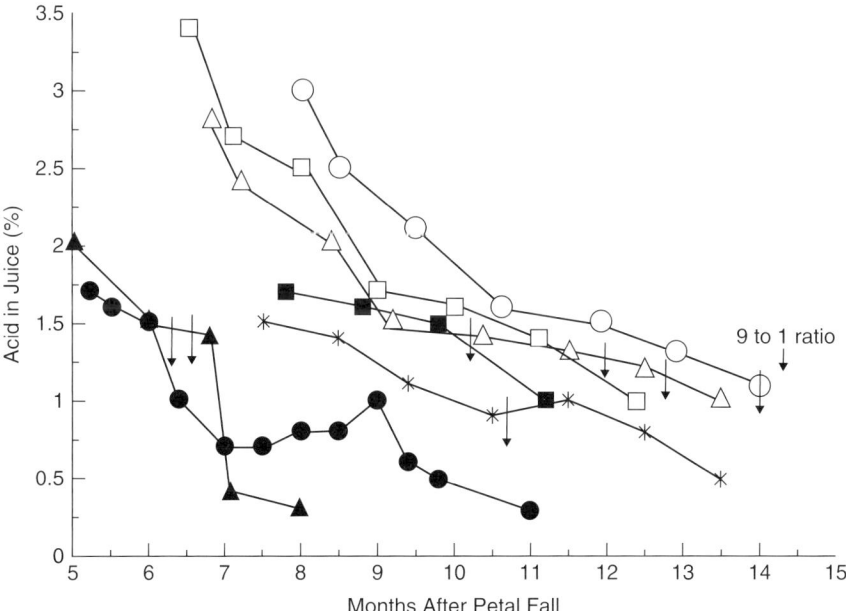

Fig. 4.11. Comparison of the trends in TA (calculated as anhydrous citric) concentration in the juice of 'Valencia' oranges in relation to advancing maturity. ▲ Cartagena; ● Palmira; ■ Medellin; ✶ Indio; □ Lindsay; △ Riverside; ○ Santa Paula (Reuther and Rios-Castano, 1969).

fruit in these areas is often insipid because of the lack of acidity. However, grapefruit attain very high quality due to reduced TA in hot, tropical regions. A similar but more protracted decrease in TA occurs in mid-elevation tropical and humid subtropical areas. Juice may also become insipid in these regions if fruit are held on the tree for extended periods. Total acid levels are generally greatest in semiarid or arid subtropical and coastal regions and decline more slowly than in other regions. This decrease in TA is primarily a function of temperature (hu accumulation) and the rapid respiration of organic acids at high temperatures, although rate of decline is also increased and minimum levels decreased by excessive rainfall or irrigation (dilution), or by selection of certain rootstocks (Chapter 3).

The ratio of TSS:TA is a primary determinant of fruit edibility and is linked to maturity standards in many citrus-growing regions (see Chapter 8). For example, attainment of a minimum TSS:TA ratio of 9:1 for 'Valencia' sweet oranges occurs in 6–7 months in lowland or mid-elevation tropical areas, but may not occur for 8–12 months in upland tropical and subtropical areas, and is slowest to increase in coastal subtropical areas, taking as long as 14–16 months (Fig. 4.12). These time frames, of course, change with cultivar, rootstock

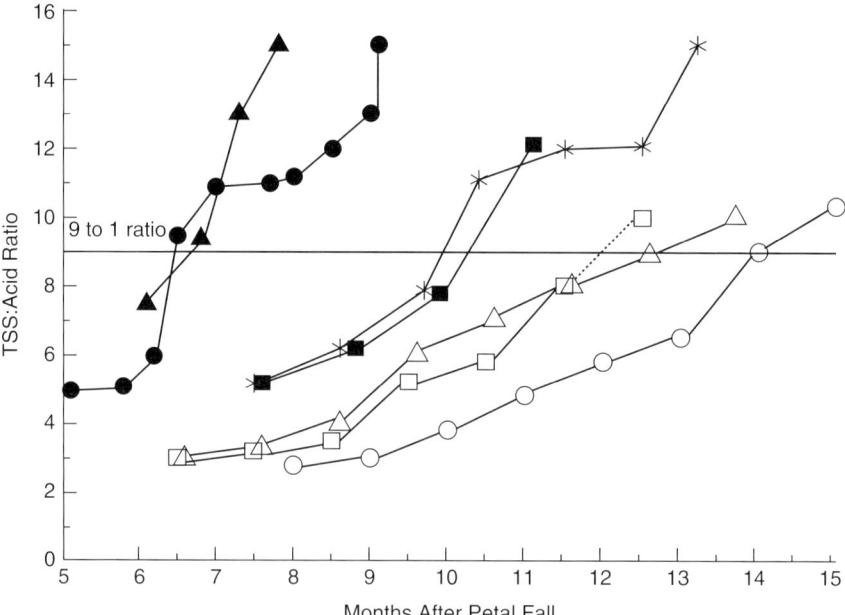

Fig. 4.12. Comparison of the trends of ratio of TSS and TA concentrations in the juice of 'Valencia' oranges in relation to advancing maturity in tropical and subtropical climates. ▲ Cartagena; ● Palmira; ■ Medellin; ✶ Indio; □ Lindsay; △ Riverside; ○ Santa Paula, ... extrapolation (Reuther and Rios-Castano, 1969).

and value for minimum ratio. However, the general trends are related to climate irrespective of cultivar.

It is quite obvious that climate has a significant effect on nearly all aspects of citrus growth and development. Failure to assess accurately the impact of climate on economic profitability of citrus is a major reason for crop losses (in the case of severe freeze-damage) or at least reduced income because of low yield and quality potential.

It is important to note that climatic factors are both limiting and coactive. For example, in most citrus regions, light intensity is not a limiting factor for tree growth but low light may reduce yields, especially in the interior of the canopy. Low temperature and water availability are certainly major limiting factors but temperature and water stress may also act together to reduce or enhance growth or productivity. Therefore, a successful citrus grower must first choose a location that ultimately optimizes climatic factors that are coactive and then reduce or limit the risk of limiting factors. Choice of a site with optimum temperature conditions for a particular cultivar is of primary concern. Factors such as irrigation and drainage can then be regulated to achieve optimum returns.

Climate Change and Potential Impact for Citrus

Glacial cycles have occurred every 100,000 to 120,000 years for at least 4 cycles. Currently the Earth is in the latter phase of an interglacial period, about 18,000 years in, with up to an additional 2000 years remaining. In each cycle from the glacial to interglacial phases atmospheric CO_2 and temperature have risen about 80 ppm (200 to 280 ppm) (parts per million) and 10°C, respectively. During the current interglacial period, industrial global warming has also occurred with an additional rise in CO_2 of 80 ppm. Over South America, atmospheric temperature is expected to rise by 2.5°C from 1955 to 2050 and an additional 4.8°C from 2050 to 2100. This translates to a current temperature increase of 1°C since 1955. Short term (60 year), sun flare (sun spot, 10 year) and ENSO cycles (3 to 7 year) are interposed on top of these major changes.

In tropical citrus areas, temperatures are fairly constant over an annual period with average minimum temperature decreasing about 3°C per 500 m elevation and average maximum temperature decreasing only about 1 or 2°C per 500 m elevation. Day/night temperature differential may be 8°C at sea level but 14°C at 1400 m elevation. The day/night differential helps citrus develop good quality in oranges and mandarins at higher tropical elevations. Global warming might be expected to change citrus production in these areas as if trees grown at 500 m elevation were growing at sea level, or an area could now become suitable for limes and grapefruit. Oranges and mandarins that can now grow and produce quality fruit at 500 m might need an elevation of 1000 m for suitable fruit quality.

In other citrus areas the effect of climate warming may be less predictable due to variables other than elevation alone. Many areas are expected to be drier than now. Water stress rather than cool temperatures may become more important in flower bud induction, but these same areas may not have adequate water for irrigation. Many Mediterranean areas may become more suitable for juice production if they have adequate water for that production. Warmer ocean temperatures may increase the occurrence of hurricanes in the Caribbean region coupled with higher late summer and autumn rainfall and humidity.

Strategies for continued production may include shifting planted areas to more suitable locations, changing citrus types more suited to the changing climate, or breeding new cultivars that have capabilities to adapt to the specific increasing stresses in the current production area. Areas with more rainfall may have increased pest and disease pressures for citrus production.

Plant Husbandry

NURSERY OPERATIONS

Establishment of a reliable source of planting material is essential to the success of a citrus industry. Dissemination of diseased or genetically inferior trees can have catastrophic effects on the productivity of the industry for years to come. Consequently, most of the major citrus-producing countries have stringent nursery regulations, and the establishment of a budwood registration programme is essential for the long-term success of developing as well as established citrus industries. Nursery operations differ from those in mature orchards because of higher tree densities and demands for resources. Nurserymen must be able to predict production trends, rootstock and cultivar demands and the availability of budwood, often 1 to 2 years in advance. Moreover, disease outbreaks like the nursery strain of citrus bacterial spot in Florida in the 1980s or severe freezes can virtually eliminate a nursery overnight. Traditionally, nursery trees were grown in the field, generally requiring 2 years from planting seed to reaching saleable size. Because maintenance of disease-free nursery trees requires protected plants, most nurseries today are located in insect-proof screenhouses or greenhouses and are able to produce marketable trees in as little as 9–15 months (Moss, 1978; Castle and Ferguson, 1982; FDACS, DPI, 2013).

Key factors in modern nursery operations to produce disease-free trees are initial production of disease-free material, nursery procedures to prevent contamination of this material and diagnostic procedures to determine that the plant material is still disease free. Seedlings for rootstocks are generally considered disease free (Mink, 1993). Scion varieties that are disease free are readily produced by shoot-tip grafting from rapidly growing scion trees (Navarro and Juárez, 2007). Young scion plants are often grown in a hot greenhouse or growth chamber to accelerate shoot growth compared to movement of virus or bacterial diseases into the new tissue. The small shoot tip is then free of disease. Mother plants for disease-free buds are grown in insect-proof houses. Careful movement of plant material and insect-proof houses maintain disease-free material. Monitoring to determine that plant material is still disease free used to be done by indexing – budding from nursery material into sensitive plants for various viruses or bacterial diseases. Now, almost all of this monitoring is done using biotechnology methods, to a large extent PCR methods (Vidalakis *et al.*, 2010). Vidalakis presents a comprehensive overview of disease management in citrus nurseries.

Various aspects of greenhouse and field nursery operations will be discussed below.

Site Selection

Site selection is extremely important for proper and successful establishment and operation of a citrus nursery. In selecting a field nursery site it is important to choose a warm location with adequate air and water drainage. Use of virgin sites minimizes the danger of soil-borne disease problems such as *Phytophthora*, pythium and nematodes, but this is less important now that trees are primarily produced in protecting structures. Requirements for greenhouse nurseries are less stringent because the environment can be controlled and artificial media rather than the local soil are used. However, availability of clean water, road access and power are very important.

Seed Selection

All nurseries must begin with a reliable, true-to-type seed source since nearly all citrus rootstocks are propagated by seed. It is important to purchase validated (true-to-type) seed from a reputable seed company or nursery, or the seed may be extracted by the nurseryman himself from rootstock mother trees maintained by the nursery. Seed should be visually inspected to ensure that it is true-to-type and properly formed and developed. The seed lot should contain uniform seeds, free of small underdeveloped material. Seed characteristics are quite distinctive for each cultivar, varying in shape, size and surface characteristics. For example, seeds of rough lemon and *Poncirus trifoliata* are small with a pointed micropylar end and a smooth seed coat. In contrast, grapefruit seeds are large and plump. Sour orange seeds have a wrinkled seed coat and a distinctly flattened micropylar end.

Seed numbers per litre and price vary considerably. For example, a litre contains about 5500 rough lemon or 'Cleopatra' mandarin seeds, 2500 sour orange or 'Swingle' citrumelo seeds, and only 2100 'Carrizo' citrange seeds. Although the standard units may vary worldwide, seed number per litre gives an indication of seed size. Price per litre also varies with availability and demand. For example, when 'Swingle' citrumelo seeds first became available and demand was high, price per litre approached US$500, whereas rough lemon seed sold for about US$35 per litre. The price per litre of 'Swingle' seed has decreased considerably since then.

Establishment of seed source trees and seed extraction are relatively specialized cultural practices. Seed source trees of the major rootstocks are often grown from seed, although many nurseries bud from good source trees on rootstocks suitable for local conditions. This avoids juvenility and fruit for seeds are produced earlier. Fruit for seed extraction are collected in the autumn after the seeds have matured. Seeds extracted before midsummer have a lower germination percentage than those harvested later (Fucik, 1978).

Seeds should be properly treated before planting. When the seed is removed from the fruit it is covered by a mucilaginous coating, which should be removed by washing or using an enzyme preparation (Barmore and Castle, 1979). As seeds are washed, off-types and underdeveloped seeds can be removed by flotation. After the mucilage is removed, seeds generally receive two fungicide treatments, in a hot-water dip (51°C for 10 minutes) for phytophthora control, followed by treatment with a registered seed treatment fungicide to control other fungi.

Proper seed storage is essential for adequate germination and seed survival. Seeds should be thoroughly dried and placed in sealed plastic bags that permit gas exchange but limit desiccation. Seeds can be stored at 4–5°C for 6 months ('Mexican' lime) to 2 years ('Troyer' citrange and sour orange) without serious losses in germination percentage (Newcomb, 1977). Freezing must be avoided, however, as seeds may be damaged and germination reduced, especially at temperatures below –4°C.

Some new rootstocks produced from widely different *Citrus* species or relatives may not produce seeds or true-to-type seeds, and in other cases, particularly for new rootstocks, available seed is scarce. These can be produced in tissue culture (Bowman *et al.*, 1997), and some nurseries in Florida are now producing rootstocks in this way (Chaires, 2014). One potential problem is how many rootstock shoots from tissue culture may have somatic variations that make the rootstock seedling not true-to-type. This is avoided to a large extent by propagating meristems to stimulate a profusion of shoots and not using callus tissue (Grosser, personal communication).

Planting and Seedbed Preparation

Greenhouse nursery

Seed planting operations in the greenhouse are different from those in the field because growing conditions can be more carefully controlled. Seeds are exposed to ambient air temperatures prior to planting, using aerated water soaks for 8–36 h at 30°C (Castle and Ferguson, 1982). Seeds are then sown into one of a variety of tray types, usually containing soilless media depending on individual grower preference and whether the plants are to be propagated entirely in the greenhouse or moved to the field nursery after budding. Most commonly used media include a mixture of peat moss or pine bark and perlite, vermiculite, styrofoam or other inert material that provides adequate drainage. Coir (coconut husk fibre) can be substituted for peat (Fornes *et al.*, 2003). Dolomitic limestone is often added to the media to raise the pH of the peat moss to between 6 and 7. Some nurserymen add controlled-release fertilizers or micronutrients directly to the media. A number of commercial media are available, although many larger operators blend their own.

Seeds are commonly sown by hand into large rectangular styrofoam trays or into plastic tubular cells with pointed, tapered ends. Biodegradable sleeves are now available that allow seedlings to be transferred easily into the final sale pot. Many greenhouse operations cull underdeveloped seeds at this stage to improve germination percentage.

Temperature, irrigation, fertilization and pest control can be carefully controlled in the greenhouse. Temperature is controlled using cooling pads with fans or louvered vents in the top of the house. Many greenhouses have automated sprinkler systems for irrigation.

Liquid fertilizer application rates and ratios are similar to those applied in field nurseries, but usually the concentration of N in the formulation is higher. Maust (1992) suggested that these levels were too high for some rootstocks, recommending rates of 15–19 mg l^{-1} (Table 5.1). Most nurseries probably over-irrigate and over-fertilize which they justify based on the value of the trees under the erroneous assumption that trees will become saleable sooner.

Pest problems can be greatly reduced with proper greenhouse management, but they can be quite severe because of the optimum environmental conditions present in a greenhouse, which are increased by high humidity from over-irrigation and over-succulent plants from excessive N. Generally, soil-borne problems are less severe in the greenhouse, but pests such as whiteflies and spider mites can be very serious if not properly controlled.

Field nursery
Although seldom used in advanced production areas, field nursery production may still be of interest in some countries. The planted area should be discharrowed and levelled prior to planting. Some nurseries also treat the site with a pre-plant fumigant to kill soil-borne organisms and weed seed. If this treatment is to be done the area should be disced and covered with plastic (if a volatile fumigant is used) after which the fumigant is applied. The site should be left fallow for 2–4 weeks prior to planting to reduce toxicity to newly-set citrus seedlings. Fumigation also kills beneficial mycorrhizal fungi which aid in uptake of nutrients, especially phosphorus (see this chapter). Consequently, planting of some rootstocks, notably sour orange, into fumigated soils requires addition of extra phosphorus for adequate growth. Some companies also market

Table 5.1. Growth data of 'Hamlin'/'Cleopatra' citrus nursery trees as affected after 30 weeks by solution N levels (Maust, 1992).

Nitrogen level (mg l^{-1})	Total scion growth			Root dry wt (g)	Shoot dry wt[1] (g)	Total dry wt (g)	Shoot:root dry wt ratio
	Length (cm)	Leaf area (cm^2)	Dry wt (g)				
0	13.7	204.5	2.8	8.4	6.9	15.4	0.83
12.5	72.2	1395.7	20.4	15.0	26.3	41.3	1.79
25	80.0	1761.6	28.2	18.9	35.1	53.9	1.86
80	89.7	1930.1	30.5	19.7	37.2	56.9	1.92
100	96.3	2081.8	32.3	19.3	38.6	57.9	2.05
200	91.6	1832.5	27.6	14.4	33.1	47.5	2.29
Significance							
Treatment	***	***	***	***	***	***	***
Linear model[2]	0.55(Q)	0.56(Q)	0.58(Q)	0.37(Q)	0.52(Q)	0.48(Q)	0.54(Q)
Linear plateau	0.77	0.76	0.73	0.42	0.71	0.65	0.74

Significant at $P = 0.001$. Includes rootstock trunk and all scion flushes.
[2]R^2 values are given under their respective models and (Q) indicates the highest order of quadratic fit. Each with four replicates per block and fertilized once or twice a week.

mycorrhizal fungi which can be added to the site following fumigation; these fungi must be added under particular environmental conditions and may not be as effective as natural populations.

Seeds of the desired rootstock are generally planted in the spring in seedbeds after the soil temperature reaches 12.5°C or above. Presoaking seeds in water usually results in more uniform germination. Seeds are planted about 0.5 cm deep at 0.5–1.0 cm apart in the row, although spacing may vary depending on the vigour of the rootstock. Spacing between rows ranges from 15 to 30 cm. It is essential to maintain optimum soil moisture conditions during this time, without over-watering. Many nurserymen have overhead irrigation sprinklers with pumping capacities of 0.2–0.6 cm h^{-1} that can also be used for freeze protection in some subtropical areas. Water should be free of *Phytophthora* propagules and low in total dissolved solids to reduce the risk of salt damage to the foliage.

Seed germination occurs 2–3 weeks after planting, depending on soil temperature and moisture. Optimum germination temperatures range from 25 to 30°C for most rootstock selections (see Chapter 3). This temperature range also provides optimum seedling growth rate. Percentage germination varies considerably with rootstock and environment but 75–80% is considered commercially acceptable (Castle and Ferguson, 1982).

Because seedlings have a very limited root system they are extremely susceptible to environmental stresses. The seedbed requires frequent irrigation over the entire growing season. Seedlings should receive frequent (once or twice a week) fertilizations with low analysis materials. Nitrogen, phosphorus and potassium should be applied at 1:1:1 ratios in most citrus regions, although phosphorus rates may need to be increased in fumigated soils where mycorrhizal fungi have been eliminated. Fertilizer rates vary considerably between nurseries, but often exceed 1000 kg of N ha^{-1} annum^{-1}, which is probably excessive.

Citrus seedlings are susceptible to damage by diseases and insects in the seedbed. Major soil-borne problems include foot rot (caused by *Phytophthora*) and damping-off (caused by *Pythium* spp.). These pathogens cause extensive tree losses particularly in humid tropical growing regions. The foliage of lemon-type rootstocks is often affected by *Alternaria*, while scab (caused by *Elsinoë fawcettii*) may be a problem for sour orange leaves. These problems can usually be controlled by fumigation, timely fungicide sprays and provision for adequate drainage (see Chapter 7). Insect problems include citrus leaf miner, aphids, orange dogs, grasshoppers, whiteflies, root weevils and leaf-cutting ants in some areas. Damage by these pests can be prevented using standard control programmes (see Chapter 6), and except for whiteflies are usually not as likely to cause problems in screened or greenhouse nurseries.

Another important problem in the field seedbed is desiccation due to wind and sandblasting. This can be serious in open areas, especially during the spring when leaves have yet to develop a thick cuticle. Cover crops of grains or tall grasses stabilize the soil and provide windbreaks that help reduce damage to leaves and twigs.

Nursery Practices

Greenhouse operations
The greenhouse nurseryman has three options for lining-out (planting-out) rootstock seedlings: (i) seedlings are transferred to the larger, final container in the greenhouse; (ii) seedlings remain in the original container; or (iii) they are lined-out into a field nursery as described in the previous section. Seedlings are generally transferred to larger containers after 3–5 months when roots have filled the original container and trunk diameter has reached 3–6 mm. Nurserymen usually cull seedlings, using only uniform, healthy material free of distorted or bench-rooted seedlings. Most nurseries cull 20–50% of their seedlings at this time. This culling allows the nurseryman to produce uniform plants which are in demand by growers. Moreover, uniform plants from the nursery will probably produce uniform trees in the field. In contrast Rabe (1991), in South Africa, found that nursery trees propagated from bench-rooted liners were as large and productive as non-bench-rooted trees after

7 years in the field. Therefore, there is some question as to whether such extensive culling is necessary. In regions where nursery trees are scarce, growers are more likely to bud and plant small or off-type trees. Seedlings are generally grown for about 3–4 months before budding. Soilless media similar to those used for seedlings are also used in the containers.

As a second option, some nurserymen plant the seed directly into large containers and grow them in the same container through budding until they are ready for field planting. This has the advantage of saving on labour and materials for repotting but reduces seedling density, makes culling more difficult, and multiple nucellar seedlings per seed need to be separated. The third option is to transplant the seedlings to the field nursery as liners and to treat them as described in the previous section.

Field operations

When the seedlings are large enough for transplanting (4–8 mm in diameter), during late summer or autumn, they are transferred to the field nursery. At this time, large (bulls) and small (runts) trees as well as trees with damaged or distorted root systems or bench roots (curved roots) are culled in most citrus regions (Castle and Ferguson, 1982). Large and small seedlings probably represent zygotic rather than nucellar seedlings and if not culled may affect subsequent size uniformity and tree performance. Growth and yields of scions budded onto zygotic seedlings may vary from that of nucellar seedlings, although this is not always the case. This process is called lining-out and the trees are called liners in some citrus regions. Some nurseries apply a fungicide dip at this time. Trees are generally transplanted at 1–2 m between rows and 15–60 cm apart within the row, depending on the size of equipment used for cultural operations. Some nurseries plant liners in paired rows with about 1 m between rows and 10 × 30 cm between pairs in a staggered pattern. Tops may be pruned to a standard height of 45–60 cm, but again this practice varies among nurseries.

Propagation

Most citrus trees worldwide are budded using the T or inverted-T method. Budwood for the scion cultivar is collected when buds are not growing. Budding may be done shortly after budwood selection or in some areas budwood is stored at 4–5°C in moist peat moss in plastic bags if cambial growth has not yet begun in the rootstock. Cambial growth (bark slipping) is necessary to permit insertion of the bud into the rootstock. Alternatively, budding may be done immediately after budwood selection even if bark is not slipping using a chip-bud. This method has a much lower percentage of bud take than inverted-T budding and is much slower.

Commercial propagators bud 1000–5000 seedling d^{-1} depending on the closeness and condition of the plants. Many budders work in groups consisting of the budder and wrappers who follow directly behind the budder and wrap the buds. Often lower shoots are removed from rootstock stems prior to budding to make placement of the bud easier. Professional budders usually have greater than 95% take and are typically paid on a piecework basis.

The T or inverted-T bud consists of making a vertical cut about 1–2 cm long followed by a horizontal cut either above (T bud) or below (inverted-T bud) the vertical cut. The inverted-T bud is preferred in high rainfall field nurseries since it serves as a rain shield for the open wound. Inverted T budding is usually used even in protected cover nurseries. The bud is then cut from the budstick and inserted in the cut. The bud is wrapped tightly with plastic tape to prevent desiccation and to allow the cambia of the scion and rootstock to unite. Buds generally 'take' 2–3 weeks after budding. Budding height varies from 5–80 cm above ground level depending on the growing region. In many lowland tropical areas, trees are budded high above ground level to permit maximum protection from soil-borne diseases such as *Phytophthora*. Some studies, however, suggest that budding too high (>30 cm above ground level) will reduce growth of the tree (Bitters *et al.*, 1981) and requires more time to produce a mature nursery tree which also is more costly to the nurseryman.

The position of the bud on the budstick affects percentage bud take. Buds located in the apical portion of the shoot have a higher percentage take than those at the basal end (Halim *et al.*, 1988). This observation may be related to the physiological maturity and age of the bud (Guardiola, 1981), but bud location is usually not considered in the budding procedure. Moreover, buds that develop during the summer–autumn months grow more rapidly after budding than those from shoots developing during the spring, possibly because summer shoots are stronger than spring shoots (Valiente and Albrigo, 2004).

After the buds have taken, the plastic tape is cut and the bud is 'forced'. There are three major methods of forcing: (i) the entire rootstock above the bud is removed (cutting); (ii) the rootstock above the bud is cut about halfway through from the bud side and the top portion is bent and tied (lopping); (iii) the rootstock above the bud is bent over and tied without cutting (bending or looping). Each method has practical advantages and disadvantages. Cutting removes a large mass of leaves and stems, permitting trees to be spaced very closely, especially in the greenhouse nursery. Cutting is also less labour-intensive than lopping or bending, requiring only a single trip through the nursery. In contrast, bending and lopping require wider spacing and at least two trips through the nursery, one to tie the trees and a second to remove the top after the bud has begun vigorous growth.

The three forcing methods are used to promote optimum growth of the new bud. If the rootstock is left intact, the new bud grows poorly due to competition between it and the canopy of the existing rootstock. Bending and lopping decreases competition for nutrients and water between the rootstock and new bud, yet still permits transport of hormones and carbohydrates to the bud. Cutting off the entire rootstock canopy completely eliminates competition with the bud but also eliminates sources of these metabolites. Some studies suggest that the canopy area above the bud can be removed if leaves are retained on the rootstock below the bud (W.S. Castle, unpublished).

Studies from Texas (Rouse, 1988) and Florida (Williamson *et al.*, 1992) suggest that bud growth is greatest if the rootstock canopy is lopped or bent compared with complete removal (Fig. 5.1). Translocation of labelled photosynthates from the rootstock leaves to the scion was greatest for bent compared with cut or lopped trees (Williamson *et al.*, 1992).

Citrus trees are grown from seed in some regions but this is rarely done on a commercial basis. 'Tahiti' lime trees were propagated by marcottage (air-layering) in Florida. A 2–4 cm bark patch is removed from the branch and the exposed wood is covered with moist sphagnum moss, after which the moss is tightly wrapped with black plastic. Callus tissue forms on the exposed surface followed by root formation. After the roots have developed, the entire branch is removed and planted. Lime marcotts often produce fruit within a year of planting.

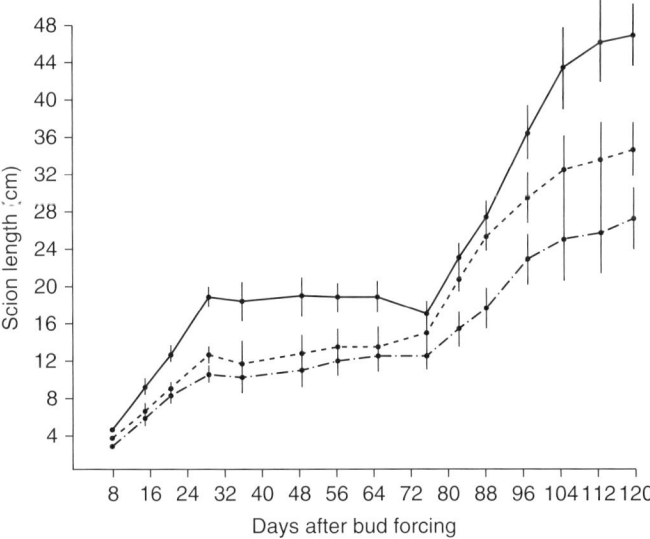

Fig. 5.1. Scion length of citrus nursery trees following bud forcing by cutting off, lopping or bending. Data points and bars represent means ± SE (n ranged from 5–22, depending on measuring date) (Williamson *et al.*, 1992). (—) Bending; (---) lopping; (–·–·–·) cutting.

Cultural Practices in the Nursery

Fertilization, irrigation and pest control practices vary from nursery to nursery. Overhead sprinkler irrigation is commonly used in the field nursery, with application rates of about 2 cm per application, 2–3 times a week, depending on soil moisture levels. Fertilizer is applied once or twice per week in soluble forms (20N-20P-20K, 8N-0P-8K or 9N-3P-6K). A survey of nurseries in Florida indicated that fertilizer rates ranged from 1000 to 3000 kg of N ha^{-1} annum^{-1} (Castle and Rouse, 1990). However, plant density is also much greater in the nursery than in a mature orchard ranging from 32,000 to 63,000 trees ha^{-1} in the field to 170,000 trees in an equal greenhouse area. Of interest, only 5–20% of the applied N could be recovered from the trees, suggesting considerable N losses due to over-fertilization. These figures contrast with 40–60% recovery of applied N under controlled greenhouse studies or in the field. In studies in Texas, weekly application of soluble fertilizers produced greater shoot growth than applications every 2–4 weeks to containerized nursery trees (Rouse, 1982).

Major pest problems in the field nursery can include citrus leaf miner, orange dogs, spider mites, mealybugs, grasshoppers and root weevils. In the greenhouse, citrus leaf miner, whiteflies, mealybugs and spider mites often are major problems.

Sales and Distribution

Citrus trees are usually sold on a contract basis directly to growers or for cash to wholesale or retail nurseries for homeowner use. Contracts are often made before trees are produced to protect the nurseryman against economic losses resulting from unsold inventory. Contracts specify cultivar, rootstock, size limitations in some cases, date of delivery and price.

Containerized trees can be loaded directly onto trucks and need not be planted immediately. In contrast, trees to be sold barerooted, balled and burlapped or wrapped must be dug from the nursery, which removes a large percentage of the root system. Digging is done by hand or mechanically by pulling a specially modified blade under and around the tree. Barerooted trees are usually misted and covered with hay and plastic or canvas during shipping and may be heeled-in if there is to be a delay in planting. Some van delivery trucks have built-in misting systems to keep the trees moist during transport to the field and until distribution for planting (contract planting). Barerooted trees should be stored in the shade if possible before planting to prevent desiccation which may affect subsequent growth of the tree. Root desiccation before planting is a major cause of poor initial tree growth and even tree death (Grimm, 1956). Recent requirements to hold mother trees and produce citrus trees under protected cover to prevent vector spread of diseases has greatly reduced availability of bareroot trees and raised the cost of ready-to-plant citrus trees to about four times the previous cost. These disease-free requirements are now eliminating field-grown citrus tree nurseries.

ESTABLISHING THE ORCHARD

Site Selection

Although citrus trees can be grown over a wide range of latitudes, climatic regimes and edaphic conditions (Chapter 4), proper site selection remains the key to successful commercial production. Several factors are important in site selection including geographic location and climate, soil characteristics, availability of water for irrigation, proximity to packing or processing facilities, availability of labour for cultural operations and harvesting, and costs associated with land and equipment purchases. Each of the above may be the limiting factor in site selection or more likely a combination of several factors may reduce the profitability of producing citrus in a given location.

Without question, low temperature is the critical factor dictating the range of citrus production worldwide. In subtropical regions where freezes occur on a regular basis it is essential to collect long-term, localized temperature information as well as regional climatic trends. For example, long-term data on freeze occurrences in Florida suggest that a severe freeze occurs on average once every 10.4 years. However, during the 1980s severe freezes occurred on average every 2–3 years. Moreover, regional data does not account for localized, microclimatic differences in temperature due to topography or proximity to oceans or lakes. It is common to observe significant differences in tree freeze-damage among closely situated orchards or those differing by only a metre in elevation in subtropical regions.

Growth and development of citrus trees is also influenced by extremely high temperatures (>50°C), although young, succulent foliage may be damaged at temperatures >40°C. This situation rarely arises except under specialized conditions such as in arid regions where leaf temperatures are typically 10°C higher than air temperatures. However, even under these conditions, leaf damage or distortion only occurs rarely. Therefore, high temperature limitations on citrus production as such usually do not occur unless water is also unavailable, thus producing extreme water stress. Certainly, high temperature greatly increases transpiration, induces stomatal closure and decreases photosynthesis (Kriedmann and Barrs, 1981).

Commercial citrus cultivars can be, and are, grown over a wide range of edaphic conditions ranging from coarse impoverished sands (sand culture) to sandy loams to moderately heavy loamy clay soils or even muck soils. Nevertheless, the greatest productivity and tree growth occur in deep sandy to sandy loam soils, provided that temperature, light and water are not limiting factors. Citrus tree growth is reduced in poorly drained soils or where impervious clay layers or hardpans are present near the soil surface. Root and tree growth are also restricted in soils having greater than 50% clay content.

Other nonstructural aspects of the soil also affect citrus growth and should be considered in site selection. Citrus trees on commonly used rootstocks generally grow best at a soil pH between 5.5 and 6.5 due to improved availability of most important nutrients, although certainly several successful exceptions are notable worldwide. A considerable amount of citrus is grown at a soil pH between 7.5 and 8.5 with no major problems provided that the appropriate rootstock is used (see Chapter 3). In many of these areas, however, micronutrient deficiencies, particularly iron, are common. Citrus trees grown in low pH soils are subject to aluminium toxicity where high levels are present in the soil. In many instances soil pH may be adjusted by addition of limestone (liming) to raise the pH, or sulfur to lower the pH. However, many citrus-growing regions have highly calcareous or acidic soils whose pH is difficult to adjust. Therefore, rootstock selection becomes extremely important for adapting citrus trees to various pH levels and soil types (see Chapter 3). In general, soil pH alone is not a major limiting factor in worldwide citrus production. Recent problems associated with huanglongbing (HLB) disease have shown that the current, commonly used rootstocks 'Carrizo' and 'Swingle' do much better at lower pH (<6.5) than previously used rootstocks like rough lemon (Anderson, 1987).

Lack of adequate quantities of good quality irrigation water limits citrus productivity in growing regions like Brazil, China, Mexico and even in many tropical areas and is the major factor limiting expansion of citrus acreage in arid regions such as Australia and Israel. Brazil traditionally had a non-irrigated citrus culture, but the need to increase yields has resulted in almost all new plantings being placed near rivers or other sources of water in order to furnish proper irrigation in dry periods. In many arid regions water quality is marginal for citrus growing due to high salinity levels (see p. 144). Moreover, even in humid subtropical regions, water may be unavailable due to inaccessibility (some mountain regions of China), lack of delivery systems (Brazil) or competition with urban areas for limited supplies of water (California).

Choosing a site with adequate drainage is equally important for successful citrus production. Citrus roots are killed or damaged by anaerobic conditions in the field (see Water Management, p. 136), and will not grow well where water has been stagnant for protracted periods. Areas with high annual rainfall and poorly drained, heavy soils like those in some of the American tropics are prone to drainage problems. Furthermore, poor drainage exacerbates problems with *Phytophthora* foot and root rot, pythium and other soil-borne diseases. Adequate drainage is equally necessary in regions using flood irrigation to prevent water from standing in low spots and causing root death.

Many new citrus plantings are being sited in isolated areas due to problems with availability of affordable land and increased urbanization plus avoiding endemic diseases. Consequently, transportation capability to packing and processing facilities may be important in site selection. In fact, inadequate transportation capability is a limiting factor to increased production and marketing in many developing world countries, or in remote locations in other regions. In developed nations, although sufficient roads or railways may be available, escalating transportation costs may limit the profitability of some citrus sites.

A final significant factor associated with site selection is the economic feasibility of purchasing and developing a citrus orchard and the long-term economic prospects for the property. Potential citrus investors must consider the initial costs for land, irrigation systems (usually necessary), trees and tree planting, and interest rates, realizing that in many cases the break-even point ranges from 6 to 10 years from orchard establishment (Ford *et al.*, 1989). About 7 to 8 years still appears to be the typical break-even point even for higher density plantings, although it is not clear if the higher cost of nursery plants is considered in these cost analyses. Moreover, the regional and worldwide supply and demand situation also is important. Recently, demand for citrus, particularly juice products, has declined, but supply of juice oranges has also declined due to climatic and disease pressures in Florida and Brazil. However, an orchard planted today will be subjected to prices 4–25 years or more in the future. The situation further varies depending where fruit is produced and for what market (fresh or processed). Changes in production in Florida, or particularly in Brazil, will have a significant effect on processed fruit prices but less impact on fresh fruit prices. For example, high production of oranges in Brazil and Florida in 1992–1993 and 2003–2004 significantly decreased prices of processed orange juice worldwide. Production changes in Spain, Italy or Israel affect fresh fruit prices in Europe but until recently have had less impact on prices in the USA due to differences in marketing channels of the primary commodities produced in the Mediterranean. Increased production of seedless mandarins in several Mediterranean countries and their expanded availability has influenced fresh citrus marketing worldwide. Tremendous expansion of citrus acreage and projected yield increases in China has the potential to decrease long-term prices and must be considered as part of any site selection decisions. At this time, China is still primarily committed to fresh cultivars but is beginning to develop some juice production. For the present, this production will probably be for internal consumption.

With the worldwide spread of HLB into traditional juice production areas, it is not clear what site selections are available to avoid this disease and still have optimum processing orange production. Selection of locations in Brazil and Florida for juice orange production may change if (i) the disease HLB cannot be controlled or (ii) climate change makes these regions no longer suitable for juice orange production. However, Spain has been able to increase its juice orange production significantly even though the climate is not optimum for such production (see Chapter 1).

Orchard Design and Planting Density

Selection of the proper orchard design and layout and planting density will have a significant impact on future yields, fruit quality, cultural operations and net returns. The overall objective of any planting design is to capture as much sunlight as possible while still allowing equipment movement through the orchard. Many different planting options exist depending on growing region and climate, topography, cultivar and rootstock and, to a lesser extent, intended use for the fruit, fresh or processed. For example, in lowland tropical regions plant density (a function of number of trees within and between rows) is generally low to moderate. Tree spacing has ranged from 7×7 (205 trees ha^{-1}) to as wide as 9×9 m (125 trees ha^{-1}) due to the excessive tree vigour in such locales, but more and more higher-density plantings now occur using less vigorous rootstocks. Even at such wide spacing it is common for trees to form hedgerows in 5–6 years after planting. Tree spacing in commercial citrus orchards in subtropical regions varies widely from 8×8 (156 trees ha^{-1}) in older Florida plantings to as close as 3×3 m (1111 trees ha^{-1}) in Japan to 1.5×3 m (2222 trees ha^{-1}) in mountainous areas of central China where all operations are by hand labour. High-density plantings are now the rule rather than the exception of 20–30 years ago. Topography, heat unit (hu) accumulation (see Chapter 4), water availability, production practice methods and disease pressure affect decisions about planting density and orchard design.

In areas where large expanses of arable land were available (e.g. USA, Brazil and Mexico) most orchards were designed in square or rectangular configurations. The term 'rectangularity' is used to describe the relationship between tree spacings between and within rows. A symmetrical orchard design with even dimensions between and within rows has a rectangularity of 1. In most citrus-growing regions orchard design is rectangular rather than square, usually resulting from decreasing within-row spacing relative to between-row spacing. The resultant increase in within-row density produces a hedgerow planting which has more trees ha^{-1} than a square configuration. Maintaining relatively wide between-row spacing allows for movement of equipment. Therefore, it is conceivable to have the same plant density with widely different spacings within and between rows, e.g. a 6×6 m and 3×12 m spacing each have 278 trees ha^{-1}, yet the rectangularities differ significantly, 1:1 vs 1:4. Considerable research has been conducted in several agronomic and tree crops, including citrus, to determine effects of varying rectangularity on yields and fruit quality. In California, varying within- and between-row spacing did not have a significant effect on yields of citrus provided that plant density did not change (Boswell *et al.*, 1975). This fact, however, does not suggest that this would apply in a climate that induces faster tree growth or that net economic returns will not vary as spacing is varied because cultural operations such as spraying and harvesting may become more difficult and costly as trees grow into a hedgerow.

Orientation of tree rows within the planting becomes important as the orchard design shifts from square to rectangular. Rectangular plantings that will become hedgerows should be oriented north–south to permit maximum light interception, particularly for plantings in subtropical areas that are considerably north or south of the equator. It is advisable to maintain the tree height at no more than twice the distance between canopy widths, again to maximize light interception, thus improving the effective fruit-bearing surface of the tree (Wheaton *et al.*, 1978). For example, trees should be no taller than 4 m if the distance between the canopy perimeters between rows is 2 m. The distance between canopy perimeters is called the 'width of the drive' or 'row middle'. This distance is considerably less than the trunk-to-trunk distance. As trees become too tall, fruit is produced primarily in the upper compared with the lower or interior portions of the tree. Additionally, total soluble solid (TSS) levels and fruit size are lower in interior and lower portions of shaded trees (Boswell *et al.*, 1970). Citrus-growing regions such as Japan, central China, and many areas in the Mediterranean region are mountainous and lack large expanses of useable, readily accessible, level land for tree crops. In these instances orchards are terraced and planted on the contour to decrease soil erosion and allow walking areas for hand labour. Cultural practices and harvesting are difficult and costly in terraced plantings; often only hand labour is practised in such locations. With terracing on hillsides, most plantings are made on the south-facing slope (in the northern hemisphere) again to maximize light interception.

High-density plantings
The concept of high-density citrus planting has been operational for many years; however, the definition of high density varies considerably worldwide depending on climatic conditions. Cary (1981) defines high density as more than 500 trees ha^{-1}. However, it may be more utilitarian to use low (<300), moderate (300–700), high (700–1500) or ultra-high (>1500 trees ha^{-1}) density categories, although these too are based on arbitrary divisions. Within each production region the overall tendency has been to increase plant density to optimize land use and most importantly increase early returns on investment by increasing yields during the early years of the orchard (Wheaton *et al.*, 1978). There is a direct positive correlation between yields and number of trees ha^{-1} during the initial phases of orchard development (Phillips, 1974; Passos *et al.*, 1977). Nevertheless, yields of ultra-high density plantings (2000 to 10,000 trees ha^{-1}) are only higher than those of low to moderate densities (300 to 700 trees ha^{-1}) for the first 5–8 years after planting (Wheaton *et al.*, 1990). This early high production may be a potential advantage where endemic diseases are likely to result in significant tree loses at an early age (Morris *et al.*, 2011). Very high density plantings may still be economically productive for 20-plus years with tree losses of 5 to 6% per year after year 1 or 2.

Several studies worldwide suggest that moderate- to high-density plantings are more profitable than low-density ones, even over a 15–20-year period, providing that they are properly managed (Boswell *et al.*, 1970; Cary, 1981; Koo and Muraro, 1982). A regular pruning or hedging and topping regime is necessary to control tree size and permits optimum light interception. It is also feasible that some trees will have to be removed within rows to improve light penetration into the canopy. In general, however, citrus growers are reluctant to remove any trees once production has begun. Currently, higher prices for nursery trees makes the cost for very high densities significant. Most new cultivars may also have royalty fees, further increasing price per tree.

Because of the tremendous variation worldwide in growing conditions no single plant density is optimum for every situation. However, long-term studies from the USA have suggested that densities of 600–800 trees ha^{-1} are economically optimum for 3–7-year-old navel orange trees under California (Mediterranean-type) conditions (Boswell *et al.*, 1970) (Fig. 5.2). This density range is intended to be a general guideline. Certainly, plant densities can typically be greater in cooler growing regions with less vigorous rootstock scion combinations, e.g. satsuma mandarin/*P. trifoliata* rootstock, such as those found in Japan (Tachibana and Nakai, 1989) and central China. Moreover, several tree-density studies using *P. trifoliata* or related rootstocks dwarfed with citrus exocortis viroid (CEV) have achieved maximum yields at densities of 1000–5000 trees ha^{-1}, with optimum long-term yields at 1500 trees ha^{-1} (Cary, 1981). In contrast, optimum plant densities are typically, but not always, lower in humid subtropical and lowland tropical areas with high hu accumulation and more vigorous tree

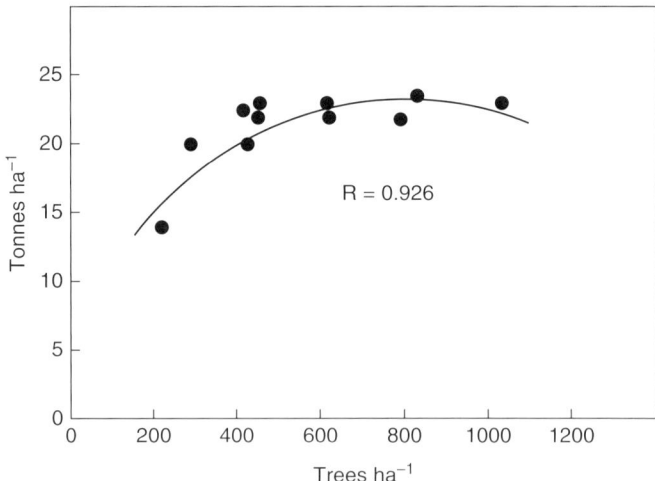

Fig. 5.2. Relation of 5-year average yield ha^{-1} to planting density of 'Frost' nucellar navel orange trees (Boswell *et al.*, 1970).

growth (see Chapter 4). Koo and Muraro (1982) found 300 trees ha^{-1} to be the economically optimum density for 'Pineapple' orange trees on rough lemon rootstock over a 15-year period in Florida, although more current studies suggest densities up to 500 trees ha^{-1} may be optimum (Castle *et al.*, 2010a,b). Tree growth rate differs among citrus-growing regions. For example, a 6-year-old sweet orange on 'Carrizo' rootstock growing in Costa Rica may be the same size as a 15-year-old tree of the same scion/rootstock combination in Japan. Moderate- to high-density plantings should be used in tropical areas only if equipment is available for pruning, hedging and topping, and preferably a moderate or dwarfing size rootstock is available.

One factor associated with the effectiveness of high-density plantings of some fruit trees like apple is the availability of dwarfing or semi-dwarfing rootstocks for controlling tree size. Although rootstocks affect tree size, few truly dwarfing rootstocks have been available in the past for citrus (see Chapter 3). Therefore, citrus tree size control was largely a matter of pruning and/or removal of trees as they grow together in a hedgerow, as in Japan. Such a system is more costly and difficult to manage than one based on dwarfing or semi-dwarfing. Lack of reliable dwarfing rootstocks has in the past been responsible for somewhat limited use of high-density plantings in many citrus-growing regions. However, introduction of dwarfing or semi-dwarfing rootstocks such as 'Flying Dragon' or 'Rangpur' × 'Troyer', 'Rusk', or several new rootstocks in Florida may be acceptable (Castle *et al.*, 2016) Alternatively, use of CEV to dwarf trees may be useful to restrict citrus tree size and is currently being used to a limited extent in California, Australia, Israel and South Africa among others.

The primary objective of a high-density citrus planting is to optimize light interception per unit land area, thus improving canopy-bearing volume and ultimately yields as rapidly as possible after planting. In traditional plantings, trees operated as individual units for many years until canopies grew together. For mature trees that have filled the space available to them, within- and between-row spacings decrease but no hedgerow has formed, and canopy leaf area per tree increases but total canopy-bearing volume is nearly constant or decreases as the trees mature and grow together (Wheaton *et al.*, 1978). In contrast, in a hedgerow system where the within-row trees are allowed to grow into a continuous hedge, a decrease of between-row spacing increases canopy surface area and most importantly total bearing volume. Almost all citrus production today is in hedgerows.

Wheaton *et al.* (1978) calculated a hypothetical change in bearing volume over a 30-year period for individual citrus trees compared with a hedgerow situation. Again, yield potential and bearing volume increased more rapidly for hedgerows compared with individual trees and as planting density increased. The production efficiency (amount of fruit produced) is initially directly related to canopy-bearing volume until the internal canopy becomes shaded. Consequently, any high-density system that increases canopy-bearing volume also has the potential to increase yields if other factors are not limiting, e.g. water or nutrients.

Potentially, high-density plantings should make more efficient use of water and nutrients than lower-density plantings; however, this supposition has not been well documented experimentally. Water use (transpiration) should be a function of the total leaf surface area of the tree. Therefore, high-density plantings may transpire more water on a per ha basis than low density, at least during early development, although transpiration would become similar as orchards mature and trees fill their allotted volume. Irrigation systems that direct water more to the individual trees or tree row may overcome some of the supposed increase in water use. This has been demonstrated for high- compared with standard-density plantings in Florida and Brazil. Nutrient requirements are a function of soil type, tree vigour and, most importantly, cropload in mature orchards. Again, during early orchard development, high-density plantings would have higher root densities ha^{-1} than lower-density plantings because there are more trees ha^{-1}. Theoretically, the denser root system would intercept more of the applied nutrients, thus increasing the efficiency of nutrient uptake. This concept applies only where nutrients are broadcast over the entire orchard floor and not when using microirrigation systems or foliar sprays that deliver nutrients efficiently to a specific area of the orchard floor or tree. Moreover, as trees form a hedgerow, radiant heat loss during freezes is reduced, thus improving orchard temperatures and possibly lessening tree and fruit damage (Wheaton *et al.*, 1978). However, cold air may collect in low spots causing more severe freeze-damage than in more open plantings.

While high-density plantings may be potentially more efficient in the use of pesticides, water and nutrients, some cultural practices are considerably more troublesome. Equipment must be adapted to operate in narrower drive middles and often spray materials do not penetrate sufficiently into dense hedgerows. Harvesting is especially difficult in hedgerows due to impaired movement of equipment and the reticence of pickers to harvest trees individually because of problems in moving ladders between trees. However, currently tree losses due to diseases provide many crossover points within a hedgerow for picker movement to both sides of the trees within the hedgerow.

Orchard Planting and Establishment

Proper planting and initial establishment of young citrus trees is essential for successful commercial development of an orchard. Although planting appears to be relatively straightforward, numerous trees have died because of a failure to follow such procedures. When planting barerooted trees it is important not to allow roots to desiccate. Grimm (1956) found that even short periods (1.25 h) of leaving trees unprotected in the field significantly reduced post-planting survival. Therefore, trees should be dug as close to planting time as possible and covered and moistened even if they are to remain in the field for less than 30 minutes. Similarly, trees in containers should not be allowed to dry out because the media will be difficult to re-wet when the trees are planted in soil due to differences in hydraulic conductivities between the soil and the media (Marler and Davies, 1987).

Rootstocks adapt the scion to various soil-related characteristics that might adversely affect a seedling tree. Sweet orange trees, for example, are extremely susceptible to collar and root rots. Therefore, by budding well above the soil level on a resistant rootstock the problem is alleviated. In many citrus-producing regions, trees are planted too deeply, thus exposing the susceptible scion to foot rot and negating the value of the rootstock. Deep planting, a problem in most citrus-producing areas, can be simply resolved through proper instruction and supervision of planting crews.

Planting operations
Planting is fairly standard throughout the world. The orchard is laid out at a predetermined row and tree spacing. In some regions, tree locations are determined visually by lining up rows with stakes at the end of the row. Many large-scale plantings are laid out using surveying equipment to ensure that rows will be straight with uniform spacing. Irrigation lines are then laid out and emitters or sprinklers installed and tested (see Water Management, p. 136) if in-place irrigation is to be used. In many areas, irrigation, particularly microirrigation, is necessary to ensure optimum production.

Tree planting consists of digging a hole that is the appropriate size for the tree, rather than pruning the roots or backfilling an overly large hole (Jackson and Tucker, 1992). Digging is done with an auger or by hand. Trees are then brought to the field either barerooted, burlapped, balled or in containers. Barerooted trees are often top-pruned in the nursery before digging because many roots are cut during digging. Burlapped or balled and container-grown trees are usually not pruned, but greenhouse container-grown trees are often topped before shipment. Trees should be planted at the same depth as they were grown in the nursery and in most cases watered-in to promote soil–root contact and remove air pockets that could cause root desiccation or the trees to settle, bringing the scion to ground level.

There is some controversy as to whether field- or container-grown trees grow better after planting. Studies from Florida (Marler and Davies, 1987) suggest that field-grown, barerooted trees grow faster than container-grown ones for the first 2 years after planting. Moreover, research from Texas indicates that these differences in growth persisted for 10 years after planting; however, containerized trees, although smaller, were easier to manage and eventually had similar yields to field-grown trees (Maxwell and Rouse, 1984). Nevertheless, thousands of containerized trees have been planted with favourable results. The slower growth observed in these studies may have resulted from inadequate wetting of the media surrounding the roots or from improper breaking-up of potbound root systems. Alternatively, container-grown trees are generally smaller in diameter than field-grown trees. Large citrus trees generally grow faster than small trees; however, these initial differences usually equalize within 5 years after planting. With more citrus nurseries shifting to greenhouse (protected) production for disease control, most trees for planting are now in containers rather than bareroot.

WATER MANAGEMENT

Proper and successful water management is often necessary to achieve commercially acceptable yields of citrus fruits. Nevertheless, most citrus and related genera are water-conserving plants capable of withstanding long periods of drought when trees are mature and healthy. Therefore, it becomes necessary to separate the physiological adaptability of citrus to drought from the commercial necessity to obtain acceptable yields of high quality fruit.

Citrus trees are water-conserving plants due to a combination of anatomical and physiological factors that limit water movement through the plant (Kriedmann and Barrs, 1981). Root hydraulic conductivity (Lp) is inherently low, possibly due to the absence of well-developed root hairs and the presence of a fairly pronounced endodermis. As discussed previously, root hydraulic conductivity is positively correlated with root temperature and also varies with rootstock (see Fig. 3.1, p. 73). As root temperature increases from 15 to 30°C, hydraulic conductivity increases (Syvertsen, 1981). Hydraulic conductivity is greater in vigorous rootstocks like 'Carrizo' citrange and rough lemon and lower in slower growing ones such as 'Cleopatra' mandarin.

The leaves of citrus trees are also adapted to conserve water. Stomata occur on the abaxial surface of the leaf. Stomatal control of water loss is quite poor in young developing leaves but becomes quite good as leaves mature (Syvertsen, 1982). Additionally, young leaves lack the structural rigidity of mature leaves and do not have as much epicuticular wax and cutin as mature leaves. Thus, young leaves are more likely to wilt during drought stress than mature ones. The waxy cuticle of mature leaves sometimes occludes stomata and is further responsible for limiting water loss from leaves. Citrus stomatal conductance oscillates throughout the day with a periodicity of 3 to 160 minutes depending on environmental conditions (Kriedmann and Barrs, 1981). This cycling is likely a function of microclimatic changes in CO_2 levels, temperature or vapour pressure deficit (VPD) (a function of temperature and humidity). VPD has a particularly pronounced effect on citrus stomatal conductance even when factors like soil moisture are not limiting. During periods of high VPD (low relative humidity [RH], high temperature) stomatal conductance decreases, thus reducing water loss from the leaf.

Citrus water use efficiency (WUE), the ratio of the amount of CO_2 fixed per amount of water transpired, is quite low compared with many other C3 plants. This factor probably results from the relatively equal contribution of stomatal and residual conductances to total leaf conductance. Residual conductance, previously termed 'mesophyll conductance', refers to the movement of water and/or CO_2 through intercellular spaces and mesophyll cells. It also applies to movement of CO_2 into the chloroplast. Therefore, the WUE of citrus leaves is relatively constant. As water stress occurs, however, the stomata begin to close, restricting both CO_2 and water fluxes but residual conductances remain similar in value to stomatal conductance and WUE changes very little (Kriedmann and Barrs, 1981). In plants where stomatal and residual conductances have highly disproportionate values either CO_2 or water fluxes are preferentially altered and WUE either increases or decreases with water stress. Under extreme water stress, however, WUE decreases as net CO_2 assimilation approaches 0, since some transpiration still occurs.

Citrus fruit, with their leathery rinds, low stomatal densities and high wax levels on the peel also contribute to overall water conservation by the tree. Classic studies from the 1930s suggest that fruits serve as reservoirs of water for the leaves during drought conditions, at least for detached shoots. Years of field observations also suggest that trees with fruit withstand drought better than those without. Some researchers, however, have suggested that fruit serve this function only in detached shoots because detachment establishes abnormal water potential gradients between the leaves and fruit. Observation of short-term wilting of leaves on mature trees after harvest suggest that the daytime supply of water to leaves from fruit is real.

Water Stress

Water stress occurs when a lack of water causes the plant to function at suboptimal levels; it is manifested in several ways including a cessation of growth, leaf wilting or a decrease in stomatal conductance, net CO_2 assimilation or root conductivity. Water stress occurs when environmental conditions cause water uptake or transport within the tree to be insufficient to replace that lost via transpiration. Severe water stress inhibits vegetative and/or fruit growth, and causes leaf thickening and eventually abscission.

Under most environmental conditions, however, citrus trees are exposed to moderate rather than severe water stress. Therefore, the key to successful water management is to minimize the amplitude and duration of the stress. Typically, as temperature increases and relative humidity (RH) decreases during the day, VPD increases (Marler and Davies, 1988). Stomatal conductance, transpiration and net CO_2 assimilation increase during early morning in response to these changes (Fig. 5.3). Transpiration increases largely as a function of increasing VPD creating tension in the xylem (water potential) and subsequent water stress within the plant as indicated by a decrease in water potential. Maximum stomatal conductance and net CO_2 assimilation

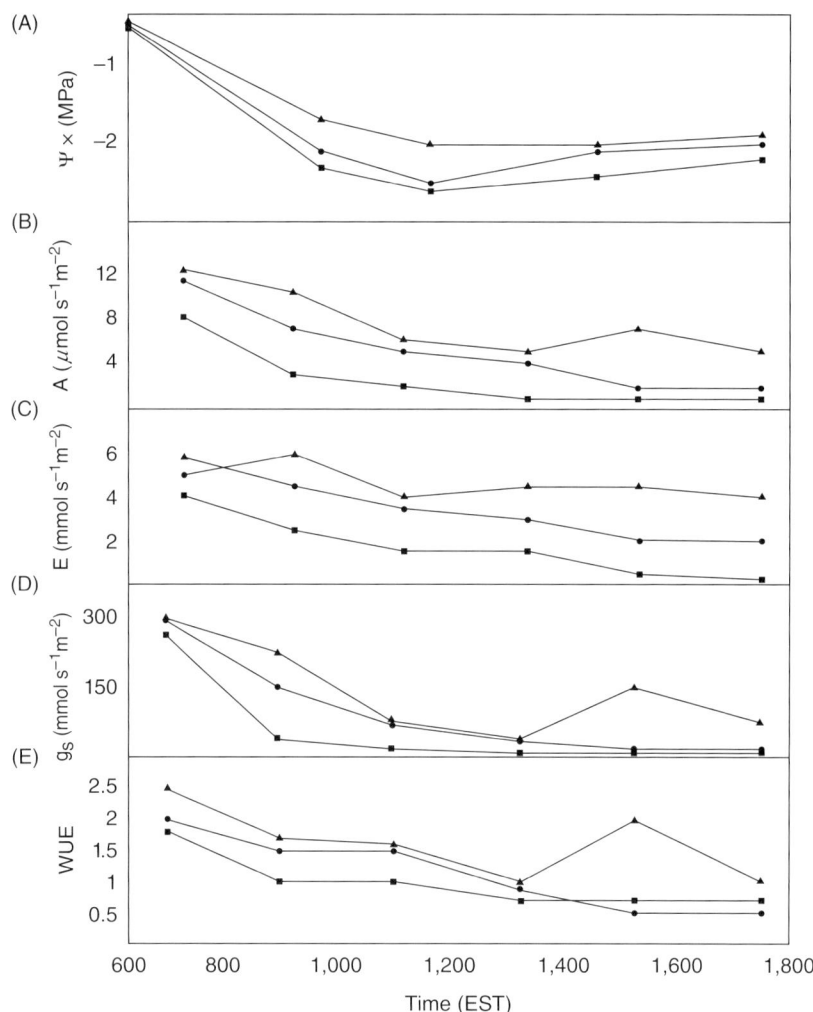

Fig. 5.3. Diurnal cycle of xylem water potential (Ψ), net CO_2 assimilation (A), transpiration (E), stomatal conductance (g_s) and water use efficiency (WUE) of young 'Hamlin' orange trees on 15 June 1987 as influenced by soil water depletion (SWD). ▲ High treatment at 20% SWD; ● Moderate treatment at 45% SWD; ■ Low treatment at 60% SWD. Mean of eight measurements ± SE (Marler and Davies, 1988).

of citrus trees generally occurs around mid-morning to midday with a characteristic afternoon decline in stomatal conductance and net CO_2 assimilation particularly at temperatures greater than 28–30°C. Temperatures above these values reduce net CO_2 assimilation primarily by reducing the activity of RuBisCo enzyme with a subsequent increase in internal CO_2 levels in the leaf. Increases in internal CO_2 levels then further decrease stomatal conductance thus limiting gas exchange.

Some degree of water stress occurs even under seemingly favourable environmental conditions. As water is lost to the atmosphere via transpiration, the tree trunk, limbs and fruit lose water in a cyclic, diurnal pattern. A moderate level of water stress does not decrease yields or fruit quality provided water potentials do not reach critical levels that severely reduce net CO_2 assimilation or tree growth. Severe water stress occurs over time as soil moisture levels decrease and water becomes progressively less available to the plant (more negative water potential). Even so, the citrus tree itself can store large quantities of water in the wood and fruit which slows the decline in water potentials to critical values. Obviously, the amplitude and durations of these diurnal changes varies with temperature, VPD, light intensity, soil moisture levels and tree factors such as cultivar and rootstock. For example, Marler and Davies (1988) observed that the amplitude and duration of the water stress is usually greatest in high VPD and low soil water conditions (20 vs 45% soil water depletion) (Fig. 5.3). Stomatal closure serves to reduce the degree of water stress, but also reduces net CO_2 assimilation. In addition, water stress is often more severe in shallow-rooted rootstocks like 'Carrizo' citrange than in deep-rooted rough lemon types (Albrigo, 1977).

The objective of any successful water management programme is to provide sufficient soil moisture to replace that which is transpired during the day. However, it is important to realize that providing adequate soil moisture is not always sufficient to prevent water stress. In many cases, soil moisture may be at optimum levels, yet stomatal conductance and net CO_2 assimilation are low and young leaves may wilt because water losses to the air occur more rapidly than water transport from the roots to the leaves. This is why irrigation regimes based on soil moisture levels alone do not always accurately reflect tree water status levels.

Irrigation

The purpose of irrigation is to minimize the deleterious effects of water stress on growth, yields and fruit quality of citrus. Irrigation is necessary to achieve maximum growth and yields on a worldwide basis over a wide range of growing conditions (Kriedemann and Barrs, 1981). In general, mature citrus trees require from 1000 (Koo, 1963) to 1563 mm (Van Bavel *et al.*, 1967) of water a year to replace that lost by evapotranspiration (evaporation and transpiration, ET), although losses due to runoff and percolation may also be large (Hilgeman, 1977). The amount of irrigation varies with growing conditions. In arid or semiarid regions such as those in Israel (Bielorai *et al.*, 1981), Arizona (Hilgeman, 1977) and South Africa (Bredell and Barnard, 1977) irrigation is essential to obtain optimum yields and tree growth, and in some cases tree survival. Irrigation, however, also improves yields and tree growth in humid subtropical regions such as Florida (Koo, 1963) and São Paulo, Brazil. In fact, lack of widespread sufficient irrigation is a major factor limiting yields in Brazil.

Furthermore, even in humid subtropical and tropical growing regions where sufficient rainfall is available for economic citrus production on a yearly basis, the seasonality and distribution of the rainfall may still adversely affect yields. Many tropical regions have pronounced wet–dry cycles and irrigation may improve yields even under these high rainfall conditions. Irrigation generally increases yields by increasing fruit size (Koo, 1963; Kriedemann and Barrs, 1981). However, irrigation also reduces physiological (June, November) fruit drop (Koo, 1963; Kriedemann and Barrs, 1981) and under extremely arid conditions improves flowering and fruit set. With extended winter drought stress in Brazil, ultra-low irrigation rates reduce the depth of drought stress and improve flowering and fruit set (Albrigo et al., 2006b; Albrigo and Carerra, 2015; see Chapter 4).

Irrigation also affects fruit quality. In general, irrigation decreases fruit TSS and total acidity (TA) by a dilution effect, but increases juice content. Thus, excessive irrigation may decrease fruit quality (Hilgeman, 1977). Irrigation may also increase problems with oleocellosis (rupturing of oil glands in the peel) during harvest and enhance the incidence of zebra skin of mandarins and stylar-end breakdown of 'Tahiti' limes, both of which render fruit unmarketable.

Irrigation scheduling
Although the advantages of citrus irrigation have been clearly substantiated worldwide, the method of scheduling irrigation is still largely unresolved in most citrus-growing regions despite years of research on the subject. In theory, irrigation scheduling should be straightforward and based on the concept of replenishing daily water losses to ET, runoff and deep percolation (Rogers and Bartholic, 1976). Several methods of irrigation scheduling exist that are based on estimation or measurement of ET from an orchard. ET is a function of solar radiation, temperature, wind speed and RH. Temperature, solar radiation levels and wind speed are positively correlated with ET, and RH is negatively correlated with ET. Estimation of ET may be made from historical data for a particular location and time of year. ET may also be measured using a USDA Class A evaporation pan. The amount of water evaporated from the pan is proportional to the ET of the orchard but must be adjusted for plant factors such as root and stomatal conductances, leaf area, ground area covered by crop, etc. that regulate water losses from the system. Thus, plant ET is always less than potential ET and ET of citrus is not more than 80% of the potential ET. Correction factors (crop coefficients), relating pan evaporation values to citrus tree ET, have been developed but still contain some measure of error and vary with tree size.

Several irrigation studies have been conducted comparing citrus tree yields at various pan evaporation levels. Yager (1977) in Israel found no differences in yields of 'Valencia' orange trees at 35, 47.5 or 60% of pan evaporation. Swietlik (1992) made the interesting observation that pan evaporation coefficients varied with tree age from 0.75 in the first year after planting to

0.20 by the fourth year in the field. Thus pan evaporation coefficients are a function of the amount of ground surface area covered by the plant canopy and they should be adjusted yearly to compensate for these changes. Similarly, Smajstrla and Koo (1984) found no yield differences for 'Valencia' orange trees irrigated at 100, 50 or 25% of potential ET. Some of this lack of differences is poor accuracy in how much rainfall and irrigation is effectively taken up by the tree. Often, as more irrigation is applied, large amounts are lost to leaching or applied outside the root zone of the tree.

A quite common method of scheduling citrus irrigation is based on soil moisture levels – the idea again being that as moisture is lost due to ET, percolation or runoff, it can be replaced. This method is fairly straightforward and relies on soil tensiometers or other similar devices to measure soil moisture content. The soil moisture content at which to irrigate is usually determined by comparing the long-term relationship between yields or tree growth and soil moisture levels for a particular location and soil type.

Therefore, a set of soil water depletion values can be derived for a particular soil type or location. Soil water depletion is a measure of how much water has been removed from a volume of soil relative to the amount of available water in the soil of the root zone. For example, a 20-year study from Arizona suggested that optimum citrus yields were obtained when soil moisture tension (as measured using soil tensiometers) remained between 60 and 70 kPa at 75 cm depth (Hilgeman, 1977) (Table 5.2). Vegetative growth increased with increasing irrigation but yields were comparable at the high and moderate levels. Other similar studies from Florida suggest maintaining soil moisture content at 33–66% of available soil moisture for optimum yields (Koo, 1963). Soil moisture depletion percentages of 30–45% (Marler and Davies, 1990) and tensions of 10–20 kPa (Smajstrla *et al.*, 1985) are recommended to obtain optimum growth of young, nonbearing citrus trees. Young trees have more limited rooting and store less water than mature trees and thus require more frequent irrigation.

Table 5.2. Effect of irrigation schedule on 'Valencia' orange growth, yields and fruit quality[1] (Hilgeman, 1977).

No. of irrigations, amount of water applied (cm^3 $year^{-1}$)	Trunk growth in 20 years (cm^2)	Canopy growth in 17 years (m^3)	Feeder roots at 0–183 cm in 17 years (gm^{-2})	Yield (20-year avg. wt tree^{-1}) (kg)	Fruit (20-year avg. wt fruit) (g)	TSS (20-year avg. Dec.) (%)
A. 15, 175	374a	107a	740a	125a	160a	10.6a
B. 10, 135	308b	76b	725a	115a	158a	11.0b
C. 15, 95	242c	66b	648b	88b	159a	11.4c
E. A Mar–July C Aug–Feb	244c	70b	697a	122a	140b	11.7d

[1]Numbers with different letters differ significantly, $P = 0.05$.

In theory, irrigation could most accurately be regulated using plant characteristics such as stomatal conductance, net CO_2 assimilation, leaf water potential or fruit or trunk growth. In practice, however, some of these are difficult to measure and may not be accurate, reliable indicators of when to irrigate under field conditions. For example, stomatal conductance and net CO_2 assimilation are affected by other factors besides plant water status such as atmospheric CO_2 levels, VPD or leaf temperature (Kaufmann, 1977). Similarly, leaf water potential is quite variable within the tree canopy during the day, although predawn water potential measurements may be used to assess tree water status. Using stem water potentials reduces this variability considerably (McCutchan and Shackel, 1992). These methods, while useful for plant science researchers, are not readily adaptable for use by growers. Measurement of diurnal fruit or trunk shrinkage to ascertain citrus tree water status is more useful to growers but is quite tedious and not commonly used.

Therefore, due to problems and limitations to each of the above methods, many citrus growers irrigate based on historical patterns and the calendar system, or on a combination of soil-based and historical considerations. New, potentially more reliable and useful methods have been tested but they often still only measure soil moisture or plant water relations at one site or tree. As yet no single method has been adopted commercially on a worldwide basis, but plant water balance methods (water applied vs water used) are more favoured. Florida citrus growers have a 'Citrus MicroSprinkler Irrigation Scheduler' available that is easy to use (Morgan, 2016).

Irrigation systems
Citrus trees have been irrigated for thousands of years, probably beginning with the use of catchment basins for rainwater in arid regions of the Middle East. Water was then diverted to citrus orchards using irrigation ditches. This flood or furrow irrigation method is still widely used in many citrus-growing regions worldwide. Berms (mounds of soil) are constructed to serve as borders along a row of trees after which water flows along the row until the entire land surface is covered. Water percolates through the soil mass to the root zone where it is taken up by the tree. In regions such as the poorly drained flatwoods areas of Florida or central China, citrus trees are grown on raised beds (ridges). Essentially, no areas of Florida are still under flood irrigation, but flood irrigation is practised in Texas, northern Mexico and limited parts of central China. Preferably the area is rapidly flooded and unpercolated water removed after 48 or 72 hours.

Historically, many citrus orchards were irrigated using permanent overhead or undertree sprinklers or by travelling guns. In the early 1960s low volume (micro) irrigation systems were developed in Israel and South Africa to conserve water without compromising growth or yields. Several studies have clearly demonstrated that microirrigation uses less water than flood or area-wide sprinkler irrigation systems without compromising tree growth or yields (Roth *et al.*, 1974; Koo, 1985; Swietlik, 1992). Roth *et al.* (1974) found yields of 'Campbell' 'Valencia' orange trees were comparable using trickle compared with flood irrigation, yet trickle irrigation used only 11% of the water applied with flood irrigation. Similarly, 'Ray Ruby' grapefruit growth and yields were comparable for trickle and flood irrigated trees, yet flooded trees received 24,410 kl ha^{-1} while the trickle irrigated trees received only 1845 kl ha^{-1} over the 4 years of the study (Swietlik, 1992).

In arid or semiarid regions such as Israel, Australia, southern California or Arizona, citrus tree roots are concentrated in areas where the irrigation is applied. In contrast, root distribution is more widespread in humid subtropical areas such as Florida where sufficient rainfall occurs to promote root growth. In these areas, it has been important to irrigate as much of the root volume as possible to achieve optimum yields, particularly on deep, sandy soils (Koo, 1985). In general, best results were obtained when more than 50% of the root system was irrigated for mature trees in well-drained, sandy soils (Smajstrla and Koo, 1984). Similarly, Bielorai *et al.* (1981) in Israel found that yields of mature 'Shamouti' orange trees were greater when 90 or 70% of the root zone was irrigated compared with 35%. Current interest is in using microirrigation systems to deliver nutrients and water as used on a daily basis (pulsed irrigation), even if smaller soil areas are wetted. This method appears to have promise without wetting all of the rootzone even on sandy soils, although roots do concentrate in the daily wetted zone. Several microirrigation systems have been developed that deliver relatively small volumes of water at fairly frequent intervals, thus minimizing the large diurnal variations in soil and plant water status that commonly occur with furrow irrigation (Swietlik, 1992). Drip (trickle) irrigation emitters typically deliver 4–8 l h^{-1} and microsprinkler emitters from 38–80 l h^{-1}. With escalating costs and lack of availability of high quality water, microirrigation is becoming widely used throughout the world and will continue to be the method of choice in many citrus-growing regions. Currently open hydroponics is being evaluated and commercially used more (daily replenishment of water used and nutrients needed).

Microirrigation systems provide an efficient method of providing water to the tree on a regular, consistent basis; however, they require more intensive management than other methods. Irrigation lines and emitters are subject to clogging by particulate matter, insects, spiders, or minerals such as calcium or magnesium which precipitate from the irrigation water. Various iron and sulfur-reducing bacteria and algae may also cause plugging of emitters (Ford and Tucker, 1975). Therefore, water quality and adequate filtration and chlorination (to control algae and bacteria) are necessary to ensure proper operation of the system. Moreover, salts (Na, Cl) may accumulate in and around emitters, especially in arid regions, in some cases causing root damage if they are not periodically leached from around the roots.

Salinity

Citrus trees are for the most part sensitive to salinity (Bielorai *et al.*, 1988; Alva and Syvertsen, 1991) so water quality is an important consideration in any irrigation system. High salinity irrigation water may be a major limiting factor in citrus-producing regions such as Israel, southern California, Australia and coastal regions of Florida. In general, growth of citrus trees is impeded when saturated extract paste levels of the soil exceed 1.4 dS m^{-1} (deciSiemens). The situation is less straightforward when quantifying levels of total dissolved solids that cause leaf damage. However, salinity damage is intensified under low humidity conditions and where irrigation water with moderate levels of salinity (e.g. 1.3 dS^{-1}) is applied regularly.

In citrus-growing regions where high salinity water is common, the trend has been to use microirrigation rather than furrow or overhead sprinklers which apply water directly to the leaves or apply large quantities of salt. Choice of rootstocks such as 'Cleopatra' mandarin tolerant to salinity (Na and Cl) also decreases problems with high-salinity water (see Chapter 3 rootstocks).

With increased competition between urban areas and citrus orchards for water, several regions use reclaimed wastewater for irrigation (Davies and Maurer, 1992; Parsons *et al.*, 2001). Secondary or tertiary treated wastewater provides an inexpensive and plentiful source of irrigation water and additionally may provide some necessary plant nutrients, thus reducing fertilizer requirements. Some reclaimed wastewater may contain high levels of Na, Cl or B, however.

Drainage and Flooding

Flooding (waterlogging) occurs when water displaces oxygen from the soil creating hypoxic (low oxygen) or anoxic (no oxygen) conditions. As oxygen is displaced by water or metabolized by microorganisms or roots, other compounds replace it as the terminal electron acceptor in respiration as the soil becomes reduced to <330 mV redox potential. Most soils suitable for growing citrus become reduced within a few days of flooding (Syvertsen *et al.*, 1983). For example, sulfides are reduced to hydrogen sulfide at a redox potential of about −150 mV by sulfur-fixing bacteria. Citrus roots are not adversely affected immediately by low soil oxygen as such, but are very sensitive to hydrogen sulfide levels in the soil (Culbert and Ford, 1972). Consequently, the presence of hydrogen sulfide in the soil is a good indicator that the soil is reduced and suggests that citrus root damage is occurring.

The production of hydrogen sulfide and ultimately the extent of root damage is a function of soil temperature, organic matter content and microbial activity. Citrus roots are killed in as little as 3 d of waterlogging at relatively high soil temperatures (30–35°C) but may survive for months at lower temperatures (<15°C) (Syvertsen *et al.*, 1983). Soils with high organic matter become reduced more rapidly than very sandy, low organic matter soils. Organic matter provides substrates for microbial reduction of the various compounds present in the soil. In contrast, citrus seedlings growing in sand culture survive months when flooded due to the absence of sufficient substrates for microbial production of hydrogen sulfide.

Citrus tree responses to flooding
Citrus trees growing in poorly drained sites are generally sparsely foliated, stunted and low yielding. In many instances trees do not die but remain marginally productive until the impediment to drainage is removed. In most cases trees are exposed to cyclic waterlogging and then drought stress related to wet–dry cycles, rather than to continuous flooding. Moreover, waterlogged soil conditions, besides debilitating the tree, are conducive to the proliferation of soil-borne fungi such as *Phytophthora parasitica* (root and foot rot) and *Pythium* spp. (damping-off). These organisms cause extensive tree death in nurseries, especially in tropical regions with high rainfall and poorly drained soils. It is advisable to grow trees on raised beds or in artificial media rather than soil in these locales. *Phytophthora* spp. in particular is a worldwide problem also causing extensive tree losses for young orchards (see Chapter 7). Symptoms for flooded compared with pathogen damaged trees are quite distinctive. Pythium causes seedlings to wilt and eventually die due to root death and girdling of the trunk. Foot or root rot symptoms include a pronounced chlorosis of the leaf midvein after root damage and girdling of the trunk. Lesions also appear on the trunk, usually near the soil level (foot rot) or roots die and slough-off (root rot). Flooding damage does not produce lesions.

Citrus trees respond physiologically to flooding long before morphological symptoms or yield reductions appear. Net CO_2 assimilation and transpiration decrease within 24 h of flooding and remain at reduced values as flooding persists (Phung and Knipling, 1976) (Fig. 5.4). Stomatal conductance also decreases as an adaptive mechanism to reduce transpiration because hydraulic conductivity of the roots decreases and thus water uptake is reduced (Syvertsen *et al.*, 1983). This survival mechanism, however, reduces net CO_2 assimilation which eventually translates to decreased shoot growth and yields.

As mentioned previously, soil temperature and texture have a significant effect on the severity and development of flooding damage in citrus. In addition, rootstocks vary in their flooding tolerance generally, with rough lemon being more physiologically flood-tolerant than sour orange, *P. trifoliata* or 'Cleopatra' mandarin. Sour orange, and especially *P. trifoliata*, are more tolerant to foot rot than rough lemon; consequently under field conditions where the *Phytophthora* organism is quite prevalent, trees on rough lemon appear more sensitive to flooded conditions than those on sour orange or *P. trifoliata* (see Chapter 3).

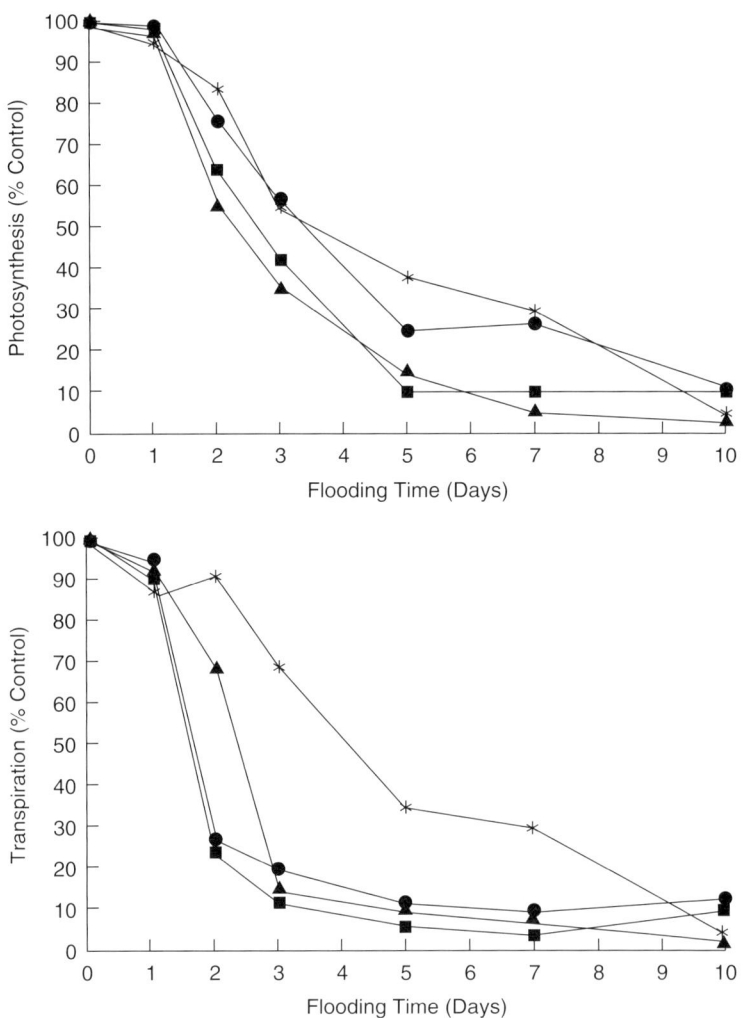

Fig. 5.4. Relative rates of photosynthesis (top) and transpiration (bottom) of four citrus rootstocks following flooding. ● Rough lemon; ■ Sour orange; ▲ 'Cleopatra' mandarin; * *P. trifoliata* (Phung and Knipling, 1976).

FREEZE-HARDINESS AND FREEZE PROTECTION

Frost- and freeze-damage is of moderate concern in many of the largest citrus-producing areas worldwide, but is of major concern in some subtropical regions such as the USA, Japan, central China, northeastern Mexico and Argentina. Crop and occasionally tree losses have occurred through much of the Mediterranean region, as in Greece in the 1991–1992 season, but at irregular intervals and with far less severity than experienced in the USA. Several economically devastating freezes have occurred in the USA, most recently during the 1980s, causing the death of over 100,000 ha of citrus trees as well as billions of dollars in crop losses in Florida. Freeze-related crop losses rarely occur in Brazil, the second largest producer of citrus worldwide, and never in low- to mid-elevation tropical growing regions.

The definition of a frost as opposed to a freeze is somewhat subjective and controversial. A frost occurs under calm, clear conditions where extensive radiation heat losses occur. Frosts are common in Mediterranean-type climates with wide fluctuations in diurnal temperatures. A freeze connotes windy (advective) conditions which often cause more extensive damage than radiative freezes. However, since citrus tissues are damaged or killed only when ice forms, it seems most appropriate to use the terms 'freeze-hardiness' or 'freeze-damage' rather than 'frost-' or 'cold-damage' since plants do not sense cold (Rieger, 1989).

Freeze-hardiness

Commercial citrus and the most closely related genera are of subtropical and tropical origins and thus are intermediate in freeze tolerance between temperate zone species which acclimate to freeze conditions based on day length and temperature cues and tropical species which have limited capacity to acclimate except through some supercooling. Citrus trees adapt to freeze conditions through a combination of freeze avoidance and freeze tolerance mechanisms (Yelenosky, 1985). Freeze avoidance involves the capacity of plant organs to supercool below 0°C without forming ice. The extent of supercooling varies with species, type of organ and extent of freeze acclimation. Flowers of commercially important *Citrus* species have the capacity to supercool to −4.3°C, fruit to −5°C, mature leaves to −7°C and stems to −8.9°C. However, young, succulent, nonacclimated expanding leaves supercool only to −2°C or less. Extent of supercooling also varies with species with leaves of *Citrus medica* (citron), *Citrus limon* (lemon) and *Citrus aurantifolia* (lime) supercooling the least (−3 to −5°C), and *Citrus unshiu* (satsuma mandarin) supercooling the most (−9.4°C) of the commercially important cultivars (Yelenosky, 1985). *Poncirus trifoliata*, when fully acclimated, has the capacity to supercool to −15°C, however, its primary use is as a rootstock and not as an edible cultivar. It does have a gene that provides cold tolerance through maltose metabolism (Ping *et al.*, 2014). Other citrus relatives also have more cold tolerance than *Citrus* sp. (Inch *et al.*, 2014).

During the 1980s the concept of ice-nucleating agents (INAs) and their potential role in citrus freeze-hardiness became popular. INAs are undesirable because they interrupt supercooling and serve as initiation points for ice formation and propagation through the plant. Although some research suggests that INAs may be important for citrus freeze-hardiness, lack of sufficient numbers of INAs under the field conditions tested and alternative mechanisms of freeze tolerance suggest that the role of these bacteria in citrus freeze-hardiness is less clearly defined than in herbaceous species (Constantinidou *et al.*, 1991). Thus, attempts to reduce INAs and increase citrus freeze-hardiness in the field have been largely unsuccessful.

Citrus trees, unlike freeze-sensitive species, tolerate varying degrees of intercellular ice formation and consequently, even if nucleation occurs cells will not necessarily be damaged or killed. Ice formation is manifested as dark-appearing areas (watersoaking) in the leaves, usually emanating from the midvein toward the periphery of the leaf. Watersoaking in the field may occur at temperatures as high as $-3°C$ and may become widespread throughout the tree and orchard, indicating that nucleation occurs at numerous sites during a freeze. Nevertheless, the extent of watersoaking within the leaf does not necessarily correlate to the amount of leaf damage. It is common to observe fairly extensive watersoaking the night of a freeze without incurring leaf damage.

The degree of damage to the citrus tree is related to the amount of freeze acclimation and the duration and minimum temperature of the freeze. Degree of acclimation of citrus is a function of soil and tissue temperature, and day length. Citrus trees will not acclimate without light and acclimate more under long- than short-day conditions. The long-day effect is probably related to increased metabolite accumulation at long compared with short day lengths and not to phytochrome-mediated changes in metabolism as with temperate zone species (Yelenosky, 1985). Maximum freeze acclimation is attained under moderate daytime temperatures ($20-25°C$) and low night-time air and soil temperatures ($<12°C$) of 2 weeks or more. These conditions cause the plant to become quiescent or to have 'non-apparent' growth. This condition is not synonymous with true dormancy, since placing citrus trees into favourable growing conditions (temperature $>12.5°C$) will cause a resumption of growth and a decrease in hardiness. Citrus trees de-acclimate much more rapidly than they acclimate, sometimes within a few days at temperatures favourable for growth.

Numerous biochemical changes occur in citrus trees as they acclimate to low temperature, some of which may be correlative only, while others appear to directly affect hardiness. Sugar:starch ratios increase during acclimation as sugar levels increase dramatically (Yelenosky, 1985). Sugars may serve to lower freezing points within the plant or act as cryoprotectants for cell membranes. Trioses in particular are known to function in this way in other plants, although their role has not been clearly defined in citrus. Increases in proline in tree sap and changes in lipid and protein composition also occur during acclimation, but their functions related to freeze-hardiness are less well understood.

The relative proportion of ice to water in the intercellular spaces and the rapidity of ice propagation through the citrus tree are also important in freeze tolerance. Differences in freeze tolerance among species may ultimately be a function of differences in the amount of frozen water tolerated at a given sub-freezing temperature (Anderson *et al.*, 1983). Therefore, both minimum temperature during a freeze and duration of temperatures below the ice nucleation point are equally important to freeze survival. Ice nucleation occurs at more sites and ice propagation occurs over a longer period for long- versus short-duration freezes.

Methods of Freeze Protection

Passive methods

Without question the most successful method of freeze protection is site selection. Choosing an area where temperatures remain above $-2°C$ is best to avoid fruit and tree losses due to freezes. It is very important to obtain long-term historical temperature records for a particular region and to assess the probability of freeze-damage. Avoidance of low spots that do not allow proper drainage of cold air (frost hollows) also reduces the risk of freeze-damage.

Other passive cold protection measures include windbreaks, clean cultivation and tree covers. Windbreaks reduce air mixing during advective freezes and thereby reduce heat losses from the orchard. They are most effective if a heat source (like orchard heaters) is also present in the orchard. Windbreaks usually consist of closely planted rows of trees planted on the north or northwest border (south or southeast in the southern hemisphere) of the orchard. Man-made windbreaks (fences or shade cloths) have also occasionally been used, but are usually too costly to construct and maintain. Natural windbreaks (trees) may decrease citrus tree vigour and yields in the area nearest the break due to shading and competition for water and nutrients. Windbreaks should not be placed in low spots in the orchard where they can impede air drainage and allow freeze-damage in that area.

Clean cultivation is another method of passive cold protection. A clean, hard-packed surface intercepts and stores more solar radiation during the day and releases more heat at night than a surface covered with vegetation or a newly tilled area. Addition of water to the cleanly cultivated area enough prior to a freeze to allow soil warming further improves heat accumulation during the day. Clean cultivation can be achieved using chemical or mechanical methods, provided sufficient time is given after tilling to allow the surface to become firmly packed.

Citrus trees have been grown for centuries by covering the trees (orangeries) to impede radiation losses and decrease freeze-damage (see Chapter 1). Orangeries have been used since Roman times and were common in northern Europe in the 15th and 16th centuries. The use of tree covers is not usually economical for large trees under commercial conditions, although some covered production occurs around Sorrento, Italy. Moreover, the Japanese grow citrus trees in greenhouse complexes with success as do growers on Jeju Island in South Korea. The greenhouse not only provides freeze protection, but also hastens fruit maturity by increasing temperatures, allowing early market fruit production. Research in Florida is evaluating protective screen enclosures for exclusion from HLB infection through the insect vector, *Diaphorina citri*. (Chaires, 2015). Some commercial plantings of this method (CUPS – citrus under protective screening) are in place in Florida, usually in 10- to 20-acre units.

Active methods
Mature trees survive moderate freezes of short duration better than young trees in most cases because of their greater canopy size and tree mass. The large canopy retards heat losses from the orchard and the greater mass requires long periods of low temperature to reach critical (damaging) levels. Moreover, young trees are typically more vigorous than healthy mature ones and thus are less quiescent and therefore less freeze-hardy in many cases.

Orchard heating was used successfully for freeze protection of mature groves for many years. The major advantage of orchard heaters was that, if properly placed and utilized, they provided effective freeze protection for trees and fruit, during both radiative frosts and advective freezes. However, heaters are costly to purchase, maintain and operate. Few growers can economically justify the use of heaters even for high value fresh fruit or during very severe freezes where tree survival is jeopardized.

Wind machines provide another method of active freeze protection for mature orchards. They operate on the principle of taking advantage of the formation of a temperature inversion layer during radiative frosts. This situation is created when the soil surface cools faster than the air at 8–25 m above the surface. Wind machines, which consist of one or two large propellers powered by a gasoline or diesel engine, are used to mix the warmer upper air with the colder air in the orchard. Wind machines are most effective during moderate radiative freezes where only 1–1.5°C of temperature increase is required. They should not be used at wind speeds above 12 ms^{-1} (advective conditions) because air mixing has already occurred and also because high winds may detach the propellers, creating a hazardous situation. Wind machines are not widely used because of high fuel and maintenance costs and damage by vandalism. In the past, wind machines were used along with heaters which provided better freeze protection than either method alone.

Various types of irrigation have been used for freeze protection of mature trees. Irrigation provides heat to the orchard as sensible heat and latent heat of fusion. In most cases the temperature of the irrigation water is 15–25°C when pumped from deep wells. As the water is applied to the orchard, heat is released to the air and trees. Another form of heat (the latent heat of fusion) is released as water turns to ice. Therefore, effective freeze protection using water is dependent on a continuous, adequate supply of water. The major reasons for failure of irrigation to supply adequate freeze protection are insufficient quantities of water or inadequate coverage of the treated tree or area. These factors are particularly important during advective freezes where evaporative cooling occurs. Evaporative cooling removes 7.5 times more energy from an irrigated area than is provided by heat of fusion – consequently, both minimum temperature and wind speed must be considered when determining how much water is needed for freeze protection. Even small increases in wind speed require relatively large increases in irrigation application rates to provide adequate freeze protection. Severe freeze-damage occurred in Florida in the 1960s and 1970s due to inadequate application rates and poor coverage using overhead irrigation at 0.25 cm^3 h^{-1}. Moreover, there was extensive limb breakage due to ice accumulation. Therefore the use of overhead irrigation was primarily limited to field citrus nurseries where high application rates (1.0–1.5 cm^3 h^{-1}) were adequate for freeze protection and economically feasible because of the limited acreage that must be covered by water. With the shift to greenhouse nurseries, even this use of overhead sprinklers for freeze protection is disappearing.

Microsprinkler irrigation has become a popular method in many areas of the world. While microsprinklers at ground level are quite effective for freeze protection of young trees, they are only marginally effective for mature trees because they do not provide sufficient quantities of water to the tree or irrigated area (Buchanan *et al.*, 1982). Generally, a temperature increase of only 1–2°C is provided to the lower canopy by a typical ground-level microsprinkler system. This is because most systems apply inadequate amounts of water. Microsprinkler irrigation generally provides no freeze protection for the fruit and upper canopy, but may protect portions of the lower trunk and leaves. However, it has been effective in protecting trees 2–3 m tall if emitters are elevated to about 1 m in the tree canopy (Parsons *et al.*, 1991). This method, when used on mature trees, provides freeze protection of the lower scaffolding, allowing rapid re-establishment of a productive tree. Continuous water coverage of lower limbs will maintain them at the freezing point of water, 0°C.

Flood irrigation is used for freeze protection in some areas of the world. In most regions, growers begin to flood the orchard the day before and during the time when minimum temperatures are expected. In Florida, water is also pumped onto the orchard the day before a freeze and then removed within 48–72 h after the freeze to minimize root damage due to anaerobiosis. Flooding an entire orchard provides 0.5–1.5°C temperature increase, mainly from sensible heat. It is a fairly cost-effective and simple means of freeze protection but is limited to areas using flood irrigation with access to a large volume of water over a short time.

Care of Freeze-damaged Trees

Although thousands of hectares of trees have been killed by freezes worldwide, historically freeze-damage generally affects only the fruit, leaves or wood to varying degrees without killing the entire tree. This leaves the grower with the major management decision of how to bring these trees back into production as soon as possible. Tree age, freeze duration, cultivar and rootstock are factors that determine how to rehabilitate the orchard.

Assessing freeze-damage
The first step in managing freeze-damaged trees is to assess the extent of damage. As every freeze is different it is very difficult to make an immediate assessment of damage. Fruit damage is estimated by making cuts through the fruit at hourly intervals in the morning following a freeze. Some ice formation in the top 0.6 cm of the juice vesicles indicates mild damage, while solid ice formation in the centre signifies severe damage and loss of a portion of the crop. Generally, 4 h or more of temperatures of $-2°C$ or below will cause some mature fruit damage. If extensive fruit damage has occurred, some fruit abscission occurs within 1–2 weeks following a freeze. High daytime temperatures following a freeze will, in particular, accelerate fruit drop and segment drying. Fruit should be harvested as soon as possible after a freeze and processed quickly to minimize decreases in juice content and yield losses due to fruit abscission. After ice in the fruit has melted, this free water is transpired through the peel rather than rehydrating the juice sacs, thus decreasing juice content. Some alternaria decay is likely to occur about 3–4 weeks after the freeze.

Leaf damage is difficult to assess during a freeze night. Water-soaked (where dark green areas occur in part or all of the leaf) or curled leaves may or may not be significantly damaged. In the morning following a freeze, leaves may be rolled up and appear dry and dull green. These leaves will probably, but not always, abscise over the next week, again depending on temperature. Freeze-damaged leaves abscise between the petiole and the lamina (leaf blade) with the petiole dropping later. Leaf abscission is usually more extensive at the top of a tree than at the base following a radiative frost because temperatures are lower in this area due to direct exposure to the sky. It is not uncommon for temperatures of exposed leaves to be $1-2°C$ lower than air temperatures reported 1.5 m above ground level. Therefore, caution should be exercised in interpreting minimum air temperature data relative to extent of freeze-damage. Within a week of a freeze the extent of leaf damage should be quite apparent. Trees can recover even from total defoliation and in some cases flowers and fruit will be produced in the next season, depending on when a freeze occurs, whether flower buds have already been initiated and the extent of wood damage.

The consequences of freeze-damage to twigs, stems and trunks is more difficult to assess than that to fruit or leaves. In general, small twigs will be damaged before larger limbs and trunks. Twig or limb dieback may not become visible for weeks after a freeze. It is common for large limbs to bud out in the spring following a freeze, only to die back in the summer or autumn due to latent freeze-damage to cambial tissues. Another indication of wood damage is when leaves turn brown but do not abscise following a freeze. This indicates more severe freeze-damage than defoliation alone, and usually indicates severe shoot damage.

Because freeze-damage to the wood is so difficult to assess, freeze-damaged trees should not be pruned earlier than late spring, at least 6 weeks following a freeze. After the extent of freeze-damage has been assessed by evaluating the extent of cambial discoloration, pruning should be done to minimize problems resulting from melanose (a fungus) harboured in dead wood. Pruning can be done by machine hedging and topping for minor damage or using chain or pneumatic saws when more selective, extensive hand-pruning is needed.

Cultural practices for freeze-damaged trees
Changes in cultural practices will probably have to be made depending on the severity of the freeze-damage. It is important to assess freeze-damage accurately before altering cultural practices. In mild to moderate damage, partial or total defoliation with no wood damage, it is important to regrow the canopy as rapidly as possible. Trees should receive recommended fertilizer rates during the winter and spring and adequate but not excessive irrigation as new leaves develop. Most water loss is through the leaves and therefore it is unnecessary to apply heavy irrigation to defoliated trees. However, adequate soil moisture is important to promote uptake of nutrients and growth of new leaves. Weed control becomes a problem because the orchard floor receives more sunlight than a fully canopied orchard. Recommended rates of pre-emergence materials should be applied.

Cultural practices should be modified when severe leaf and wood damage have occurred. In this case, the size of the canopy and roots has been reduced and the tree requires less water and nutrients. For example, if canopy size is reduced by a third, fertilizer, irrigation and herbicide rates should be reduced by that amount. Trees should receive more frequent applications of water and herbicide because of reduced tree size. There is also less demand for nutrients because a crop is not present, but it is essential not to neglect these trees since new canopy development is important. Buckhorned trees (those cut back to large scaffold limbs) may require hand-pruning to reshape the tree. Water sprouts that can lead to multiple trunks should be removed to lessen future management problems. Buckhorned trees may not require whitewashing to reduce trunk temperatures if pruning is done in late spring, allowing for regrowth of enough new foliage that will protect the trunk from high temperature damage. Trees buckhorned in the summer, however, require whitewashing to reduce trunk temperature and heat stress that may retard regrowth.

MINERAL NUTRITION

Nutrient Elements

Mature citrus trees require 12 elements besides carbon, oxygen and hydrogen which are readily abundant, to attain adequate growth and yields (Smith,

1966a). Elements required in large quantities (macronutrients) include N, P, K, Mg, Ca and S. Micronutrients include Mn, Cu, Zn, B, Fe and Mo. Citrus trees consist mainly of carbon compounds and water with nutrients comprising a small percentage of the total fresh weight (Chapman, 1968). Nevertheless, nutrients are essential to proper metabolic functioning of the tree and consistent commercial production.

Abundance of macro- and micronutrients varies considerably with soil type and citrus-growing region worldwide. Some soils have inherently high levels of a particular nutrient or nutrients as well as a high cation exchange capacity. For example, little P is required for mature citrus trees in the central ridge area of Florida due to high levels in the soil. Similarly, many high pH calcareous soils are high in Mg and Ca. Conversely, compounds such as K and particularly nitrate are commonly deficient due to rapid leaching from the soil and uptake by the tree. Therefore, application of N as urea, nitrate (NO_3^-), ammonia or a combination and K are essential parts of nearly all fertilizer programmes worldwide. Some studies suggest that as much as 40–50% of the applied N is not available to the tree due to leaching, denitrification (conversion of nitrate to N_2) and volatilization. Other essential elements are applied based on the inherent fertility of the soil and proportional to cropload.

Macronutrients
Nitrogen is a component of amino acids and proteins and is particularly important for proper growth and development of citrus trees. Adequate amounts of N are necessary to attain commercially acceptable growth and yields. Inorganic N is present in the soil solution primarily as N_2, NO_3^- or NH_4^+. Nitrate and ammonium are both taken up by citrus trees, although NH_4^+ absorption is greater at high pH and NO_3^- absorption at low pH (Kato, 1986). Several N fertilizers are available that provide these forms of N, none of which has been shown to produce superior yields compared with the others, at least under Florida growing conditions (Leonard *et al.*, 1961). Nitrate is particularly mobile in the soil solution and may be leached from the root zone by excessive rainfall or irrigation. Moreover, both forms of N may be denitrified by bacteria to N_2O and N_2 (gases) which diffuse to the atmosphere. Nitrate is taken up actively (a process requiring energy) or in the transpiration stream and translocated to the canopy in the ionic form, while NH_4^+ is converted to amino acids, primarily glutamate, in the root, after which it moves to the canopy in the transpiration stream (Kato, 1986).

Nitrogen uptake and translocation are affected by several factors including soil temperature, root and tree vigour, and soil oxygen levels. For example, N uptake as well as reduction and assimilation were 10% lower for satsuma mandarins growing in the winter (low soil temperature) compared with the summer (high soil temperature) (Kato, 1986). Winter chlorosis or winter yellowing which occurs in many areas of the world may also be the result of reduced N uptake at low soil temperatures (<12°C). Chlorosis is most severe on mandarin hybrids such as 'Orlando' tangelo. It is also related to translocation of N metabolites at low temperatures. Kato *et al.* (1982) found that acropetal translocation of ^{15}N in the winter was only 0.1% of summer rates. These differences are likely due to reduced transpiration rates at lower temperatures as well as to direct temperature effects on active transport of N.

Tree vigour and rootstock also influence N uptake. Trees on rough lemon, a very vigorous rootstock, had greater N uptake rates than those on sour orange, 'Carrizo' citrange or 'Cleopatra' mandarin (see Chapter 3). Castle and Krezdorn (1973) also observed differences in leaf nutrient levels related to rootstock selection (see Chapter 3).

Optimum N levels for vegetative growth and yields are usually determined based on previous yields or through leaf analysis. Nitrogen-deficient leaves are distinctly light green to yellow in colour. The entire leaf becomes pale compared with the interveinal chlorosis which occurs with other nutrient deficiencies. Optimum levels of leaf N generally range from 2.5 to 2.7% for most cultivars. Ranges for leaf N for mature sweet oranges from several citrus regions are given in Table 5.3. These ranges are intended as general guidelines and may not apply to every situation and cultivar (Smith, 1966b). Leaf N concentration is often much higher for young, nonbearing trees than for mature trees, particularly just after transplanting from the nursery.

Table 5.3. Standards for classification of the nutrient status of orange trees based on concentration of mineral elements in 4- to 7-month-old, spring-cycle leaves from non-fruiting terminals (Smith, 1966b).

Element	Dry matter basis	Deficiency	Low range	Optimum range	High range	Excess
Boron (B)	mg l^{-1}	20	20–35	36–100	101–200	260
Calcium (Ca)	%	1.5	1.5–2.9	3.0–4.5	4.6–6.0	7.0
Chlorine (Cl)	%	–[1]	–[1]	<0.2	0.3–0.5	0.7
Copper (Cu)	mg l^{-1}	3.6	3.7–4.9	5–12	13–19	20
Iron (Fe)	mg l^{-1}	35	35–49	50–120	130–200	250?
Lithium (Li)	mg l^{-1}	–[1]	–	<1	1–5	12
Magnesium (Mg)	%	0.2	0.2–0.29	0.3–0.49	0.5–0.7	0.8
Manganese (Mn)	mg l^{-1}	18	18–24	25–49	50–500	1000
Molybdenum (Mo)	mg l^{-1}	0.05	0.06–0.09	0.1–1.0	2–50	100?
Nitrogen (N)	%	2.2	2.2–2.4	2.5–2.7	2.8–3.0	3.0
Phosphorus (P)	%	0.09	0.09–0.11	0.12–0.16	0.17–0.29	0.3
Potassium (K)	%	0.7	0.7–1.1	1.2–1.7	1.8–2.3	2.4
Sodium (Na)	%	–[2]	–	<0.16	0.17–0.24	0.25
Sulfur (S)	%	0.14	0.14–0.19	0.2–0.39	0.4–0.6	0.6
Zinc (Zn)	mg l^{-1}	18	18–24	25–49	50–200	200

[1]Indicates lack of information regarding value
[2]These elements are not known to be essential for normal growth of citrus.

Phosphorus is essential for proper functioning of cell energy systems and as a structural component of cells; however, citrus trees require low levels of P. Phosphorus is present in the soil solution primarily as PO_4^{-3}, $H_2PO_4^{-2}$ or H_3PO_4 in the pH range of 6 to 7. Phosphorus is very immobile in the soil because it forms insoluble compounds with metals such as Al, Ca or Fe and tends to accumulate particularly in mature orchards. It is also leached and metabolized much more slowly than NO_3^- or K. Therefore, annual P application is not needed in many mature orchards. Phosphorus deficiency symptoms rarely occur on mature trees but include a pronounced reduction in bloom and poor production of mostly small fruit. Because of P immobility, leaf and soil analysis may be used to determine if P application is needed. Ranges of leaf P for sweet oranges in selected growing regions are given in Table 5.3 and are intended as general guidelines (Smith, 1966b).

Potassium is necessary for regulating ionic balances in the cell and for developing adequate fruit size and regulating peel thickness. Leaf K has little effect on vegetative growth of citrus within the range 0.35 to 2.0%. However, it has a significant effect on fruit quality. Low leaf K results in small fruit size and reduced peel thickness which predisposes the fruit to splitting, plugging (tearing of the peel around the calyx during harvesting) and creasing. Overapplication of K produces large, coarse fruit with thick peels. Potassium, like NO_3^-, is readily leached from the soil and is usually applied as a ratio of the N content either as muriate or sulfate of potash. For example, a 1:1 N/K ratio might be used where a thick peel is desirable, while a 1:0.5 ratio would be more desirable where fruit with a thin peel is more in demand. In cooler climates the ratio of N to K is important for proper balance of yield to fruit size (DuPlessis and Koen, 1988). Higher N favours more fruit but smaller size, while higher K favours fruit size over yield. Potassium is naturally high in some soils and may not have to be applied. Ranges of K levels in sweet orange leaves in selected growing regions are given in Table 5.3 and are intended as general guidelines (Smith, 1966b). Potassium deficiency symptoms include production of small fruits with thin peels. Visual leaf symptoms are not as distinctive as those of N or Fe, therefore deficiencies are usually detected through leaf analyses.

Magnesium is essential to many enzyme reactions in the tree, but is not limiting in many orchards especially if dolomitic limestone (which contains Mg) is used to regulate pH. Magnesium deficiency decreases yields, tree vigour and freeze-hardiness; however, within acceptable levels Mg has little effect on vigour or yield. Deficiency symptoms include tip and margin interveinal chlorosis with a greener zone at the base of the blade. Sectional yellowing without distinct delineation between veins and interveinal areas of the older leaves as is observed with Fe or N deficiency is often present. Symptoms are more likely to occur in seedy cultivars or on calcareous soils that have received treatment with calcitic rather than dolomitic limestone. Magnesium deficiencies may also be discovered via leaf analysis (refer to Table 5.3 for general ranges of Mg) and are correctable using soil-applied MgO or $MgSO_4$, although leaf symptoms may not be corrected for over a year after soil application. Foliar sprays of Mg $(NO_3)_2$ (7% Mg) are also used to correct deficiency symptoms. Magnesium is most readily taken-up by the tree at pH values between 7 and 8.5, but this is too high for best uptake of essential heavy metals. In most calcareous soils, Ca competes with Mg for uptake sites on the root and limits uptake of Mg.

Calcium is the most prevalent element in a mature citrus tree, comprising over 20% of the elemental content (Chapman, 1968). Calcium is important for enzyme functioning and is an essential component of cell wall structure and in transport of metabolites. In most citrus-growing regions Ca deficiency is rare since Ca is either abundant in the soil or added in lime to increase soil pH. Under controlled experimental conditions Ca deficiency decreases growth but effects on yields are not well-documented. Calcium deficiency symptoms are not distinctive but include a decrease in root and canopy vigour along with interveinal chlorosis similar to Fe or Mn deficiency symptoms. Calcium sources are most important to citrus culture for their role in increasing soil pH via liming. With recognition that HLB affected trees, particularly on 'Carrizo' or 'Swingle' rootstock, need to be kept at a lower soil pH, Ca levels may be less than adequate and use of $Ca(NO_3)_2$ may be warranted.

Sulfur is an essential component of proteins and enzymes but is rarely deficient for citrus trees under commercial conditions (Smith, 1966b). Sulfur is usually provided as sulfate in many commercial formulations of fertilizer. These fertilizers are used more commonly now to help keep soil pH lower. Sulfur deficiency symptoms in leaves are very similar to those of N deficiency symptoms, but N deficiency is much more common.

Micronutrients

Micronutrients are necessary for proper enzyme functioning and, as the name implies, are present in small quantities in the tree. They are applied either to the soil or more commonly as foliar sprays. Soil-applied micronutrients become complexed in the soil and may be taken up slowly, requiring foliar application. Many growers apply micronutrients annually whether leaf deficiency symptoms are present or not. Long-term research from Florida has suggested that regular applications of micronutrients are unnecessary and that growers should wait until leaf symptoms or leaf analysis dictate that micronutrients are needed (Wutscher and Obreza, 1987). In citrus areas with HLB, recent studies suggest that Mn and Zn may need to be elevated in the soil-applied fertilizer to overcome some of the tree decline problems associated with the disease (Handique *et al.*, 2012 and current research in Florida, unpublished). The ranges of micronutrient concentrations for sweet oranges in selected growing regions are given in Table 5.3.

Manganese deficiency symptoms are common on young, expanding leaves, appearing as yellow patches between veins. Such deficiencies do not require correction unless they become quite severe. Usually Mn is not limiting in acid soils but may become complexed and unavailable in high pH, calcareous soils. In this case, Mn should be applied as a foliar spray. Higher Mn levels may be warranted for HLB affected trees.

Copper deficiency symptoms are rare but do occur in newly planted trees on virgin soils in the nursery, particularly for 'Swingle' citrumelo rootstock and occasionally in mature orchards. Symptoms include abnormally large leaves, limb dieback, gum pockets under the bark of young wood, brownish gumming on fruit, twigs and leaves, and multiple buds. Sufficient Cu is often applied in fungicidal sprays to prevent deficiencies, but Cu may be applied when necessary as a foliar spray along with other micronutrients or as a component of mixed fertilizer to the soil. Copper toxicity may occur where regular Cu sprays have been applied over the years. Copper becomes readily available to trees growing in acid soils and may become toxic when more than 100 kg ha^{-1} has accumulated in the top 30 cm of soil (Koo *et al.*, 1984). Therefore, soil testing becomes a very important means of detecting excess Cu levels. Toxicity symptoms in the tree include reduced yields, production of small, dull green leaves, iron deficiency symptoms and stubby roots. If toxicity symptoms appear, soil pH should be maintained above 7 by applying lime to decrease the solubility of Cu in the soil. Better tree growth at lower pH in HLB areas may make balancing soil Cu levels more of a management dilemma for growers, particularly for grapefruit, which traditionally received frequent Cu fungicide sprays, or if citrus canker is present which also calls for frequent Cu sprays.

Zinc deficiency symptoms are common worldwide and are readily distinguishable as distinct interveinal chlorotic regions. Zn is a co-factor for several important enzymes and is complexed by a phloem protein in citrus blight affected trees (Taylor *et al.*, 1996). Leaves are generally smaller than normal and may occur in rosettes. Nevertheless, Zn levels alone rarely affect citrus growth or yields. Zinc deficiency is not readily corrected using soil applications, but is generally remedied by foliar application of ZnO or ZnSO$_4$ during the post-bloom period. Zinc deficiency symptoms commonly occur on trees affected by citrus blight, citrus variegated chlorosis and greening (see Chapter 7) on leaves of freeze-damaged shoots and in sweet orange trees on 'Carrizo' citrange rootstock, especially in high pH soils. Zn uptake from soils may be improved by lower soil pH and application in a restricted wetted zone in the tree row.

Boron level regulation in mature trees may be problematic since there is a narrow range between deficient, adequate and toxic levels (Koo *et al.*, 1984). Deficiency symptoms include corkiness of leaf veins and cork spots in the albedo of the fruit (Smith and Reuther, 1949). Toxicity produces dull green leaves with brown surface pustules. Boron should be applied either in the fertilizer at 1/300 of the N rate or as a foliar spray. Boron should not be applied both by ground and as a foliar spray in the same year as toxicity may occur.

Molybdenum deficiency is manifested as interveinal yellow spots, usually in late summer. Deficiency symptoms are rare in most orchards, especially in high pH soils. Molybdenum deficiency is more severe on acid compared with neutral soils. Thus, deficiencies can often be corrected by liming to raise the pH or preferably by foliar application of sodium molybdate.

Iron deficiency is a common problem and can have many causes. Chlorosis punctuated by a fine network of green veins is symptomatic of Fe deficiency. Severely Fe-deficient leaves may become light yellow to nearly white in colour. Iron deficiency, unlike other micronutrient problems, is not easily and readily corrected by foliar application of Fe. For soils with a pH greater than 7.0 (lime-induced chlorosis) FeDDHA iron chelate is applied to the soil or as a foliar application. On acid soils FeEDTA chelate is used. Chelated forms of Fe are not readily complexed into insoluble or unavailable forms. Deficiency symptoms are usually alleviated 6–8 weeks after soil application. Finely ground sources of chelated Fe, which are very expensive, should be mixed with an inert material and carefully applied to prevent burning of the fruit and leaves by drifting material.

Soil and Leaf Analysis

Soil analysis

Soil analysis can be used in conjunction with field observations, yield and leaf analysis in determining fertilizer and limiting rates or in correcting nutrient disorders. Soil analysis is most useful in monitoring levels of relatively immobile nutrients like Mg, Ca, Cu and P or for determining pH. Levels of these compounds in the soil are usually expressed in kg ha^{-1} with low, medium and high ranges presented. Low levels of Mg and P can be corrected by applying these minerals as part of the fertilizer mixture. Soil N and K analyses are usually not representative of levels in the tree because of the mobile nature of these ions.

Proper soil sampling and extraction are necessary to ensure reliable results. Sampling is done using a soil auger or shovel to remove soil from a depth of up to 30 cm around the dripline of the tree. Usually samples from 16 locations within an area no greater than 8 ha are combined into a single composite sample (Koo *et al.*, 1984). This type of sampling is adequate if the soil in the orchard is uniform; however, composite sampling can be very misleading in variable soil types. With restricted wetted zones with most current irrigation systems, soil samples from the wetted zone may be more representative of the active soil area where the tree gets most of its soil nutrients. The sample is sent to a soil testing lab where it is dried, sieved and its pH determined. A sample of the remaining soil is extracted using a variety of acid mixtures, although the double acid (Mehlich) method is commonly used. The double acid method may cause excessive extraction of nutrients and does not correlate well with leaf analysis (Anderson and Albrigo, 1977). The extraction slurry is shaken for 5 minutes, filtered, and the extract analysed using atomic absorption (AA) spectrophotometry, inductive plasma spectrophotometry (ICAP) or a colorimetric method. It is important to be aware of the extraction technique used and how well its value correlates with citrus requirements for a particular soil type. Values are typically expressed in mg l^{-1} or kg ha^{-1}.

pH

Availability of some nutrients, notably Cu, Fe, P, K and Mg, is affected by soil pH. It is advisable to maintain pH at 5.5 to 7.0 if possible, although many orchards worldwide are productive at pH values considerably lower and higher than these. Liming of acid soils using calcitic (or preferably dolomitic limestone because it also contains Mg) will raise soil pH. Liming was a regular part of the production programme in the past but has become less important in recent years. In contrast, some orchards are planted in high pH, calcareous soils. In these soils Zn, Fe and Mn deficiency symptoms can become quite severe. Through a costly and relatively slow process, pH can be decreased using soil additions like S, S-containing compounds and acid-forming fertilizers. Deficiency problems can be somewhat managed by foliar applying chelated materials, but this is relatively expensive. With increased knowledge of soil chemistry and tree disease interactions it is now known that lower soil pH is better for citrus trees with HLB, particularly if on 'Swingle' or 'Carrizo' rootstocks and bicarbonate levels in irrigation water are part of the adverse pH situation (Graham and Morgan, 2018).

Leaf analysis

Leaf analysis is generally preferred to soil analysis for most nutrients. The leaves integrate uptake of nutrients over an extended period and the tree is the most accurate soil extraction system to indicate adequacy of each nutrient. A yearly programme of leaf analysis may be useful to citrus growers in assessing the overall nutritional health and fertilizer requirement of an orchard. Sampling at least every other year is needed to be sure that leaf nutrient levels remain in acceptable ranges. As with soil analysis, proper leaf sampling is important. Samples of 100 leaves should be collected over at least 20 trees representative of the general orchard condition in not more than 8 ha units (Koo *et al.*, 1984). Fertilizer and cultural practices should be consistent over the sampled area.

Several extensive studies have been done on effects of leaf age, location on the shoot and fruit load on nutrient levels (Smith, 1966b). Generally N, P and K levels decrease from 1 to 4 months of age, while Mg and Ca levels increase (Fig. 5.5). Levels plateau for leaves between 4 and 7 months, with values decreasing for all elements as leaves age and as nutrients are translocated to newer, developing tissues. Similarly, leaf N and K levels increase with leaf age within a shoot, with newly developing leaves having lower levels than older ones. Conversely, levels of P, Mg and Ca remain fairly constant along the shoot. Leaf nutrient concentrations are also higher for non-growing compared with growing shoots, with the exception of P concentration which is relatively constant.

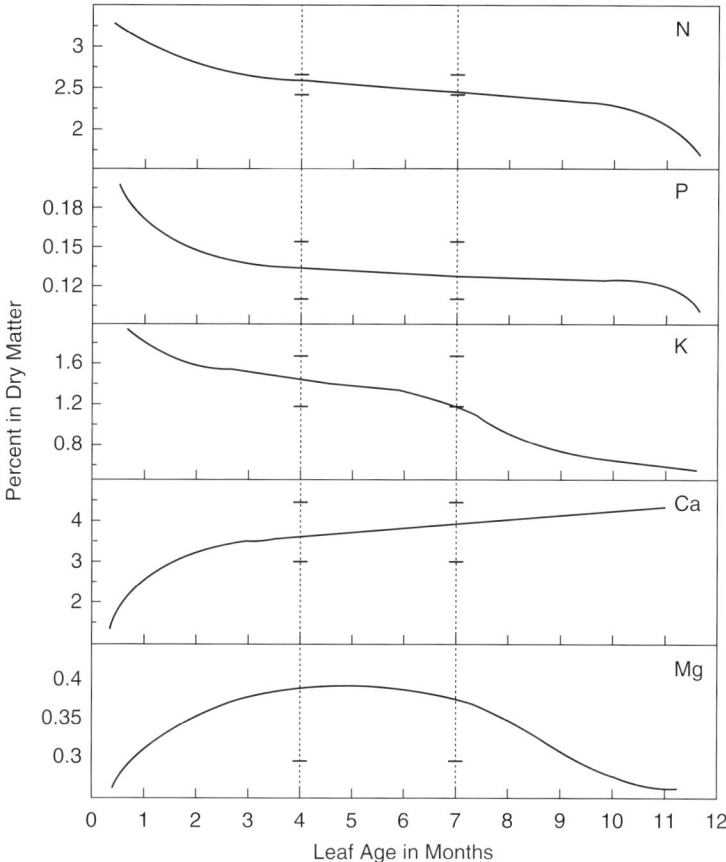

Fig. 5.5. Schematic diagrams of the change in concentrations of the macronutrient elements in orange leaves on non-fruit-bearing shoots as they increase in age. The dotted vertical lines indicate the age bracket of 4 to 7 months commonly used in taking indicator samples. The optimum ranges of concentration for leaves in this age bracket are indicated by cross marks for each element. The trends may vary somewhat under different conditions but these are fairly typical of the published results (Smith, 1966b).

The presence of fruit on a shoot significantly decreases leaf nutrient concentrations of N, P, K, Mg and Ca in most instances (Embleton *et al.*, 1963; Smith, 1966b). This is expected since leaves compete for nutrients with developing fruits. This is evidenced by leaf N and K levels being inversely related to cropload (Rasmussen and Smith, 1961). For example, leaf N ranged from 1.97 to 2.31% in heavy crop years to 2.55 to 2.75% in years with light crops, with differences in K varying nearly twofold between heavy and light croploads. Therefore, leaf samples are usually collected from nonfruiting, fully expanded, 4–7-month-old leaves in most citrus regions, with the exception of California and South Africa where samples are taken from fruiting shoots. If micronutrient or fungicidal Cu sprays have been applied it is very difficult to remove contamination, even with acid washing, and analysis will be inaccurate. Leaf samples should be washed three or four times with deionized water before air-drying, and shipped in paper or plastic bags to an appropriate laboratory for analysis. Typical ranges of leaf analyses for sweet oranges in several citrus producing areas are given in Table 5.3 (Smith, 1966b). Values may differ slightly for grapefruit, lemons and some other commercial cultivars (Embleton *et al.*, 1978). Leaf nutrient classifications include deficient, low, optimum, high and excess. These ranges refer to yield and fruit quality responses.

Fertilization

Citrus trees are grown on a wide range of soil types. Consequently, availability and inherent levels of nutrients may vary widely. In most areas, supplemental nutrients are needed to obtain commercially acceptable growth and yields. The type of nutrients required and the amounts are naturally a function of soil type, growing region and cropload. In impoverished soils with low cation exchange capacity (CEC) all the major macro- and micronutrients may be required to attain adequate growth and yields. In most soils suitable for citrus, however, primarily supplemental N and K, which are readily leachable, are needed. Growing region also affects the amount of supplemental nutrition required (Chapter 4). Regions with high rainfall and temperatures are more likely to lose nutrients from the soil due to leaching or volatilization. Cropload is also closely tied to fertilization programmes for mature trees since many nutrients are removed by harvesting the fruit. For example, a tonne of 'Valencia' oranges contains 1.31 kg N, 0.19 kg P and 1.8 kg K (Smith and Reuther, 1953). Therefore, an orchard producing 50 tonnes ha^{-1} would require 66, 9.5 and 90 kg ha^{-1} of N, P and K to replenish nutrients lost during harvest alone. This does not account for losses due to leaching or volatilization or utilization of nutrients for vegetative growth. Mature trees need about 166 kg N/ha for maintenance (Morgan *et al.*, 2006).

Fertilization of young, nonbearing trees

Fertilization practices for young, nonbearing trees differ from those required for mature, bearing trees because nutrients are not harvested with the fruit and the trees are much smaller. The objective in a young tree fertilization programme is to produce fruit as soon as possible by growing the tree as rapidly as possible. Productivity is highly correlated to canopy-bearing volume. Therefore, young trees receive high levels of nutrients, particularly N, to promote vigorous growth. In fact, N levels on a per treated hectare may approach 1000 kg because of the relatively small area that receives fertilizer. Several studies suggest that this rate is excessive.

A number of studies have been conducted on fertilization practices for young trees (Rasmussen and Smith, 1961; Marler *et al.*, 1987; Willis *et al.*, 1990, 1991; Swietlik, 1992; Albrigo and Syvertsen, 2000). In general, young trees (nonbearing trees less than 4 years old) are fertilized more frequently than mature trees, 4 to 6 times per year vs 2 to 3 times, using smaller amounts of material for each application. This is due to the relatively limited root zone, particularly of a newly planted tree. Usually, low analysis fertilizer is used to reduce the chance of salt damage due to over-fertilization. Fertilizer rates are based on tree age or size, soil type and growing conditions and may vary considerably among citrus regions.

Nitrogen is the most important element regulating growth of young citrus trees as during this time trees are growing very rapidly. It is not uncommon for tree trunk diameter to increase by 100 to 200% or more during the first and second growing seasons (Jackson and Davies, 1984; Davies and Jackson, 2009), and canopy volume to increase tenfold in the humid subtropics. Other macronutrients, particularly P and K, are usually added to the fertilizer mix but have less impact on tree growth than N. In fact, studies from Florida indicate that leaf macronutrient levels from the nursery are often in the high to excess range. Similarly, leaf micronutrient levels are usually relatively high when the trees leave the nursery and thus supplemental sprays or soil applications are not needed unless indicated by leaf analyses or nutrient deficiency symptoms.

Traditionally, fertilizer has been banded or broadcast in a granular form; however, more recently liquid fertilizer is being applied through irrigation lines (fertigation) in many areas of the world. Some studies suggest that frequent application of low levels of liquid fertilizer improves tree growth over less frequent applications (Dasberg *et al.*, 1988; Willis *et al.*, 1991) – the theory being that the concentration of nutrients is maintained at a constant level in the soil solution thus allowing continual uptake. Other studies, however, indicate that fertigation frequency or the use of granular compared with liquid fertilizer has no effect on tree growth, possibly because trees take up and store nitrogen as amino acids to be used for subsequent growth (Willis *et al.*, 1991; Swietlik, 1992). It is likely that these regional differences in growth response to fertigation are due to edaphic, rootstock and environmental differences. For example, more frequent fertigation improved growth of 'Hamlin' orange trees on 'Carrizo' citrange rootstock but did not affect growth for trees on sour orange rootstock (Willis *et al.*, 1991). In earlier work, fertigation appeared to have more effect on tree growth in areas such as Israel and Spain than in more humid regions like Florida and Texas. Recent studies suggest that highly managed daily replacement fertigation may be successful in more climates and soils (Kadyampakeni *et al.*, 2013b). More effective and better-managed irrigation resulting from fertigation may also be involved in growth differences.

Controlled-release (slow-release) fertilizers may be useful and economical for young citrus tree management. Controlled-release materials reduce the number of fertilizer applications needed per year and losses due to leaching

(Marler *et al.*, 1987). This may be very useful for replants in mature orchards in order to minimize travel to fertilize the small number of replants in an established planting. Several products are available that have potential, especially in isolated reset (replant) situations. The major drawback to acceptance of controlled-release materials is the high cost per unit of N.

Fertilization of mature citrus trees
As stated previously, fertilization of mature, bearing citrus trees is necessary to replenish nutrients lost during harvest and leaching or volatilization, to maintain tree vigour and to obtain optimum yields (Smith, 1966a). Nitrogen has the greatest effect on tree growth and yields of all elements. For example, increasing the N rate for mature grapefruit trees in Florida initially produced a linear and significant increase in yields; however, at rates above 200–250 kg ha^{-1} increasing N had a far smaller effect on yields. This law of the minimum, or the Mitscherlich effect, suggests that while N is necessary to obtain optimum yields, excessive N cannot be justified economically (Smith, 1966a). Nitrogen increases yields primarily by increasing fruit number rather than fruit size. Moreover, trees receiving optimum N are more densely foliated and produce more flowers than N-deficient trees.

Excessive N is not only economically unjustified but also may contaminate groundwater. Nitrate is highly water-soluble and moves rapidly through the soil profile into groundwater. High NO_3 levels in drinking water may be a health threat, especially to infants. Nitrate competes with oxygen in the bloodstream and in some instances 'blue baby syndrome' is caused; however, this problem occurs very rarely and is usually not of major concern.

Several studies have been conducted worldwide to determine optimum rates and timing of fertilizer application with, as expected, variable results depending on location, cultivar and intended use of the fruit (fresh or processed). Long-term studies by Smith (1969) in Florida suggested that about 900 g N tree^{-1} annually was necessary to achieve optimum yields for mature grapefruit, and that multiple compared with single applications had little effect on yields or fruit quality. Similarly, Mungomery *et al.* (1978) found that 900 g N tree^{-1} was the annual optimum for mature navel orange trees growing in Australia. They also observed that increasing N rates from 900 to 1350 g tree^{-1} did not increase yields, supporting Smith's previous findings (1966a). Twenty years of research from Brazil has suggested that yields of mature 'Bahianinha' navel orange increased as N rates increased from 0 to 250 g of N tree^{-1} but no further increase was observed at 500 g tree^{-1} (Rodriquez and Moreira, 1969). These results were for non-irrigated trees, which may explain the lower optimum than in previous studies. It was also observed that application of optimum levels of N and P together gave better results than either individually. DeVilliers (1969) conducted a 6-year study on 'Washington' navel orange in South Africa, finding that about 1300 g N tree^{-1} annually produced optimum yields. Leaf mineral composition, growth and yields were compared over 9 years for

'Valencia' oranges in Arizona (Sharpies and Hilgeman, 1969). Ammonium nitrate (granular) at 9 annual rates from 0 to 3630 g of N tree^{-1} was compared with foliar urea sprays (230 g tree^{-1}). Little effect was observed on trunk growth related to rate or source; however, optimum yield occurred at about 600 g N tree^{-1}. Studies from Spain on 'Navelate' oranges again suggested optimum N rates in the 1000 g tree^{-1} range (Legaz *et al.*, 1981). These values should be lower if more trees are planted per hectare. Fruit weight tended to decrease with increasing N rate probably due to increased cropload. In this study, split application of fertilizer and fertigation were superior to single or granular applications. Studies from Israel also support this viewpoint, yet research from California (Jones and Embleton, 1967) and Florida (Smith, 1966a) suggests no yield advantage to split applications or use of fertigation for mature trees. Undoubtedly, climatic and edaphic differences between regions account for some of these conflicting results, although certainly optimum levels of N are similar worldwide. These rates no doubt also reflect losses that might be reduced if leaching was minimized by more frequent applications of lower rates per application. Work by Syvertsen and Smith (1996) has suggested that half to two-thirds of these previously determined optimum rates may be sufficient if leaching and volatilization losses are better controlled.

Citrus trees require relatively low levels of P to attain adequate growth and yields. The citrus fruit itself contains rather low levels of P which are removed during harvest and vegetative growth is not affected by increasing P levels provided that leaf levels are >0.08% dry matter (Smith, 1966a). Therefore, P deficiency in citrus is rare in most citrus-growing regions with the exception of some regions of South Africa. In fact, in the central Ridge area of Florida soil levels are naturally high (phosphate is mined there) or have been raised to high levels after years of fertilization. Thus, little P is added in the fertilizer. Sources of P for citrus fertilization include superphosphate or triple superphosphate.

Potassium fertilization is very important in many citrus-growing regions, particularly in fresh fruit producing areas where fruit size is an important concern. Nevertheless, leaf K levels vary considerably even in Florida soils (Kadyampakeni *et al.*, 2013a) and studies suggest that vegetative growth is not significantly affected within a 0.35–2.0% range. Potassium is usually applied as potassium chloride, sulfate or nitrate.

Trees with HLB disease show considerable leaf deficiency symptoms, particularly Zn, and they look weak often with chlorotic zones in many leaves. In Florida, growers (notably M. Boyd) have found that multiple foliar applications of minor and major elements reduce these symptoms (Rouse *et al.*, 2010; Spann *et al.*, 2011). Coupled with correction of other stress factors such as high soil pH and bicarbonates in the irrigation water, mature trees can be much more thrifty and productive for several years when infected with HLB (Rouse *et al.*, 2010). However, growers in Florida are currently relying more on soil applied nutrition for HLB affected trees, usually adding slow release fertilizer often with sulfur and higher rates of Zn and particularly Mn.

Interactions of elements
There are several noteworthy interactions among the various nutrient elements. Nitrogen levels influence those of most other elements (Smith, 1966a). Nitrogen and P leaf levels are inversely related, with N levels having a pronounced effect on P levels. The N interaction with K has been widely studied. In general N and K levels are inversely related. However, the ratio between N and K has a significant effect on yields and fruit size (DuPlessis and Koen, 1988). Maximum yield for 'Valencia' orange trees in South Africa was obtained at a N:K ratio between 2.4 and 3.0. Both N and K had to be at levels greater than 2.1 and 0.8% dry weight, respectively. In contrast, optimum fruit size was attained at a ratio between 1.6 and 2.2 with N greater than 1.8% and K greater than 0.9%. Most importantly, greatest income was achieved at relatively low N levels. This situation occurs only for areas producing fresh fruit, such as South Africa, where fruit size and quality are directly related to income. Where citrus processing is more important than fresh fruit production, N levels should be in the moderate range to attain maximum yields. The ratios used in this study may differ from those in other citrus areas of the world due to climatic factors and because leaf nutrient levels in South Africa are determined for fruiting rather than nonfruiting shoots. Nitrogen and Mg levels in the leaf are positively correlated and synergistic, while conditions which produce high Ca leaf levels generally depress N levels. Potassium interacts with other elements besides N. Potassium and Ca displace one another in the soil and thus are mutually antagonistic. Thus, concentrations of K and Ca in the leaf are inversely correlated. Similarly, NH_4^+ and K compete for sites in the soil. Potassium in addition is a strong antagonist of Mg, although Mg has only a moderate effect on K levels. There are several other interactions among elements but these have less effect than those mentioned above. For more detail on these interactions consult the review by Smith (1966a).

Nutrition and fruit quality of citrus
Nutrition and fertilization practices also influence fruit quality. Thus, varying fertilizer rates and the ratio of elements is used commercially to change fruit quality depending on market demands. Nitrogen, P and K have the greatest influence on fruit quality, provided that other elements are not severely limiting (Embleton *et al.*, 1978). The general influence of N, P and K on fruit quality also varies with species, with sweet oranges showing more response to N and P than lemons. Nevertheless, the general effects of changing leaf N levels from 2.0 to 3.0% dry weight for 'Valencia' and navel sweet oranges are given in Fig. 5.6 (Embleton *et al.*, 1978). As leaf N levels increase fruit size, juice content and ascorbic acid levels decrease. Effects on TSS and TA are inconsistent. In addition, increased N levels increase peel thickness and coarseness of peel texture and decrease peel colour. Therefore, leaf N levels in the excessive range adversely affect fresh fruit quality. These same trends also occur for navel oranges. Increasing leaf P levels from 0.10 to 0.21% dry weight generally

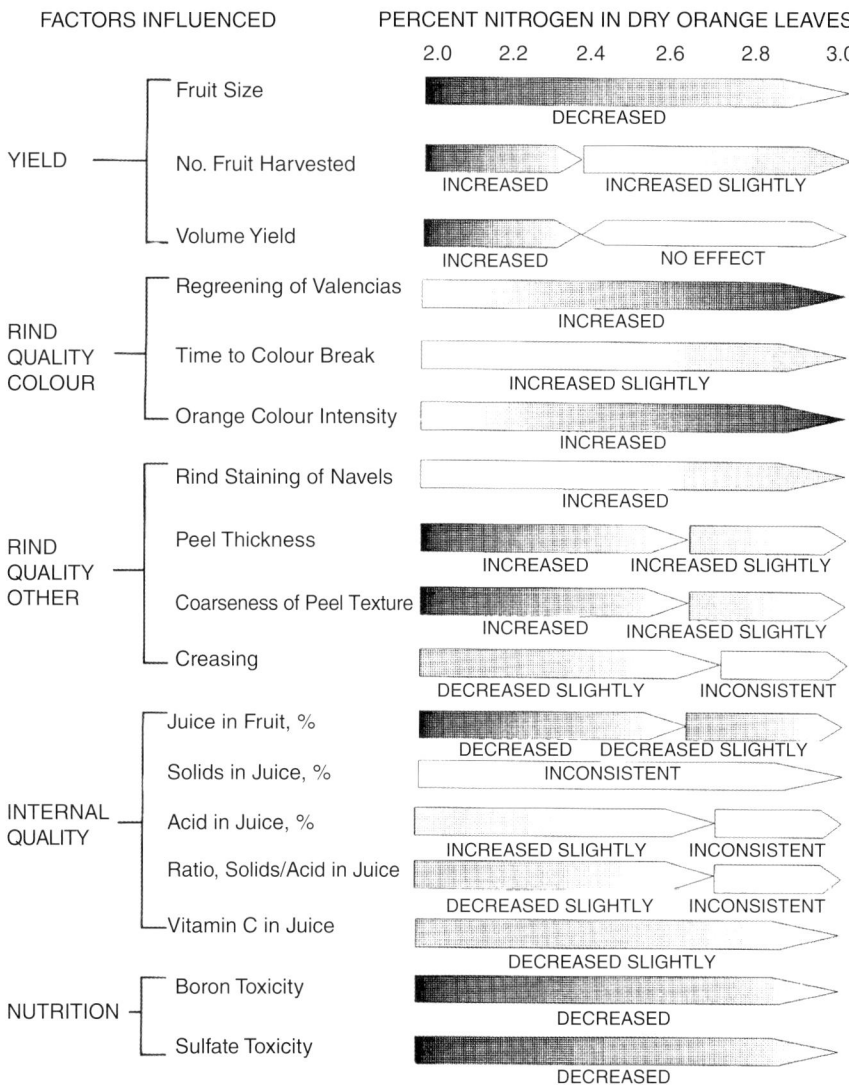

Fig. 5.6. Influences on yield, quality, and B and S nutrition resulting from changes in the percentage of N in 5- to 7-month-old, spring-cycle orange leaves. The greater the intensity of stippling in the graph, the greater the effect on the factor indicated (Embleton *et al.*, 1978).

slightly decreases fruit size, TSS, TA, ascorbic acid, and peel thickness and coarseness. However, P levels have less effect on fruit quality than N or K. Potassium levels in the leaf have a pronounced effect on fruit quality as they increase from 0.3 to 1.9% dry weight. Fruit size, peel thickness and coarseness increase with increasing K levels and juice content decreases slightly. Increased K levels also reduce the amount of fruit creasing (see Chapter 8).

TREE SIZE CONTROL

Mature citrus trees can attain heights of more than 10 m and widths of nearly 10 m. Over time the trend has been to considerably decrease tree size and spacing within rows, with a smaller decrease between rows. Higher density plantings have the advantage of earlier returns and higher yields, but also become hedgerows within as little as 5 years after planting. As trees grow together, lower limbs become shaded and fruit production occurs primarily toward the outside and the top of the canopy in standard size scion/rootstock citrus trees. The interior of the tree becomes twiggy, limb dieback may occur and fruit size decreases. Yields and fruit quality decrease as shading continues. Standard-size trees that are allowed to grow unrestricted are difficult to harvest and spray. Consequently, as planting density has increased, interest in tree size control has also increased. In many citrus regions, trees have grown into canopy orchards and extensive and expensive rejuvenation pruning has been necessary to increase production and improve fruit quality. The most commonly used methods of size control include detailed pruning and mechanical hedging and topping, although there is considerable interest and recent success in finding dwarfing or semi-dwarfing rootstocks or in the use of viroids such as exocortis to limit tree size (see Chapter 3).

Pruning

The two major types of pruning cuts are heading-back and thinning-out. Heading-back removes a portion of a branch and promotes lateral bud break, thus altering the morphology of the citrus tree. Heading-back cuts are usually made nonselectively using mechanical hedging and topping machines and are by far the most commonly used type of pruning for citrus trees in mechanized citrus-growing regions such as the USA, Brazil, etc. Thinning-out cuts selectively remove entire branches or limbs with a few cuts. Large portions of the tree can be removed, significantly changing light penetration into the canopy. Thinning-out cuts are generally used in conjunction with hand-pruning programmes. They also remove dead or unwarranted branches.

Selective pruning
Meticulous selective hand-pruning is used to control tree size and shape trees in citrus areas where labour is relatively plentiful and fruit are grown primarily for the fresh fruit market such as China, Japan and Spain. Hand-pruning is

also common in these countries because most orchards are small and often family-owned and operated. The system of pruning varies but generally involves hand-clipping small limbs and removal of dead or spindly wood from the interior of the tree, thus improving fruit distribution there (Iwagaki, 1981). Tree canopies (skirts) are raised above ground level to minimize fruit damage from soil-borne problems, such as *Phytophthora*. In some areas of Japan, limbs are clipped immediately after fruit are harvested to moderate the next season's crop and improve fruit size. In Spain, pruning and shaping usually begins when the tree is 2–3 years old. Scaffold limbs are selected around the trunk and other weak or poorly placed shoots are removed (Zaragoza and Alfonso, 1981). Interestingly, these researchers conducted a pruning study on 'Salustiana' sweet orange trees. They found that yields were the same whether trees were pruned annually or every 2, 3 or 6 years. Moreover, there was no effect on fruit size or quality. Thus, they questioned whether frequent, costly pruning was necessary, although it is commonly practised in Spain. However, seedless mandarins probably need pruning with other practices in order to space out limbs for maximum sun exposure and reduced crop load to help improve fruit size to meet market requirements.

Selective pruning may be necessary to remove damaged or dead limbs, particularly after a severe freeze in subtropical areas. Pruning is done selectively using hand or pneumatic pruners or loppers, or chainsaws. Hand-pruning involves heading-back and thinning-out cuts of moderate to large limbs that cannot be pruned using mechanical hedging equipment. Selective pruning is also used to skeletonize (removal of a large portion of the canopy) or buckhorn (cutting back to major scaffold limbs) trees when they are to be top-worked or for rejuvenation of mature, nonproductive trees.

Rejuvenation pruning is used primarily where trees have grown into a densely foliated canopy orchard and yields have become severely limited, but where tree structure is basically sound. The entire canopy is removed back to four to six major limbs to promote the production of vigorous new growth from latent buds. Buckhorning is often used after severe freeze damage. Rejuvenation eliminates production for about 2 years after pruning, after which yield will gradually increase. However, on a long-term basis the orchard will be more productive than if left unpruned. Pruning of this type is costly and produces large amounts of brush that must be removed from the orchard. Furthermore, severe pruning exposes the orchard floor to light and promotes weed growth.

Various types of pruning paints have been applied to pruned limbs over the years with varying levels of success. During the 1950s and 1960s, black asphalt-based paints were used but were found to result in excessive heat build-up of the wound area which delayed healing. White latex-based paints were developed to reduce limb temperature and promote healing and have been used by citrus growers for many years. Years of scientific study with forest trees, however, indicate that pruning paints do not improve wound healing and, in fact, may delay it by preventing natural wound periderm formation.

Certainly, hedging and topping practices produce millions of freshly cut limbs, most of which heal without painting the cut surfaces.

Hedging and topping
Hedging may be required on a regular basis by the fifth or sixth year after planting, depending on location, initial tree spacing and rootstock/scion combination. Hedging is typically accomplished using large circular saws vertically mounted on an adjustable mechanical arm. The cutting angle of the arm may be adjusted from 90° for routine hedging to 65–80° from horizontal for a combination of hedging and topping, commonly called 'Christmas tree' topping. Hedging may be done any time during the year but is most effective if just before the spring flush, or after the threat of severe freezes has passed in subtropical areas (Phillips, 1980; Zaragoza and Alfonso, 1981). Hedging during this dormant period allows improved light penetration into the canopy during the spring growth flush, reduces canopy volume and promotes growth of lateral buds. Hedging at this time will not remove flowers (although flower buds may be removed) or fruit for most cultivars, with the exception of 'Valencia' or 'Natal' oranges (which require greater hu accumulation for fruit maturation than other citrus cultivars), and occasionally grapefruit. Nevertheless, it has been shown that moderate, consistently timed hedging, even of those cultivars, will not significantly decrease yields. Fucik (1977) found hedging and topping of 'Ruby Red' grapefruit trees in Texas decreased yields in the following year compared with unpruned trees. Nevertheless, average yields over a 7-year period were similar and hedging and topping improved fruit quality and increased fruit numbers in the inside and bottom of the canopy. Similarly, Zaragoza and Alfonso (1981) compared manual pruning and hedging and topping with no pruning for navel orange trees in Spain. All pruning methods decreased yields compared with unpruned trees in the pruning year; however, yields were similar for all treatments in the next season when no pruning occurred. Most importantly, the average annual yield tree^{-1} was the same over a 6-year period for all treatments. Hedging of 20-year-old 'Valencia' orange trees in Cuba increased fruit size and yields over unpruned trees during all 4 years of the study (Borrel and Diaz, 1981). Hedging also decreased external blemishes on the fruit by allowing improved penetration of spray materials and increased air movement through the tree. An elaborate hedging and topping experiment was conducted over a 3-year period on lemon, mandarin and sweet orange trees. The treatments included: (i) hedging a single side of the tree in year 1, the other side in year 2 and topping in year 3; (ii) hedging both sides in year 1 and topping in year 2; or (iii) no hedging or topping. In general, moderate hedging and topping did not reduce yields of any of the cultivars but did improve fruit quality.

Several different patterns of hedging have been used, none of which proved to be superior to the others. Some growers hedge only a single side of the tree in a given year, while others hedge both sides along the row in the same year. In the first case hedging costs are less on a yearly basis, but fewer advantages of hedging, namely improvement of regrowth and increased light penetration into the canopy, are realized. In the second instance, tree width is controlled in both directions, further improving light penetration into the canopy. In areas where orchards are not bedded (ridged), trees may also be cross-hedged to limit size in the row, but this practice is seldom or not used at all currently. Most effective hedging programmes will remove only the outer 15–20 cm of the canopy, thereby cutting limbs no greater than 0.5–1.0 cm in diameter. A routine maintenance programme avoids having to cut large limbs and provides for growth renewal each year and reasonable cropping on all sides. In contrast, trees allowed to grow together between rows over a number of years will become shaded, unproductive and require more severe pruning into older wood. Generally, the more severe the cuts, the more vigorous and vegetative the regrowth will be, often requiring 2 years or more to return to normal crop levels on that surface. By then, the surface will require severe cuts again. Light annual hedging also tips many previous year spring or summer shoots. If the hedging is done in mid-autumn, some buds below the cuts can be induced to flower for the following spring flush, when otherwise they would not flower (Chica and Albrigo, 2011).

Topping is also done with mechanical saws mounted on adjustable arms. Topping improves light penetration into the upper canopy and decreases tree height to reduce the cost of harvesting and spraying operations. It also improves spray coverage. As stated previously, tree height should be no greater than twice the width between the outer perimeter of the tree canopies between rows for optimum interception of light (Wheaton *et al.*, 1978). Limiting tree height becomes increasingly important in high-density plantings with very narrow between-row spacings. Generally, tree height is maintained at about 4–5 m.

Trees are topped in different patterns. Some growers prune straight across the top of the tree in a 'flattop'. Others set the cutting angle of the saws between 10 and 25°, creating a 'rooftop' effect. These patterns also remove varying amounts of wood from the sides of the tree. Top edges usually grow more than the centre top, thus angle topping tends to compensate for this extra growth resulting in an even top when topping is again needed. It is preferable to re-establish required tree height by making smaller cuts over several seasons rather than cutting into large wood. As with hedging, removing large limbs produces vigorous, vegetative shoots from few growing points, thus defeating the purpose of topping. This is especially true for inherently vigorous scion cultivars such as lemon or for scions on vigorous rootstocks such as rough lemon.

GROWTH REGULATORS

Growth regulators have been used on a limited scale worldwide since the 1940s (Davies, 1986b) primarily to improve size and quality of fruit for the fresh market but also to prevent sprouting of small trees (Coggins, 1981;

Wilson, 1983). The growth regulators, 2,4-D and gibberellic acid (GA_3), are applied routinely to navel oranges in California (Coggins, 1981) to improve peel quality and prevent fruit drop. Although the same materials have proved effective in other countries, they are not widely used because their effectiveness may vary from season to season and margin for error is small. Growth regulators are generally applied at very low concentrations (mg l^{-1}) and if improperly used may cause leaf, fruit or tree damage. Therefore, few citrus growers use growth regulators as part of their cultural programme. Although some increase in use of growth regulators may occur, it is unlikely that their use will become universal.

Most mandarin hybrids and some mandarins differ from other citrus cultivars in their requirement for cross-pollination to attain commercially acceptable yields. Characteristics of a good pollinizer (tree) cultivar include overlapping bloom periods, adequate flowering, production of saleable fruit, similar freeze-hardiness and cultural requirements, attractiveness to bees (the agents for cross-pollination) and cross- or self-compatibility with the main cultivar. Pollinizer placement varies from orchard to orchard from a 1:3 to 1:1 ratio (pollinizer:main cultivar). In some cases, pollinizers are interplanted within the row every third to fourth tree. This pattern may cause problems during harvesting, however, due to mixing of fruit. Solid rows of the pollinizer make harvesting much easier, particularly if the pollinizer has good market value.

Growth regulators were used on a very limited basis to improve fruit set of mandarin-hybrids such as 'Orlando' tangelo, and 'Nova' and 'Robinson' tangerines. These hybrids are weakly parthenocarpic and without sufficient cross-pollination do not usually produce a commercial crop. GA_3 applied at bloom increases fruit set and yields and produces seedless fruit which is especially desirable for most fresh markets (Krezdorn and Cohen, 1962; Agusti *et al.*, 1982).

Many mandarins and mandarin-hybrids produce large crops of small fruit followed by small crops of large fruit. This problem, termed 'alternate bearing', may cause marketing and management difficulties for growers because fruit less than a minimum size are unmarketable. Sprays of GA_3 just before flower buds start growth can revert some buds back to a vegetative state, thus reducing flowering in the 'on' year (Monselise and Halevy, 1964; Chica and Albrigo, 2015). Dilute sprays of aphthaleneacetic acid (NAA) or ethephon after initial fruit set at the beginning of the 'physiological drop' period remove some of the fruit, allowing the remaining fruit to grow to a more favourable size (Wheaton and Stewart, 1973; Stover *et al.*, 2002). Ethephon and/or NAA have been used successfully to thin mandarin crops in Japan, Israel, Florida and Australia (Coggins, 1981; Davies, 1986b). This also helps even out the crop load from year to year. It is important to thin the crop only during 'on' years and to avoid spraying trees of low vigour or that are stressed as excessive fruit removal often occurs (Wheaton, 1981). Light hedging of mature trees after the postbloom drop period is another method for adjusting crop load in the 'on' year.

Citrus fruits held on the tree past their normal harvest time typically soften, become off-coloured, and are prone to drop, thus reducing yields. Extension of the harvest season allows the grower to market fruit over a longer period and to take advantage of high prices for late-season fruit. Application of GA_3 to delay peel softening and colour development and 2,4-D to delay fruit drop has been used with success to extend the harvest season of grapefruit in Texas, Florida (Ferguson *et al.*, 1982) and South Africa and navel oranges in California and Australia, etc. (Coggins, 1981; Davies, 1986b). These materials also delay peel senescence and onset of peel breakdown known as 'black eye' of 'Minneola' tangelos stored on the tree. Spray application of 2,4-D alone delays premature fruit drop of 'Pineapple' oranges (Gardner *et al.*, 1950) and decreases the severity of summer fruit drop of navel oranges in Florida (Davies, 1986b). It is important to consult current spray guides concerning limitations and label restrictions before using 2,4-D. Some nearby crops such as peppers and grapes can be damaged by very low levels of 2,4-D. Similarly, GA_3 applied at high rates defoliates citrus trees and causes fruit burn. Therefore, caution should be exercised in applying these materials.

WEED CONTROL

Competition between weeds and mature citrus trees for nutrients and water is less severe than for younger trees. Nevertheless, severe weed pressure may reduce yields (Jordan, 1981), impede harvesting operations and clog drainage ditches, requiring special control (Vandiver, 1992a, 1992b). Vines are a particularly acute problem because they may cover the entire canopy, thereby reducing light interception and hampering harvesting operations (Wondimagegnehu and Singh, 1989). Control of weed growth is a major source of expense, especially in tropical and humid subtropical areas like Florida, where high temperatures and rainfall support excessive weed growth.

An economically successful weed control programme depends on proper identification of weed species and selection of the most efficient control practices (De Barreda, 1977; Ito *et al.*, 1981). Variation in programmes occurs from area to area depending on orchard structure, soil conditions and irrigation system.

Classification of Weeds

Accurate identification of weed species is necessary before selecting and implementing a control programme (De Barreda, 1977; Tucker and Singh, 1992).

Weeds are classified on the basis of their structure, life cycle and seasonality. Grasses (monocotyledons with strap-shaped leaves) that cause problems in citrus orchards are generally perennial and reproduce by seeds and rhizomes (underground stems) (e.g. Bermuda and torpedo grass), or vegetatively by tubers, stolons (above-ground stems) or bulbs. Sedges are similar morphologically to grasses but differ in having triangular rather than flat or round stems. Sedges may be annual or perennial and reproduce by seeds, rhizomes and tubers also. Yellow and purple nutsedge are particularly troublesome to control because they are perennial and have an extensive system of rhizomes which produce underground tubers. Control measures for grasses and sedges may differ, especially in choice of herbicides. Therefore it is important to distinguish between these species prior to selecting a control practice (Wondimagegnehu and Singh, 1989).

Broadleaf weeds are dicotyledons characterized by a branching growth habit and broad, vein-netted leaves. Flowers of broadleaf weeds are generally more brightly coloured and more distinctive than those of grasses and have easily distinguishable parts (sepals, petals, ovaries, stamens). Broadleaf and monocot weeds may be annuals, biennials or perennials. Annuals complete their life cycle from seed within a season. Summer annuals begin growing in the spring and produce seed and die during the late autumn or winter. In contrast, winter annuals germinate at cooler temperatures in the autumn and grow throughout the winter, usually dying during the summer as temperatures increase. Biennial species develop from seed during the autumn and produce roots and leaves during the first season. Plants flower, produce seed and die by the autumn of the second year. Perennial weeds live more than 2 years, either initiating growth in the spring (torpedo grass, nutsedge) or growing during the autumn and becoming dormant during hot summer months (Florida betony). Many perennials accumulate reserves in rhizomes, tubers or stolons during their respective growing seasons, becoming dormant and overwintering or oversummering as these structures. Other perennials are woody and overwinter on stored carbohydrates or are evergreen.

More than 100 species of weeds have been recorded in citrus orchards in the USA, although fewer than 20 are of economic importance in mature orchards. Specific problem weeds depend on location and history of the site (Table 5.4). In the Mediterranean region (De Barreda, 1977; Lo Giudice, 1981b) and Japan (Ito *et al.*, 1981; Suzuki, 1981) many of the problem weeds are of the same genus but different species than those found in the USA.

Weed Control Practices

Mechanical weed control is commonly used worldwide. Disking and clean cultivation was commonly practised for many years in the Ridge area of Florida. Some growers felt that regular disking not only controlled weed growth, but also chopped tree roots, driving them deeper into the soil, thus improving drought tolerance. However, this so-called 'dust mulching' actually damages surface roots and causes soil compaction even in sandy soils. Moreover, regular movement of equipment can cause mechanical damage to limbs and remove fruit from the lower parts of the tree. There is no evidence that regular disking improves drought tolerance or root growth and, in fact, it may lessen tree vigour. Orchards kept weed-free with chemical herbicides are warmer than those having a sod cover during radiation freezes because of more heat flux out of the warmer soil. Disking is not used as a method of weed control in bedded orchards because bed maintenance requires stable soil. Also, roots are located close to the soil surface and subject to mechanical damage from even shallow tillage. Clean cultivation may also be disadvantageous because of loss of habitat for beneficial insects necessary for biological pest control.

Mechanical mowing is commonly used as an effective method of weed control worldwide, particularly where soil erosion is a problem. A number of different types of mowers are available that operate either by direct drive or more commonly by a power take-off (PTO) from the tractor. Specially modified mowers are used for bedded orchards to permit mowing of the ditches and sides of the beds. These are called 'bat wing' mowers because they are hinged in the middle allowing for flexibility of motion. Mowing of perennials removes tops and reduces stored reserves, thus weakening the plant and reducing subsequent growth. Annual weeds should be mowed before they go to seed.

Chemical control
Chemical weed control is a common practice in many citrus-growing regions. While chemical control is very effective if used properly, improper selection of materials, variations in environment and improper application may result in either poor weed control or damage to citrus trees. Most important to any herbicide programme is to understand the chemical properties and limitations of the product. Product labels should be read and the instructions followed and carried out by the operator during application. Materials should be applied within the recommended ranges on the label. The concept of 'if a little is good, a lot is better' is always dangerous when applying herbicides or other pesticides. Herbicides are applied before weed emergence (pre-emergence), or after the weeds emerge (post-emergence). Some materials are applied before the trees are planted, but most are used after planting. Soil fumigation controls weed seed germination and may be considered a preplant application. Currently, pre-emergent soil residual herbicides are used much less in mature orchards. These products should be applied before weeds have emerged or just as they are emerging in cases where some post-emergence capability exists. Pre-emergent materials should be applied to a clean soil surface and require rainfall or irrigation to become effective. Ultraviolet rays from the sun may inactivate soil residual herbicides if they are not incorporated to a shallow level. Deep disking following application of residual herbicides defeats the purpose of chemical

Table 5.4. Weeds commonly found in citrus orchards in the USA. Many of these species are also found worldwide, but this is by no means a complete list.

Common/Family name	Scientific name	Life cycle	Characteristics
Grasses			
(*Gramineae*)			
Paragrass	*Panicum purpuresceus*	Perennial	Reproduce by stolons
Alexandergrass	*Brachicria plantaginea*	Annual	Reproduce by seeds
Guineagrass	*Panicum maximum*	Perennial	Reproduce by stolons
Vasseygrass	*Paspalum urvillei*	Perennial	
Bahiagrass	*Paspalum notatum*	Perennial	Seed propagated
Bermudagrass	*Cynodon dactylon*	Perennial	Seed, stolon, rhizome propagated
Torpedograss	*Panicum repens*	Perennial	Reproduce by rhizomes
Foxtail	*Setaria* sp.	Summer annual	Seed propagated
Texas panicum	*Panicum texanum*	Summer annual	Seed propagated
Sandspur	*Cenchrus pauciflorus*	Summer annual	Seed propagated
Maidencane	*Panicum hemitoman*	Perennial	Semiaquatic, hollow stems
Johnson grass	*Sorghum halapense*	Perennial	Reproduce by rhizomes
Brown top panicum	*Panicum fasciculatum*	Summer annual	Reproduce by seeds
Sedges			
(*Cyperaceae*)			
Yellow nutsedge	*Cyperus esculentus*	Perennial	Reproduce by seeds, rhizomes, tubers, triangular stem
Purple nutsedge	*Cyperus rotundus*	Perennial	Reproduce by seeds, rhizomes, tubers, forms tuber chains
Compositae			
Spanish needles	*Bidens bipinnata*	Summer annual	Flowers with yellow centre, white petals
Horseweed	*Erigeron canadensis*	Annual	Small yellow flowers, tough stems
Ragweed	*Ambrosia artemisiifolia*	Summer annual	Hairy and slender stems
Dogfennel	*Eupatorium capillifolium*	Perennial	Purplish stems, thin leaves
Sowthistle	*Sonchus* sp.	Winter annual	Prickly leaves, yellow flowers

Plant Husbandry

Gourds (*Cucurbitaceae*)			
Balsamapple	*Momordica charantia*	Perennial	Vine, fruit orangish berry
Morning Glory (*Convolvulaceae*)			
Field bindweed	*Convolvulus arvensis*	Perennial	Thick vine, pink seed pods
Morning glory	*Ipomoea* sp.	Perennial and annual	Vines, funnel-shaped flowers in red, purple, white
Nightshades (*Solanaceae*)			
Brazilian nightshade	*Solanum seaforthianum*	Perennial	Vine, red berries
Milkweed (*Asclepiadaceae*)			
Milkweed vine	*Morrenia odorata*	Perennial	Vine, milky sap, seed pods with silky hairs
Vervain (*Verbenaceae*)			
Lantana	*Lantana camara*	Perennial	Woody shrub, pink, cream, purple and orange flowers
Pokesweed (*Phytolaccaceae*)			
Pokesberry	*Phytolacca americana*	Perennial	Reddish-purple, large taproot, purple fruit
Pigweed (*Amaranthaceae*)			
Redroot pigweed	*Amaranthus retroflexus*	Summer annual	Small green flowers, black seeds
Cashew (*Anacardiaceae*)			
Brazilian pepper tree	*Schinus terebinthifolius*	Perennial	Shrub or small tree, bright red fruit

weed control by breaking up the herbicide layer and bringing weed seeds to the surface as the herbicide is turned under.

Post-emergence herbicides are applied after weeds have emerged and act either by contact or root uptake. Many post-emergence materials are contact herbicides and thus are only effective if they come into contact with foliage or stems. They require good coverage because they are not translocated, and for this reason they also have no activity in the soil. Some are systemic and translocated to other plant parts including the roots. A list and application information for the various kinds of herbicides approved for use on citrus in each country should be available from a local advisory group.

Herbicides may be broadcast over an entire area, banded in a specific area, or directed to a particular region. Broadcast, also called trunk-to-trunk, herbicide application, is a common practice in many orchards, with chemical weed control replacing disking as a means of maintaining a clean orchard floor. Trunk-to-trunk programmes virtually eliminate all weed growth, permitting ease of harvesting and spray operations. Amount used and cost of herbicides, problems of soil erosion and reduction of alternate habitat for biological control agents are major disadvantages of this system. Its future may be limited by the potential for groundwater contamination. Banded herbicide application in the tree row is used in some areas such as Florida and Brazil. A sod cover is maintained between rows with mechanical or chemical mowing. This system reduces herbicide costs, soil erosion and provides an alternative plant canopy for beneficial insects needed for biological control but provides less freeze protection and requires regular mowing. Directed or spot sprays are used primarily to control outbreaks of difficult weeds in particular areas of the orchard. In places such as China, intercrops like cabbage, wheat or asparagus are grown between tree rows to maximize land use. Weed control is then primarily by hand-hoeing.

Establishment and maintenance of a weed control programme is an essential part of orchard practices. Well-designed programmes rely on a combination of methods and materials, rather than a single strategy. Programmes vary with location and severity of weed pressures and are more difficult in high rainfall and temperature areas (tropics). Because of increased herbicide costs in these areas, more hand cultivation is practised and weed control is frequently inadequate. Hand cultivation does not usually kill the reproductive root or stem parts of persistent grasses which quickly become re-established. Hand cultivation is not practical on ditch banks and in reservoir ponds typical of high rainfall, bedded orchard areas as found in southern Florida, Belize and other lowland tropical areas. Several herbicides are available for control of aquatic weeds (Vandiver, 1992a, 1992b; Langeland *et al.*, 2006). Proper application is essential when applying herbicides to waterways or reservoirs because of the potential for environmental damage.

Many factors including temperature, rainfall, wind, soil type and stage of development of weed species influence herbicide efficiency (Wondimagegnehu and Singh, 1989). Temperature affects weed growth rate and uptake of herbicides. Low soil temperatures decrease uptake of residual materials and low air temperatures slow uptake and translocation of foliar applied materials. Soil residual herbicides usually require water from rain or sprinkler irrigation for activation and uptake. Low soil moisture induces water stress in the weed species, thereby decreasing foliar uptake and translocation of foliar-applied herbicides. Rainfall within 2 to 4 h of application of contact herbicides will wash off and greatly reduce or eliminate the effectiveness of these materials. Application of herbicides during windy conditions is inadvisable because of the risk of tree injury from spray drift, but also because of uneven distribution to the target weeds.

Soil type has a pronounced influence on the effectiveness of soil residual herbicides. Organic matter may complex herbicides, reducing their effectiveness. In contrast, herbicides are readily available in the soil solution of coarse sandy soils and this may result in phytotoxicity to the tree. Also, residual materials are readily leached from these soils resulting in noneffective levels of the herbicide while the leached material can contribute to surface- and groundwater pollution. Effective weed control using soil residual materials is especially difficult in orchards with highly variable soil types as found in the flatwood, bedded orchards in Florida.

Stage of development of the target weed species has a major influence on performance of the herbicide. Most pre-emergent materials must be taken up by weed seedlings as they begin to grow and these materials become ineffective at later stages of weed development. Some herbicides require clean soil conditions to be effective, because weed cover intercepts the material before it reaches the soil. Initiation of a chemical weed control programme should start with a contact herbicide or clean cultivation before applying a pre-emergence soil residual material. Most contact herbicides are more effective on young, rapidly expanding tissues than older tissues that have developed thick, protective cuticles or pubescence. Some difficult-to-control perennials like brambles should be treated with herbicide during the autumn when materials are actively translocated to the roots. Application of herbicide is much less effective if made during the flowering period of the weed: Biological or chemical weed control materials should be applied to vines before emergence when the stems are not yet woody. Once vines have grown into the tree canopy, chemical control becomes almost impossible and they must be physically removed, a costly and difficult means of control. For these reasons, shallow cultivation or mowing before application of herbicides to heavy weed populations is often helpful.

Herbicide injury
Nearly all herbicides are potentially hazardous to citrus trees if not properly applied. Symptoms of herbicide injury, however, are not always obvious and may be confused with nutrient disorders, diseases or mechanical damage.

The type of injury depends on the amount and kind of material applied. Contact herbicides produce localized damage to a particular part of the tree, usually near ground level where spray comes into contact with tree leaves. Rarely will symptoms be observed in the upper canopy of the tree unless a major error has been made. Attempts to kill vines with 2,4-D sprays to the canopy may lead to canopy injury. Applying glyphosate to freshly cut tree stumps to prevent sprouting can lead to adjacent tree damage through xylem transport across natural root graphs in hedgerow plantings (Futch and Singh, 2009). Damage may appear as small yellow or necrotic specks on leaves or as dieback of the entire limb if heavier contact is made. Damage from contact sprays is normally randomly distributed in an orchard since the herbicide applicator will not usually apply the material to every tree. Young row-end trees may be injured if the applicator does not turn off the spray machine as the turn is made. Large trees with thick bark are not usually killed by improper application of contact herbicides to the trunk, but the outer bark is still living and heavy application, particularly of systemic herbicides, on the trunk even at 5 to 10 years of age can girdle the tree if using the highest rates of more toxic herbicides. Possible adverse effects of glyphosate through uptake and damage to weakened roots of HLB affected trees have been reported, but no definitive results have been reported beyond usual direct contact effects. Concentrated sprays to stumps of cut trees have caused damage to adjacent trees, apparently through transport through natural root graphs down the tree row in close-planted tree rows, therefore phytotoxicity seems possible for excessive root uptake or reduced vigour of disease affected root systems.

Slight over-application of residual herbicides is not easy to detect as leaf symptoms are not specific. However, in more severe cases, distinctive leaf chlorosis and even abscission may occur. Trees appear stunted and new leaves may be strap-shaped or lanceolate. Damage from residual herbicides generally occurs in blocks of trees or in entire rows, often associated with changes in soil type, irrigation practices or an application rate error. This is why it is extremely important to follow label instructions carefully and to adjust rates based on soil conditions. Some soil-applied fungicides that improve root growth such as Ridomil may also increase uptake of residual herbicides causing severe tree damage (Graham *et al.*, 2013).

Sprayer calibration
Although many innovations in herbicide sprayer design have been introduced, proper and repeated calibration of herbicide sprayers is still the most important operation to achieve effective weed control, limit damage to trees and the environment from over-application and save money. In any herbicide spray programme the goal is to apply a specific amount of material per treated area. Ground speed, spray nozzle operating pressure and pattern of output regulate the amount of herbicide applied (Matthews, 1992).

Pattern of spray application also affects output. Spray nozzles are manufactured to apply material in a number of different patterns and are rated in 1 (or g) min^{-1} at a given pressure. Commonly used nozzles include flat fan, wide-angle fan and even spray patterns. The particular characteristics of a nozzle are stamped on it and additional information on output at various pressures is available from the manufacturer. The nozzle angle is chosen to allow adequate overlap of the spray at a given boom height and nozzle spacing. The optimum height is determined by the minimum tree skirt height, weed height and pressure to direct the spray the distance required along the outside of the spray pattern. Booms usually allow for independent shut-down of 1 or 2 nozzles near the tractor so that narrower strips can be treated for younger trees. Nozzles will wear with time, causing an increase in output. For this reason they should be recalibrated at regular intervals and replaced when they exceed specified output by more than 5%.

Calibration should be done regularly after determining type of nozzle, operating pressure and boom height. Calibration may be done by spraying a prespecified area of ground or by operating the sprayer in a stationary mode at a carefully controlled rate and pressure while collecting the spray output. Detailed information on sprayers, spraying and calibration should be obtained from in-depth presentations on these subjects.

An alternative method of application is to deliver the herbicides to the irrigated area under the tree as done in Israel (Oren and Israeli, 1977) and Florida (Wondimagegnehu and Singh, 1989). Herbigation is practised in conjunction with use of microsprinkler irrigation. This system works well in a dry climate like in Israel where weeds tend to grow and compete for fertilizer and water only in the irrigated area under the tree. In high rainfall areas such as Florida, supplemental herbicide treatments must be applied to the remaining ground area that has significant tree root exploration. The herbigation system may be better applied to young trees in their first 2 or 3 years of growth when the tree root zone approximates the wetting zone of a sprinkler. This system does not work well if weeds are allowed to develop any height since they will interrupt the sprinkler spray pattern.

Biological control
Biological methods may have potential for future integrated control of weeds. A current list of weed biological control agents is available from the USDA Forestry Service (Winston *et al.*, 2014). *Phytophthora palmivera* has been shown to control milkweed vine in Florida if applied at the early seedling stage (Tucker and Singh, 1992). Allopathy may also have a role. For example, several compounds in extracts from *Lantana camara*, a weed itself, suppress the growth of rye grass plants (Singh *et al.*, 1989). In Florida, this allopathic effect is problematic when allowing *Lantana* to develop preferentially in the field compared with easier-to-control weeds. Growing *Lantana* in another area and applying an extract to specific problem weeds controlled by *Lantana*'s chemicals might

be an alternative. Insects that feed on specific weeds like Lantana may also have promise for future biological control (Habeck, 1977).

PLANT HUSBANDRY SUMMARY

Citrus nursery practices are almost all in protected cover, providing disease-free trees that are much more costly. Almost all plantings are high density to more quickly recover costs and provide economic returns before pests and diseases reduce productivity of the plantings. Due to serious diseases, especially HLB, costs of production are much higher than in the past. Pest and nutritional sprays plus ground fertilizers are now the major costs of production, with weed control usually the third most costly practice. Please see Chapters 6 and 7 for details about various pests and diseases of citrus.

Arthropod Pests

Pest management in citrus has been significantly transformed due to the introduction of citrus greening disease or huanglongbing (HLB) (Wang *et al.*, 2017). The insect vector of greening, the Asian citrus psyllid (*Diaphorina citri* Kuwayama), was first detected in Florida in 1998, and rapidly established throughout the state (Grafton-Cardwell *et al.*, 2013). There is no current cure for citrus greening and the impact of the disease on Florida citrus production has been substantial since 2005 (Spreen *et al.*, 2014). The disease limits citrus production by reducing yield and decreasing fruit quality, and can eventually kill trees (Bovè, 2006). Citrus canker (caused by *Xanthomonas axonopodis* pv. *citri*) is another severe disease currently affecting citrus industries worldwide. Like greening, canker is closely associated with an insect pest, the citrus leafminer (CLM) (*Phyllocnistis citrella* Stainton), present in Florida since 1993 (Gottwald *et al.*, 2007). While not a true vector of citrus canker, CLM larvae damage young citrus leaves when feeding, making them highly susceptible to canker infection (Hall *et al.*, 2010). The threat of these two diseases has required much more intense chemical pest management in Florida citrus than in previous industry history (Qureshi *et al.*, 2014). In areas where these disease complexes affect citrus production, all other insect pests are considered secondary and are often acceptably controlled by treatments targeting *D. citri* and *P. citrella*.

Other mites, insects and nematodes that attack citrus number in the hundreds (Talhouk, 1975). Many of them only occasionally cause economic damage. However, there are pests that feed on the roots, trunk, stems, leaves or fruit and injure the tree enough to cause economic loss. In Florida, more than 21 different mites or insects are reported to have caused economic injury for fresh fruit production (Albrigo, 1978). Many of the insects that are reported to require control because of economic damage are listed in Table 6.1, and are of concern in most major citrus production areas of the world. One reason for this wide distribution is man. As people have moved citrus from the areas of origin or older production areas to other parts of the world (see Chapter 1), pests and diseases have often been moved also. Because of large numbers of pests, chemical control is widely practised by citrus producers. Many individual orchard pest outbreaks are related to chemical upsets of natural enemies. Biological control, which is cost-effective when it works and helps meet environmental and health concerns regarding use of pesticides on food products, has historically been a major emphasis of entomological research. Biological control has been practised in citrus for many years (Browning, 1992). Much of the current biological control has been developed from indigenous or unplanned introduction of control agents and researchers have learned to manage these systems efficiently. Many of the natural enemies were probably imported with citrus just as the pests were. Purposeful importations of predators and parasites have occurred also (McCoy, 1985). Most of the useful biological control agents are found in the areas of origin of the plant host (Chapter 1). The reader should refer to the 'World Crop Pest Series' (Elsevier) or similar current reviews for detailed information concerning many of the mites and insects affecting citrus and plants in general.

ASIAN CITRUS PSYLLID

The Asian citrus psyllid (ACP), *Diaphorina citri* Kuwayama, can transmit three species of bacteria belonging to the genus *Candidatus* Liberibacter. The African strain of greening is usually associated with the African citrus psyllid, *Trioza erytreae* (Del Guercio). *Ca.* Liberibacter asiaticus (CLas) is the most likely causal agent of citrus HLB in the USA although Koch's postulates have not been fulfilled (Tylor *et al.*, 2009; Pelz-Stelinski *et al.*, 2010). First discovered in Florida in 1998 (Grafton-Cardwell *et al.*, 2013), ACP (Fig. 6.1) quickly became established, making its eradication impossible. Currently, ACP

Fig. 6.1. The Asian citrus psyllid (ACP). (A) Nymphs feeding on a young shoot; (B) Close up of nymph stage; (C) Adult stage of ACP. Images from Kirsten Pelz-Stelinski, University of Florida.

can be found in all southern US states. In 2005, plants with HLB infection were detected in Florida; within 4 years, all 32 citrus growing counties in the state had HLB-infected citrus (Anonymous, 2008). HLB is also present in both California and Texas, where positive trees are detected in about 30% of the groves. HLB is considered the most destructive disease of citrus crops in the world. All known citrus cultivars are susceptible to HLB (Folimonova et al., 2009) and prevention of disease transmission has proven difficult worldwide (Bové, 2006). Without intervention, infected mature trees drop foliage, weaken, and may die within 5–10 years of infection (Roistacher, 1996). Direct effects of HLB may be exacerbated by secondary affliction(s), including infestation by Diaprepes root weevil (*Diaprepes abbreviates*) and indigenous root weevils (e.g. *Compsus* sp., *Epicaerus* sp.), which may lead to root girdling and infection by *Phytophthora* spp. Infected mature trees produce significantly less fruit for the fresh or processed market; some fruit are small, misshapen, with uneven colour development, or remain green.

Given efficient pathogen acquisition and transmission capabilities of ACP (Pelz-Stelinski et al., 2010), HLB spreads rapidly within and between groves by short and long-range dispersal capabilities (Tiwari et al., 2010). Nymphs are primarily responsible for acquisition of the pathogen, which may have been brought to the shoot by an infected adult that in most cases emerged from an infected nymph (Lee et al., 2015). Thus, infection of the tree as well as the next generation of ACP may take place within a single vector generation (Lee et al., 2015). Once the ACP acquires the bacterium, a latent period ranging from 7–25 days may exist before the ACP is able to transmit bacteria into another host plant (Pelz-Stelinski et al., 2010). On the other hand, an infected ACP can transmit the bacteria into another host plant by continuous feeding for only 5–7 h (Xu et al., 1988). Time periods involved in acquisition, latency and transmission provide a window of opportunity for controlling infected ACP and preventing disease spread. Gravid ACP females lay eggs only on new flush, and hatched early instar nymphs feed on new flush and go through five instars before becoming adults (Halbert and Manjunath, 2004; Grafton-Cardwell et al., 2013). This results in rapid and very high buildup of ACP populations during flushing periods that are capable of acquiring and transmitting HLB-causing bacteria. Furthermore, within perennially growing citrus in subtropical climates, multigenerational and season-long occurrence of arthropod pests, such as ACP, is common.

Soil and foliar applied insecticides play a vital role in suppressing ACP populations that are infected or capable of acquiring and transmitting HLB bacteria (Qureshi et al., 2014). There are currently only two modes of action (neonicotinoids and cyantraniliprole) that are available, as soil-applied, systemic treatment for control of ACP. The neonicotinoids have been a particularly critical tool for young tree protection and are the most widely used insecticides for protecting newly planted groves or young tree resets (Langdon et al., 2018a, 2018b). Reflective mulch covering in the tree row also benefits young trees by disorienting ACP flight to the trees above the reflective mulch (Stansly et al., 2016). The above modes of action insecticides serve as contact and feeding poisons for all life stages of ACP. Insect growth regulators (IGRs) such as diflubenzuron, targeting the egg and nymph stage only, are also used. IGRs serve as important rotational insecticides to limit resistance development and can contribute to overall ACP population decline (Tiwari *et al.*, 2012). Vector control is carried out with the goal of maintaining ACP population densities as low as possible (Qureshi *et al.*, 2014). The threshold for a vector of a disease that causes significant fruit drop, may kill trees, and/or render fruit unsuitable for consumption is extremely low, if it exists at all, although economic injury levels appropriate for different conditions are demonstratively attainable (Monzo and Stansly, 2017). Consequently, intense use of insecticides has been seen as necessary with up to 12 applications per year for ACP in Florida, with even more sprays routinely applied in São Paulo, Brazil. However, this use has already historically caused widespread indications of insecticide resistance in ACP populations throughout Florida (Tiwari *et al.*, 2011) and requires refinement through use of realistic thresholds to eliminate unnecessary or poorly timed sprays on mature trees.

CITRUS LEAFMINER

The CLM, *Phyllocnistis citrella* Stainton (Lepidoptera: Gracillariidae), is a worldwide pest of citrus (Rutaceae) whose larvae burrow into epidermal cells of young leaves and create serpentine mines that damage leaves and stunt growth (Achor *et al.*, 1997; Heppner and Fasulo, 2010; Richardson *et al.*, 2011). CLM is multivoltine with as many as 13 generations per year in Florida (Heppner and Fasulo, 2010). Small populations are thought to overwinter without diapause, reproducing on the few leaf flushes available during winter (Lim and Hoy, 2006).

CLM has become a major pest of citrus throughout Florida (Heppner and Fasulo, 2010) and many other citrus-producing countries. Most of this dispersion occurred over a 2-year period (https://www.cabi.org/isc/datasheet/40831). The emerging larvae feed and develop within galleries just beneath the leaf surface (Fig. 6.2). The damage can stunt growth of young trees, and it has been well documented that leaf wounding caused by CLM prevents photosynthesis in the damaged area (Schaffer *et al.*, 1997) and renders leaves more susceptible to infection with citrus bacterial canker (Hall *et al.*, 2010). Essentially, the leaf mine is an open wound that can be readily colonized by the bacteria. The larvae eventually form pupae within the leaves, and following metamorphosis, new adults emerge to re-start the cycle. In humid subtropical areas CLM usually does not build up until late spring and to an economic level until the second leaf flush at the beginning of the summer (Rogers *et al.*, 2018).

Insecticides remain the most important tools for control of this pest. The majority of current insecticides that specifically target the leafminer are designed to affect the early larval stage as it emerges from the egg and begins to feed. These insecticides can be toxic to the mining larvae or prevent their normal

Fig. 6.2. Larval feeding tunnel and larva of a citrus leaf miner (CLM) on the underside of a leaf. The serpentine tunnel is often filled with the citrus cancer organism, *Xanthomonas citri*.

feeding and growth, killing them through starvation or by preventing normal development (Stelinski *et al.* 2010).

Male moths of *P. citrella* are attracted to a blend of two pheromone components emitted by the female: (Z,Z,E)-7,11,13-hexadecatrienal (triene) and (Z,Z)-7,11-hexadecadienal (diene) released in a 3:1 ratio (Lapointe *et al.*, 2006). Dispersing the natural pheromone blend or a component of it in citrus groves interferes with the ability of male moths to find traps baited with synthetic pheromone lures that act as surrogate females (Stelinski *et al.*, 2008; Lapointe *et al.*, 2009; Stelinski *et al.*, 2010). Reduced capture of male moths in such traps provides a proxy measurement of mating disruption. Also, early investigations from small plot trials (<30 trees per plot) suggested that damage to leaves by *P. citrella* could be reduced by application of mating disruption (Stelinski *et al.*, 2010). Several methods have been used to release pheromone components of *P. citrella* into citrus groves, including elastomeric dispensers that gradually release the triene over a period of weeks (Stelinski *et al.*, 2010). Mating disruption is an effective and environmentally friendly, but more expensive, management tool for CLM populations than insecticides (Lapointe *et al.*, 2015).

ROOT WEEVIL COMPLEX

With the exception of ACP and CLM, the root weevil, *Diaprepes abbreviatus* (Linnaeus) (Coleoptera: Curculionidae), is the most destructive arthropod to citrus trees in Florida (Campos-Herrera *et al.*, 2015) and also in many citrus production areas, particularly in Central America and the Caribbean. *D. abbreviatus* was first introduced into Florida in 1964 (Beavers and Selhime, 1975). Over the past 40 plus years, it has significantly contributed to the spread of disease and damage to citrus, ornamental plants and other crops. *D. abbreviatus* is a native economic pest of the Caribbean where at least 19 additional species are known within the genus (Weissling *et al.*, 2002). *D. abbreviatus* has spread over a large area of central and southern Florida where it causes approximately $70 million in damage annually (Lapointe, 2000). The initial area of infestation was an estimated 6500 acres in Apopka, Florida and has now increased to an estimated 164,000 acres over 20 counties in central and southern Florida (Weissling *et al.*, 2002).

D. abbreviatus has a wide host range, attacking approximately 293 different plant species including citrus, sugarcane, vegetables, potatoes, strawberries, woody field grown ornamentals, sweet potatoes, papaya, guava, mahogany, containerized ornamentals and non-cultivated wild plants (Simpson *et al.*, 1996, 2000). *D. abbreviatus* damage to the vegetative portion of plants is most often seen as a characteristic notching of young leaves (Fennah, 1940). This is a key trait of *D. abbreviatus* infestation. Adults continue to feed on foliage (Fig. 6.3) and lay eggs between older leaves (Fennah, 1940; Schroeder, 1992). However, the greatest damage is caused by larvae feeding below ground. Upon hatching, the larvae fall to the soil and make their way to the roots of plants where later instars feed and develop (Schroeder, 1992). This feeding can girdle the taproot and rhizosphere causing damage that prevents the plant from taking up water and nutrients resulting in plant death (Schroeder, 1992). This type of damage also facilitates secondary infections by *Phythophora* fungus species (Graham *et al.*, 1996). Young hosts can be killed by a single larva while several larvae can result in serious decline of older, established hosts (Weissling *et al.*, 2002).

In the tropical Americas, root weevil larvae cause extensive damage to citrus, including death of young trees because of girdling of the trunk or primary roots from root bark feeding. These insects are also reported to cause injury of economic significance in Brazil and Florida (Futch and McCoy, 1992). Adults feed on and notch leaf edges, but overall have little direct impact on the tree except in young trees fed on by unusually large numbers of adults. The adults lay egg masses between two leaves cemented together or in other protected areas such as the fruit calyx area in the case of the Fuller rose beetle (*Asynonychus godmani* Crotch). Small larvae hatch in about 7–20 d and fall to the soil where they initially feed on fibrous roots. As they enlarge, larvae often move to larger roots and can girdle these. The larval stage may feed for only 35–40 d (*Artipus*) compared to 6–24 months for most other species. Damage from these insects is particularly severe in Florida and the Caribbean where the Fiddler beetles (*Prepodes* spp.) and Diaprepes root weevil predominate.

Two features of the *D. abbreviatus* life cycle have made it difficult to control in citrus. First, its life stages are active in the field throughout the majority of the annual season. Second, adults and larvae occupy separate habitats (above and below ground); therefore, each life stage must be targeted separately (Georgis *et al.*, 2006). Given that adults continuously emerge from soil to produce offspring, which in turn return to the soil, control methods that target only adults

Fig. 6.3. (A) Diaprepes weevil adult; (B) Damage to root from larval feeding; (C) Leaf notching from adult feeding.

or larvae will only sporadically reduce the pest population density. Because persistent insecticides (e.g., dieldrin and chlordane) are no longer available, a combination of non-persistent tactics timed to kill both life phases of the population is a strategy often used by growers (Georgis *et al.*, 2006). Entomopathogenic nematodes have been investigated and implemented for management of *D. abbreviatus* larvae in Florida citrus for almost two decades. These nematodes vector a symbiotic bacterium into the body cavities of insects. The life cycle of entomopathogenic nematodes consists of these major steps:

(i) penetration into the body cavity of the potential host; (ii) release of bacteria; (iii) development of mature adults; (iv) mating and reproduction of infective juveniles; and (v) emergence of infective juveniles in search of a new host. The infective juvenile is a third-stage juvenile and is morphologically and physiologically adapted to remain for extended periods without ingesting food (Poinar, 1990). Infection with entomopathogenic nematodes can result in death of their insect host within 48 h. Early investigations focused on *Steinernema glaseri*, *S. carpocapsae*, and *Heterobditis bacteriophora* for control of the weevil (Downing *et al.*, 1991; Schroeder, 1992; Bullock and Miller, 1994). Currently used formulations containing *S. riobrave* and *H. indica* have become adopted commercially for *D. abbreviatus* management. Of the several species evaluated in laboratory bioassays and greenhouse trails, *S. riobrave* and a Florida isolate of *H. indica* were the most effective against the Diaprepes root weevil, and reproduction of *H. indica* in the weevil exceeded that of other species (Shapiro-Ilan and McCoy, 2000a, 2000b). *S. riobrave* and *H. indica* are currently the only two nematode species commercially marketed for the Florida citrus industry. *H. indica* is formulated as a paste and *S. riobrave* can be obtained in liquid or water dispersible granular formulations. In 1999, approximately 20% of the ha infested with this weevil were treated with nematodes (Shapiro-Ilan *et al.*, 2002).

The weevil complex is very difficult to control. A number of biological control agents and cultural practice strategies exist that together can give some level of control. Spray programmes have been fairly ineffective as discussed earlier. In the 1940s, it was found that citrus trees could be planted on mounds and then the primary roots exposed so that weevil larvae could not inhabit the area at the base of the trunk to girdle it and the small major roots. This strategy has been combined with the use of a tight mesh black polyethylene screen over the mound with roots not exposed. When the larvae fall to the ground from the hatched eggs, they cannot penetrate the soil directly under the tree but instead take up residence on lateral feeder root areas at the edge of the canopy. This allows the young tree to become established since water can still penetrate the mesh. Adults are often collected by hand as they feed on the leaves. Entomophagous fungi, *Beauveria bassiana*, *Metarrhizium anisopliae* and other species attack the larval stage in the ground. Other predators and parasites beyond entomopathogenic nematodes also exist but they have not provided consistent control. Reflective mulch used for ACP disorientation on young trees also prevents the weevil larvae from reaching much of the citrus tree root system (B. Adair as reported by Giles, 2018).

MITES

Mites can be pests of citrus throughout the world. All mite damage results from feeding by piercing and sucking mouthparts. The smallest mites, *Eriophyidae* spp., cause some of the most pronounced injury. The citrus bud mite (*Aceria sheldoni* Ewing) causes malformed twigs, leaves and fruit from feeding on flowers and buds. Populations build to injurious levels more often on lemons than other citrus and malformed fruit often reach maturity. Under southern California conditions, the life cycle requires from about 10 d in the summer to 1 month in the winter. Mites must be observed with a hand lens during population increases in order to verify the mite as a cause of injury. Citrus bud mite causes problems on fresh citrus in the Mediterranean region, southern Africa and California.

The citrus rust mite (*Phyllocoptruta oleivora* Ashm.) occurs worldwide and can be a pest of fresh fruit production in most humid climate production areas. Populations of this mite usually increase faster on fruit and can develop densities of 70–100 mites cm^{-2} on the fruit surface before injury occurs (Fig. 6.4) (McCoy and Albrigo, 1975). However, some hybrid cultivars like 'Sunburst' mandarin are a preferred host and citrus rust mites can increase to higher populations on leaves than fruit. This cultivar may be more sensitive to injury as leaf injury can develop into severe blister-like lesions on the leaves and young stems (Albrigo *et al.*, 1987). Because mites have small mouthparts, feeding injury only extends to the epidermal cells and requires several feeding probes per cell over a short period in order to kill cells. Subsequent damage development follows formation of a wound periderm under the injured cells and break-up and some sloughing of the injured layer. In late injury, no wound periderm forms and the injured cells remain intact but dark brown and the surface remains shiny (Albrigo and

Fig. 6.4. Citrus rust mite (inset) is very small but large numbers of them can cause significant scarring of the fruit surface.

McCoy, 1974). Severe early damage, greater than 50–75% surface injury, can lead to reduced fruit growth, excess drop and fermentation off-flavour development in late harvested fruit (Allen, 1978; McCoy, 1988)

In warm, humid subtropical areas where rust mite is the most severe problem, reproductive cycles require 7 to 14 d depending on temperature. Mite numbers increase in late spring or early summer and damage often occurs before the rainy period when the entomophagous fungus, *Hirsutella thompsonii*, develops, invades the mites and suppresses the population (McCoy, 1981). Sufficient populations may again increase in the autumn to injure fully developed fruit. Chemical control is usually required for fresh fruit production. In a dry Mediterranean climate, citrus rust mite control was adequate after the introduction and mass rearing of predaceous mites and the judicious use of selective miticides instead of sulfur, which greatly reduces populations of natural predators (Cohen, 1975). Under similar climatic conditions in California, selective chemicals integrated with other practices have successfully controlled citrus rust mite (http://ipm.ucanr.edu/PMG/r107400511.html). Some other Eriophyoids such as the pink citrus rust mite (*Aculops pelekassi* [K.]) and the citrus blotch mite (*Calacarus citrifoli* K.) occasionally also cause fruit damage.

Other mites that cause injury similar to that due to the Eriophyoids are the broad mites, *Hemitarsonemus latus* Banks, *Brevipalpus californicus* Banks or *Brevipalpus phoenicis* Geijskes. *Brevipalpus phoenicis* is common and able to develop larger populations in the warmer climate of central Brazil (25.7°C max. and 12.8°C min. during the coldest 4 months) than in central Florida (23.7 max. and 11.7°C min.). *Brevipalpus* occasionally reaches injurious populations in south Florida. These small mites reproduce in 3–7 d. *Brevipalpus phoenicis* is also the carrier of the causal pathogen for the disease leprosis, which occurs in Brazil but not in Florida (see Chapter 7, p. 223).

The spider mites (Family Tetranychidae) are broadly distributed, with the two common pest species being the citrus red mite (*Panonuchus citri* McGregor) and the Texas citrus mite (*Eotetranychus banksi* [McG.]). These species and the carmine spider mite (*Tetranychus cinnabarinus* Boise) rapidly develop injurious populations in early spring, damaging leaves and in some cases causing defoliation. These mites have not been observed to cause russet injury to fruit (Albrigo *et al.*, 1981). The six-spotted mite (*Eotetranychus sexmaculatus* [Riley]) only injures leaves from small colonies on the underside of the leaf. Most spider mites feed to a depth of about 100 μm and completely remove cell contents so that no oxidized cytoplasm can develop. These areas appear translucent due to the cells void of cytoplasm. New cells are formed from a more central periderm and these cells displace the empty cells. When damage is severe, injury is not repaired and in conjunction with water stress, mesophyll collapse can occur (Albrigo *et al.*, 1981). Development of this disorder first appears as chlorotic zones in the leaf blade, followed by leaf browning and finally premature leaf drop. Citrus red mite densities under 20 leaf^{-1} did not appear to have any adverse effect on tree productivity in California (Hare and Phillips, 1992) suggesting that leaf cell replacement rates are adequate to compensate for cell losses from feeding by populations of that size.

Spider mites, like citrus rust mite, reach higher populations on lemon, but all citrus are fed upon. One generation of a spider mite species requires about 2 weeks at optimum spring temperatures, 30°C max. and 19°C min., with 10 to 12 generations year^{-1}. They overwinter as adults and populations usually decline during summer rainy seasons in marginally subtropical climates like Florida due to biotic and abiotic factors. Although predaceous mites, such as *Euseius hibisci*, and lady beetles, *Stethorus picipes*, attack spider mites, the current predator complex in California and Florida does not usually control citrus red mite or Texas citrus mite, particularly when spray programmes of organic phosphates or sulfur upset natural control. Economic losses appear to be rare from spider mite feeding in field trees. Therefore, little additional work on biological control has occurred.

Selection of an appropriate miticide depends on the intended target of control. With the current emphasis on Asian citrus psyllid and CLM control, it would be wise to choose a miticide that may also have some activity against one of these two pests, such as diflubenzuron (Micromite 80 WGS) or spirotetramat (Movento). With the exception of petroleum oil, no miticide should be applied more than once per season to avoid development of resistance. For more detailed information on the latest miticides and recommended rates see Chapter 10 of the Florida *Citrus Pest Management Guide* on mite control at http://edis.ifas.ufl.edu/cg002.

Fruit appearance is a much greater priority in fresh-market fruit and thus fresh fruit will require more intensive management to prevent mite damage. Fruit growth and abscission will not be affected until 50% to 75% of the fruit surface is injured by rust mites, so there is less need for chemical control where fruit is destined for processing. Groves producing fresh-market fruit may be treated with miticides 3 to 4 times annually, typically during April, June, August and October depending on level of population. However, groves producing fruit for the processing market may receive 0 to 2 applications for rust mites annually, usually with spray oil unless other pests are also present.

SECONDARY INSECT PESTS

Many insects affect citrus around the world. Scientific names, common names and locations that have reported them are presented in Table 6.1.

Thrips are widely distributed but appear to be primarily a problem for the fruit blemishes they cause. They have modified mouthparts for rasping and sucking. The early flower activity of the western flower thrips (*Frankliniella occidentalis* Pergande) or citrus thrips (*Scirtothrips citri* Moult.), as occurs in California, often results in a ring of scarred peel tissue around the stem end of the fruit. Occasionally, population increases of greenhouse thrips (*Heliothrips*

Table 6.1. Major citrus pests identified in various citrus production areas worldwide.

Order/family	Species	Common name	Distribution
Acarina			
Eriophyidae	*Aceria sheldoni* Ewing	Citrus bud mite	A, CAL, MED, SAF, SAM
	Phyllocoptruta oleivora Ashm.	Citrus rust mite	A, MED, NAM, SAF, SAM
Tetranychidae	*Eutetranychus anneckei* Meyer	Texas citrus mite	FL, IND, ISR, SAF, TEX
	Panonychus citri Mc G.	Citrus red mite	A, CAL, FL, MED, SAF, SAM
	Tetranychus spp.	Carmine spider mite	A, CAL, MED, P, SAF
Phytoptipalpidae	*Brevipalpus californicus* Bks.	Citrus flat mite (Z), false spider mite	A, CAM, MED, NAM, SAF
	(*Brevipalpus australis* Tucker)	(USA)	
	Brevipalpus phoenicis Geijskes	Broad mite, reddish black flat mite	A, B, CAM, MED, NAM, SAF
Thysanoptera			
Thripidae	*Frankliniella occidentalis* Perg.	Western flower thrips, grass thrips	ARIZ, CAL, SAM, SPAIN
	Heliothrips haemorrhoidalis Bché.	Greenhouse thrips	A, CAM, FL, MED, SAF, SAM
	Scirtothrips aurantii Faure	Orange (citrus) thrips	SAF
	Scirtothrips citri Moult.	Citrus thrips	ARIZ, CAL, CAM, P
	Thrips tabaci Lind.	Onion thrips	A, CAL, CAM, MED, SAF, SAM
Hemiptera-Homoptera			
Aleyrodidae	*Aleurocanthus spiniferus* Quaint.	Orange (citrus) spiny whitefly	A, CAM
	Aleurocanthus woglumi Ashby	Citrus blackfly	A, CAM, FL, SAF, TEX
	Aleurothrixus floccosus Mask.	Flocculent (wooly) whitefly	A, CAM, MED, SAF, SAM
	Bemisia citricola Gomez-Clemente	Whitefly	A, MED
	Dialeurodes citri Ashm	Citrus whitefly	A, CAM, CHILE, FL, MED
	Dialeurodes citrifolii Morg.	Cloudy winged whitefly	B, FL, IND

Arthropod Pests 195

Family	Species	Common name	Distribution
Aphididae	Aphis gossypii Glov.	Cotton aphid	A, CAM, MED, NAM, SAF, SAM
	Aphis spiraecola Patch	Spirea aphid	CAM, MED, NAM, SAM
	Myzus persicae Sulz.	Green peach aphid	A, CAM, MED, NAM, SAF, SAM
	Toxoptera aurantii Boy.	Black citrus aphid	A, CAM, MED, NAM, SAF, SAM
	Toxoptera citricida Kirk.	Brown citrus aphid	A, CAM, SAF, SAM
Margarodidae	Icerya purchasi Mask.	Australian fluted scale, cottony cushion scale (USA)	A, CAM, MED, NAM, SAF, SAM
Pseudococcidae	Planococcus citri Risso (Pseudococcus commonus)	Citrus mealybug (USA), common mealybug	A, CAM, MED, NAM, SAF, SAM
	Pseudococcus adonidum L.	Long-tailed mealybug	A, B, MED, MEX, SAF
	Pseudococcus citriculus Green	Citrus mealybug (Japan)	A, EGY, ISR
	Pseudococcus comstocki Kuw.	Comstock mealybug	A, B, CAM, EGY, ISR
	Ceroplastes floridensis Comst.	Florida wax scale	A, AUS, B, FL, MED, MEX
Coccidae	Ceroplastes destructor (Newst)	White wax scale	AUS, SAF
	Ceroplastes rubens Mask.	Pink (red) wax scale	A, AUS, ITALY
	Ceroplastes sinensis Del Guer.	Chinese wax scale	AUS, B, MED
	Coccus hesperidum L.	Brown soft scale	A, AUS, CAM, MED, NAM, SAF, SAM
	Coccus viridis Green	Green coffee scale	A, CAM, SAM
	Saissetia hemisphaerica Targ. (S. coffeae Walk.)	Hard brown scale, coffee helmet scale	A, CAM, MED, SAM
	Saissetia oleae Bern.	Black scale (USA), brown olive scale (A)	AUS, CAM, IND, MED, NAM, SAF, SAM
Diaspididae	Aonidiella aurantii Mask.	California red scale	A, SAF, MED, SAM, CAM, NAM, AUS
	Aonidiella citrina Coq.	Yellow scale	AUS, CAL, FL, IND, JAPAN, MEX
	Chrysomphalus aonidum L	Florida red scale (USA), circular scale	A, AUS, CAM, MED, NAM, SAF, SAM

Continued

Table 6.1. Continued.

Order/family	Species	Common name	Distribution
	(Chrysomphalus ficus Ashm.)	purple scale	MED, SAM
	Chrysomphalus dictyospermi Morg.	Dictyospermum scale	
	Lepidosaphes beckii Newm.	Purple scale	CAM, MED, NAM, SAF, SAM
	Lepidosaphes gloverii Pack.	Glover scale (USA), Long mussel	B, CAM, JAPAN, MED, NAM, SAF
	(Insulaspis g.)	scale (Z)	
	Parlatoria pergandii Comst.	Chaff scale	A, CAM, MED, NAM, SAF, SAM
	Pinnaspis aspidistrae Sign.	Aspidistra (fern) scale	A, FL, SAM
	Selenaspidus articulatus Morg.	Rufous scale	CAM, SAM, SEA
	Unaspis citri Comst.	Snow scale	AUS, CAM, CHINA, FL, SAM
Heteroptera			
Pentatomidae	Rhynchocoris humeralis Thunb.	Citrus green stink bug	A, MED
Coreidae	Leptoglossus phyllopus L.	Leaf-footed bug	FL, MEX, SAM
Coleoptera			
Cerambycidae	Chelidonium gibbicolle White	Roundheaded borer	CHINA
	Melanauster chinensis Forst.	Black and white citrus borer	CHINA, JAPAN
Curculionidae	Otiorrhynchus cribricollis Gyll		
	Diaprepes spp.	Sugarcane root stalk borer	CAM, FL, SAM
	Pachnaeus citri Mshl.	Citrus root weevil	CAM, FL
	Pantomorus cervinus Boh.	Fuller rose beetle	NAM, SAF, SAM
Hymenoptera			
Formicidae	Acromyrmex octospinosus Reich.	Leaf-cutting ant	A, CAM, SAM
	Atta cephalotes L.	Leaf-cutting ant	B, CAM, PERU

	Atta sexdens L.	Leaf-cutting ant	CAM, SAM
	Solenopsis invicta Buren	Imported fire ant	CAM, FL
Lepidoptera			
Lyonetiidae	*Phyllocnistis citrella* Staint.	Citrus leaf miner	A, AUS
Hyponomeutidae	*Prays citri* Mill	Citrus flower moth, bud moth	A, MED
Metarbelidae	*Indarbela tetraonis* Moore	Litchi stem borer	A
Papilionidae	*Papilio demoleus demoleus* L	Citrus leaf-eating caterpillar	A
	Papilio memnon L.	Pastor swallowtail	A
	Papilio polytes polytes L.	Citrus butterfly	A
Diptera			
Trypetidae	*Anastrepha fraterculus* Wied.	South American fruit fly	CAM, SAM
	Ceratitis capitata Wied.	Mediterranean fruit fly	AUS, CAM, MED, SAF, SAM
	Dacus dorsalis Hend.	Oriental fruit fly	A
	Dacus tsuneonis Miyake	Chinese citrus (Japanese orange fruit fly)	CHINA, JAPAN

Abbreviations: A, Asia; ARIZ, Arizona; AUS, Australia; B, Brazil; CAL, California; CAM, Central America and Caribbean; EGY, Egypt; FL, Florida; IND, India; ISR, Israel; MED, Mediterranean; MEX, Mexico; NAM, North America; P, Pakistan; SAF, southern Africa; SAM, South America; SEA, Southeast Asia; TEX, Texas.

haemorrhoidalis Bche.) may occur on grapefruit in Florida, and injury may result in superficial scarring at the boundary where two fruit touch, leaving a ring of injured peel, similar to injury that occurs from citrus rust mite feeding. Life cycles range from 27 to 10 d for temperatures of 16 to 30°C. Citrus thrips are known to scar fruit in late spring during the second generation in California (California IPM Manual Group, 1984). Orchards with ground cover and minimum spray programmes that avoid organophosphates tend to have fewer thrip problems. Predaceous mites, *E. hibisci* and *Anystis agilis*, and the minute pirate bug, prey on thrips as do natural enemies found in ground litter. A single spray at petal fall will usually control thrips if other factors are favourable for a minimum population increase.

Whiteflies and blackfly are most serious in nursery tree production but build-up on young flushes is significant in that it leads to sooty mould development. Whiteflies have sucking mouthparts and feed on leaves. Cloudy winged (*Dialeurodes citrifolli* Morg.) and other whiteflies have become a problem in nursery trees, particularly greenhouse production in Florida and Brazil. These and other whiteflies are difficult to control in the greenhouse because frequent sprays are required. Citrus and woolly whiteflies are controlled under Mediterranean climates by parasitic wasps such as *Cales noacki* and *Eretmocerus* spp. This control is easily disturbed by ants protecting this source of honeydew. Dust or insecticide treatments such as sulfur also kill these wasps.

The citrus blackfly (*Aleurocanthus woglumi* Ashmead) was a serious pest in Mexico and moved into Texas in the early 1970s and Florida in 1976. Blackflies feed on the leaves and produce honeydew that contributes to a severe accumulation of sooty mould and a decrease in tree vigour. Successful biological control of blackfly occurs when the two parasitic wasps (*Encarsia opulenta* and *Amitis hesperidum*) are present. These parasites are easily killed by sprays of organophosphates or sulfur, while the blackfly is difficult to kill with insecticides. Best control is maintained by introduction of the parasites and avoidance of these chemicals (Knapp and Browning, 1989).

Aphids also have piercing-sucking mouthparts and are phloem feeders but are usually not economic pests as such. They are important vectors of citrus tristeza virus (CTV) and possibly other viruses. The brown citrus aphid (*Toxoptera citricida* Kirk.) is the most efficient vector and is capable of developing populations that can cover a spring flush on young trees and probably reduce tree growth (Lee *et al.*, 1992). These populations increase very fast on both leaves and stems of the spring flush, with reproduction time as short as 4 d in the warm spring but up to 10 d in cooler periods. A number of parasites, predators and fungi have been observed associated with aphids in South and Central America, some of which may control the brown citrus aphid. Biological control agents usually result in decreasing the levels of a pest rather than eradication. For an efficient virus vector like the brown citrus aphid, this level of control would probably only slow down the spread of CTV (see Chapter 7).

More kinds of scale insects attack citrus than any other family of insects. Because scale insects are sessile, small and often inconspicuous, they have been spread widely on citrus plants and are now present in most citrus production areas of the world. Some scales are mobile throughout all stages of development, but many become immobile under soft or armour covers after the crawler stage (Fig. 6.5). Eggs are protected under the cover or the body of the female. A few to a few thousand eggs are produced by each female. Winged males are found in most species, but reproduction is parthenogenic in others. Two to 10 generations are produced each year depending on the species. From crawler to adult there are as many as three stages in females and the long stylet mouthparts of all stages can be found in the plant tissues below the resting adult. Most efficient scales penetrate to phloem cells for long-term feeding (Albrigo and Brooks, 1977).

Fluted scales like the cottony cushion scale (*Icarya purchasi* Mask.) produce a copious waxy deposit over the body. This is the most common and widely spread of the fluted scales. Up to 1000 eggs may be carried in the waxy mass. Dense colonies can occur on branches and the trunk. Infested small branches and fruit stems are often weak and fruit abscission can occur.

Mealybugs have an appearance similar to the fluted scales but with less wax. The citrus mealybug (*Planoccocus citri* Risso) is the most widespread species. This insect inhabits the stylar cavity of navel oranges and the area around the calyx which can lead to yellowing of the fruit and subsequent abscission.

Fig. 6.5. Scale insect hard covers on leaf surface, Florida red scale. When on fruit, the scale covering often does not easily wash off.

The soft scales include several citrus pests like *Ceroplastes* spp. (white wax scale) and *Saissetia* spp. (black scale and brown soft scale). Generations per year range from two or three for the lowest fecundity to four to ten generations per year for brown soft scale (*Coccus hesperidum* L.). In all cases, more generations are produced in tropical climates and in warmer years. Maritime locations are also more favourable because of moderated climates. Young crawlers can be found on all parts of the canopy, but adults tend to settle on twigs and to a limited extent in the protected area near the midrib of the leaves. Preferred plant part varies with species. These insects also produce honeydew which attracts ants and causes sooty mould development.

Scale insect pests of citrus are generally under some of the best biological control of any insects (McCoy, 1985; Browning, 1992). Soft scales like brown soft scale and black scale are parasitized by *Metaphycus* spp., but the host parasites are usually specific and pest species often differ by location. Ladybird beetle larvae and parasitic flies also contribute to soft scale control. The most common cause of upset of these control equilibria is the use of broad spectrum pesticides that kill predators and parasites as well as the target pest. Programmes with spray oil and specific target pesticides are essential for effective IPM programmes.

Armoured scales are protected by their wax-based covering. California red scale (*Aonidiella aurantii* Mask.) is an important and damaging member of this family. The first generation crawlers rapidly disperse even to small fruit causing noticeable distortion. Infested young leaves become yellow and small twigs may die. Two more generations are produced during the early and late summer, but infestation of more mature fruit does not lead to their being misshapen. Many successful parasitic wasps attacking scale insects are of the genus *Aphytis*. In California, *Aphytis melinus* is an important parasite of the California red scale (California IPM Manual Group, 1984, http://ipm.ucanr.edu/PMG/selectnewpest.citrus.html). Combined with the effects of *Comperiella bifasciata* (Howard), another effective parasite, moderate biological control is obtained. The Florida red scale (*Chrysomphalus aonidum* L.) and the dictyospermi scale share similar biology to the California red scale, but they are much less severe pests. Florida red scale is controlled by *Aphytis holoxanthus* DeBach, which was purposely introduced (Bullock and Brooks, 1975). In South Africa, the predators *Cilocorus negrita* and *Cilocorus distigma* combined with *Aphytis* spp. give good control of California red scale.

Glover, chaff and citrus snow scale are less well controlled biologically, but the first two are seldom a problem if repeated sulfur sprays are avoided and some ground cover is maintained to provide a good habitat for predators. Chaff scale (*Parlatoria pergandii* Comst.) is characterized by reservoir populations on limbs and trunks, but it can be widely distributed on the tree and feeds on leaves, stems and fruit tissue which sometimes leads to fruit abscission. This is a problem in some citrus-growing locations such as Brazil where clean cultivation is practised which eliminates the desirable higher humidity habitat for alternate hosts for the predators. Heavy use of inexpensive, broad-spectrum pesticides like sulfur also kills predators and parasites, exacerbating the problem. Purple (*Lepidosaphes beckii* Newm.) and glover (*Lepidosaphes gloverii* Pack.) scales are very similar having two main reproductive cycles per year. These scales feed on all plant parts. Excessive feeding on leaves results in chlorosis. In Florida, purple scale is controlled by *Aphytis lepidosaphes*, a chance introduction.

Citrus snow scale (*Unaspis citri* Comst.) is also widely distributed but is primarily a limb and trunk feeder. Large populations can build up and turn the trunk and primary scaffolds white (males). The related species, *Unaspis yanonensis* Kuwana, referred to as arrowhead scale, is more severe on fruit and leaves and results in delayed colour development. This scale is a prominent problem on mandarins in Japan. *Aphytis lingnanensis* Compere was introduced to Florida for snow scale control, but it has been largely unsuccessful. Some predators (ladybird beetle from China being tested in Australia) and parasites appear to be better biological control agents for this difficult to spray scale insect.

Plant bugs (there are several species in Heteroptera) will feed on mature fruit, particularly if their natural weed host population is disturbed. This commonly occurs when disking or chopping is employed to prepare the orchard floor for harvesting (Albrigo and Bullock, 1977). Year-round availability of alternate hosts such as nightshade, guava, citrons, etc. should be avoided. Only the adults feed on fruit, and they penetrate into the juice vesicles with their long stylets. Generally there is little external damage, and most losses are due to decay development in the feeding punctures between oil glands. Feeding punctures through oil glands appear to be sterilized by the toxic oils. Sometimes small oil-induced necrotic spots are visible. Dry juice vesicles at the outer surface of the segments and a reddish reaction of these vesicles during alkaline peeling for sectioning are sometimes the only indication of feeding.

Wood borers have not been a problem for citrus production in Europe or the American continent, but borers infest citrus in Japan, China (Talhouk, 1975) and Jamaica. Borers cause damage during the larvae stage using chewing mouthparts and usually bore into the trunk and primary scaffolds. They usually attack trees that are already weakened from other problems. Invasion of wood by secondary decay fungi often occurs.

Leaf cutting ants (*Acromyrmex octospinosus* Reich and *Atta* spp.) can cause severe leaf loss to the point that growth of young trees is inhibited, and trees become weak and are susceptible to foot rot and other fungi. Leaf pieces are removed and used as substrata on which to culture fungi in the ant colonies for food. Ant beds can cover large areas of the orchard. These ants are a serious problem in Brazil and South-Central America in general. Imported fire ants (*Solenopsis invicta* Buren) have become a serious problem in new plantings in Florida. The ants colonize the soil near the trunk and feed on the tender trunk bark, particularly under freeze protection wraps. In some cases girdling occurs, but more often the trees are lost to infection by *Phytophthora* spp. Similar losses occur on recently cleared land in tropical America where the colonies develop in the feeder root area. Argentine ant species attack blossoms and transport and protect aphids, mealybugs, scales, whitefly, etc. from natural enemies. This increases honeydew production and associated sooty mould. Most baits and sprays have short residual effects and re-infestation occurs requiring continual monitoring and re-treatment where ants are a problem.

Termites (*Reticulitermes flavipes* Kollor) also can cause tree losses under circumstances similar to those which promote fire ants. Young trees on recently cleared land are good hosts for these disrupted termite colonies (Stansly *et al.*, 1991). Direct tree death from girdling occurs, but some losses are due to phytophthora invasion through the feeding injuries. Even phytophthora resistant rootstocks can be lost when termite or root weevil girdling of roots reduces their natural resistance.

Grasshopper populations can become large enough to cause defoliation of nursery stock or young trees in the orchard. Eastern lubbers (*Romalea microptera* Beauv.) and the American grasshopper (*Schistocera americana* Drury) are pests of minor economic significance some years in Florida.

Lepidoptera such as leafrollers and butterflies are generally not economic pests in most parts of the world. In some locations in Asia, species in this order are pests. The caterpillars are similar and are often referred to generically as orange worms. Occasional leaf or fruit damage occurs by leaf miners, leafrollers and butterfly caterpillars in most growing regions. In Southeast Asia, Australia and very recently in Florida, citrus leaf miners have become a serious pest. More damage is usually observed on young than mature trees. The orange dog butterfly (*Papilio cresphontes* Cram.) can be an economic problem in nursery production where so much new foliage is present and one or two caterpillars can defoliate a seedling or newly-budded tree. Proportionally more tender flush, ideal for feeding, is usually available on very young compared with mature trees in the orchard and economic damage has usually been associated with young trees. Various Lepidoptera attack citrus, usually on an inconsistent basis, making it difficult to develop any standardized control practices (California IPM Manual Group, http://ipm.ucanr.edu/PMG/selectnewpest.citrus.html). Several predators and parasites exist and can be promoted by sound pesticide management so that these minor pests are even less of a problem.

A number of fruit flies (Diptera: Tephritidae) attack citrus throughout the world. Adults oviposit in mature fruit and the larvae feed and develop in the fruit pulp. Except in isolated instances, fruit losses are usually not high but fresh marketing is difficult because infested fruit cannot be easily graded out. Quarantines against importation of fruit from areas with various species of fruit flies significantly restrict movement of fresh fruit in world markets. The Mediterranean fruit fly (*Ceratitis capitata* Weid.) is widely distributed in Europe, Africa and South America. It has a wide host range and will deposit eggs in relatively immature citrus. The South American fruit fly (*Anastrepha fraterculus* Weid.) and the oriental fruit fly (*Dacus dorsalis* Hend.) are widely spread within each respective continent. The Caribbean fruit fly (*Anastrepha suspensa* Loew) has been a problem for Florida grapefruit export shipments but rarely deposits eggs in grapefruit until the fruit is overmature. Other *Anastrepha* spp. exist with potential to harm citrus, especially in Central and South America. Current control strategies or shipping from fly-free zones are not very satisfactory from a long term point of view.

Nematodes

Nematodes are microscopic, cylindrical worms, many of which are parasitic to plants and feed on fibrous roots. Three nematodes account for most of the problems experienced in citrus production around the world, but six genera are reported to attack citrus (Duncan and Cohn, 1990). The citrus nematode (*Tylenchulus semipenetrans* Cobb) is widespread and the most common nematode on citrus. It is a common problem in Florida and South and Central America, but also occurs throughout the Mediterranean Basin (Lo Giudice, 1981a). At least four biotypes exist worldwide. The nematode partially penetrates the feeder roots and establishes a permanent feeding site. Eggs are laid and remain near the posterior end of the adult. The larvae hatch in 12–14 d and migrate to new feeder roots nearby so that natural spread from tree to tree is fairly slow. Adults survive until normal root death which is probably not more than 1–2 years. Trees are generally weak without specific symptoms even though they may be on a good nutrition and irrigation programme. Under Florida conditions, this nematode causes more severe symptoms in bedded orchards. This may be due to the shallow root systems that are very sensitive to any stress.

The burrowing nematodes (*Radopholus citrophilus* Huettel, Dickson & Kaplan) and *Pratylenchus* spp. are endoparasitic. They enter the root and can complete their life cycle there. Reproduction and growth are optimum at 24°C, but the nematode is active from 12 to 32°C. Larvae move to new roots in search of new food sources. Generally, the damage by an individual nematode is more severe than from a citrus nematode. The burrowing nematode is primarily a problem in Florida where it leads to spreading decline, primarily on deep sandy soils. This may be due to the more severe damage to deep rather than shallow roots. Preferential feeding may occur due to more favourable soil temperatures below 30 cm or to less competition between nematodes and other soil-borne organisms at this depth. Nematode feeding leads to decline and dieback. All trees in the nematode zone decline and trees on the border of the affected area develop symptoms as the nematodes spread. *Pratylenchus* spp. are more widespread and some other nematodes may be of isolated importance in citrus areas. Most site-to-site spread is by man on new plant material (often ornamentals) or with cultivation equipment. Best control is by avoiding contamination of planting sites and complete preplant sterilization if required. Some rootstocks are tolerant of some nematodes, e.g. 'Milam' to burrowing nematode (see Chapter 3). Furthermore, we now understand that HLB is also a root disease and thus the root zone of infected trees is rapidly weakened (Johnson *et al.* 2014). This has likely exacerbated the impact of sting nematode (*Belonolaimus longicaudatus* Rau) as an important factor limiting citrus production in localized regions of the central Florida sandridge citrus production area, prompting research efforts into development of resistant rootstocks for this pest (Grosser *et al.*, 2007a).

PEST MANAGEMENT

Several management options are available for ACP and CLM in Florida and continue to evolve for protection of young, non-bearing, as well as, mature citrus. Integrated pest management (IPM) involves employing multiple methods of pest control in a coordinated action plan. The goal is sustainable pest control that also returns an economic gain. Individually, these methods may not be enough to achieve needed pest reductions and not all methods are applicable in all environments. However, when used in concert, the benefits of an integrated approach are realized. For example, conventional insecticides play an integral role in integrated systems available for commercial production systems, but may not be suitable for urban areas or organic production. Biological control, on the other hand, may play a larger role in these areas and, in general, will contribute more to overall pest suppression in groves employing an integrated approach rather than sole reliance on insecticides for pest management. IPM also focuses on pest identification and monitoring, prevention and decision-making based on economic injury levels so that insecticide use is prescribed when it is necessary. In the case of pathogen vectors, such as ACP, application thresholds are exceedingly low since very few individuals can cause disease outbreak. Therefore, many current research projects for citrus production are either complete vector exclusion or by reducing vector populations sufficiently to maintain economically viable yield. For example, growing citrus under protective screening (CUPS) and therefore excluding the vector from the production system completely is an idea gaining momentum for fresh fruit production based on ongoing investigation (A. Schumann as reported by P. Rusnak). Furthermore, yield improvements have been demonstrated by reducing ACP populations in standard mature groves with HLB as compared directly with counterparts where vector populations are not well managed and allowed to re-innoculate trees (Stansly *et al.*, 2014). These results suggest that vector control is still important even when nearly 100% of a block is already expressing symptoms of HLB.

Pest Monitoring

Regular scouting for pests is important for making need-based decisions regarding applications of insecticides for pests in order to minimize unnecessary investment in control measures and potential collateral damage to beneficial insects and mites. At present in HLB affected citrus, most spray applications are made for ACP. The adults spread the pathogen bacteria causing HLB from tree to tree and grove to grove. Therefore, ACP should be monitored. Sticky traps, sweep nets, suction devices and stem tap sampling are all options as tools for estimating ACP populations. Stem tap sampling is easy to conduct compared to other methods and provides instant data for making management decisions. A tap sample is made by using a white clipboard (22 × 28 cm) or laminated white paper sheet held horizontally under randomly chosen branches. One should strike the branch three times with a length of PVC pipe (https://edis.ifas.ufl.edu/pdffiles/IN/IN111600.pdf). Psyllids can be counted as they fall onto the clipboard. It is also possible to monitor other insects with this method, such as CLM, weevils and beneficial insects such as ladybeetles, spiders and lacewings, if desired. Complete elimination of ACP is not possible without exclusion cages, and a threshold is needed to use such monitoring data for making spray decisions. Investigations in Florida in HLB infected blocks have revealed that in blocks where 0.1 to 0.2 adults are counted per infected tree, yields can be maintained with proper nutrition. These numbers translate to 10 or 20 ACP adults per 100 tap samples. Eggs and nymphs of ACP and similarly larvae of CLM are also important to monitor, however, they are hard to see, but small predators and parasitoids are able to reach their locations for control where they develop.

Cultural Control

Preventing ACP from colonizing citrus, particularly non-bearing trees, is paramount to minimize HLB and maximize chances for these trees to enter production. Growing citrus under screens offers a possibility of complete protection from ACP and HLB. This technique is currently under investigation and shows promise. However, there may be some risk from other pests such as thrips, mites and CLM, which otherwise are manageable. Eventually, biological control might be established in screen enclosures for other pests not excluded by the screen. Growers have already begun establishing these protected structures for growing citrus by physically excluding vector and pathogen. Significant progress has also been made toward reducing psyllid access to young citrus by planting seedlings on UV reflective mulches. Success has been demonstrated with reduction in pest and disease incidence and increased profits (Croxton and Stansly, 2014).

Chemical Control

Chemical control has evolved from the use of basic inorganic materials like the Bordeaux mix of lime–sulfur and copper sulfate applied by hand-sprayers to currently using sophisticated application machinery and organic pesticides, some with very specific pest ranges and some that control specific development stages of the insect, i.e. those chemicals that specifically alter development by growth regulation (Knapp, 1992). Moults from one developmental stage to another are inhibited so that the life cycle cannot be completed (Henrick, 1982). Specific chemicals are recommended in most major production areas, and information on these is published frequently since new chemicals, new rates and cancellations of chemicals occur frequently.

Insecticides are critical components of management programmes for ACP and CLM. However, it is important to understand the class of product being used and proper timing of application. During the period between November and February, most mature trees go through rest and do not produce new growth required by psyllids for reproduction. Therefore, psyllid populations are mainly comprised of adults during this period and generally in comparatively lower densities because reproduction is not taking place widely. CLM is also not generally reproducing at this time for the same reason and overwinters as adults. Beneficial insects are also less common due to lack of prey. Therefore, broad-spectrum organophosphate (Chlorpyrifos, Imidan, Dimethoate) and pyrethroid (Mustang, Danitol, Baythroid) insecticides are good candidates for use during this time. These products are comparatively less expensive than other chemistries and are effective adulticides for ACP. If populations of ACP adults are effectively killed off during this time with these broad-spectrum chemistries, it can take many weeks for populations to re-establish (Qureshi and Stansly, 2010). One application during November–December and a second before the spring flush in January or early February can be sufficient to significantly reduce populations of ACP (Qureshi and Stansly, 2010). If these sprays are applied in a coordinated fashion over a large contiguous area, it may take psyllid populations 4–6 weeks (or more) to effectively recolonize that area (Qureshi and Stansly, 2010). A third spray may be needed if temperatures are high and if trees should flush during November–December (Qureshi et al., 2014). The best timing for the last winter spray is at budbreak in the spring to minimize adult ACP just before feather flush is available for egg laying. The products in the two classes of insecticide above are hard on beneficial insects and mites and therefore restricting their use to this time of the year is best for integrating chemical and biological control because natural enemy populations begin to increase during the spring. These dormant sprays are very important and are known for providing significant reductions in psyllid populations into the growing season demonstrated through individual and coordinated programmes such as Citrus Health Management Areas (Qureshi et al., 2014). If adults are controlled effectively during the dormant season, then most flush can be maintained with much lower infestation during spring. Heading off a major population increase directly prior to bloom is important because of restrictions imposed on insecticide use during bloom when pollinators are present.

Selective insecticides, such as horticultural crop oils and soaps, are less hazardous to beneficial insects and mites and safe for organic production.

They are better choices for use during the growing season when beneficial organisms are common and contributing to control of multiple pests (Qureshi and Stansly, 2007). These products are not adequate if more control of ACP is desired. Selection of products should be made according to the labels and guidelines which provide details on product use rates, timing and expected efficacy against multiple pests in order to minimize unnecessary applications. The University of Florida provides an example of these guidelines (http://www.crec.ifas.ufl.edu/extension/pest/PDF/2017/ACP%20and%20Leafminer.pdf). Psyllids generally colonize the borders of blocks in greatest abundance and become more abundant directly after unexpected flush events as seen after hurricanes. Therefore, supplemental border sprays throughout the growing season or additional cover sprays following an unexpected flush event with an organophosphate insecticide may be warranted.

Soil applied systemic neonicotinoids all belong to the same insecticide chemical class and thus share the same mode of action (MOA 4A). These have been used for well over a decade to protect non-bearing citrus from ACP and CLM. Pest populations can evolve resistance to insecticides following repeated exposure to the same MOA. This has been documented with certain populations of ACP in Florida for several products that have been extensively used in spray programmes, particularly the neonicotinoids commonly used as soil drenches. Diversifying and strictly rotating MOAs can allow populations of resistant pests to reverse back to susceptibility and this practice is very useful for ACP (Chen and Stelinski, 2017). It is extremely important to integrate sprays with soil drenches of a different MOA to conserve effectiveness of neonicotinoids. Programmes are being expanded for additional area-wide monitoring of ACP resistance in order to maintain efficacy of available tools.

For CLM, insecticides that kill the larvae are currently the most important tools for control of this pest. The early larval stage is the most susceptible and best target to avoid injury as larvae emerge from the egg and begin to feed on flush. Young trees are often the most susceptible to CLM damage. Soil applications of neonicotinoid insecticides can provide multi-week control of CLM in non-bearing trees and should be applied up to 2 weeks prior to a leaf flush in order to allow lethal concentrations of insecticide to accumulate in the foliage. However, it is important to remember that all neonicotinoid insecticides share the same mode of action, so back-to-back applications may hasten development of insecticide resistance in both CLM and ACP, which must be avoided.

Insecticides recommended for leafminer control can be found in the IFAS Florida *Citrus Pest Management Guide* (web link above). Proper timing of foliar sprays to coincide with flushing cycles is critical to optimize CLM management. The goal is to kill larvae as soon as they begin mining. Few adults overwinter so little larvae activity occurs in the first spring flush. Control becomes more important in the second flush of adult and non-bearing trees. This is usually the first summer flush of adult trees and a late spring flush, after flowering, on non-bearing trees. Although broad-spectrum insecticides targeting ACP adults may kill adult CLM, typically populations rebound in groves soon after foliar sprays of organophosphate or pyrethroid insecticides and may even increase under intense ACP management due to suppression of natural enemies. Applications of foliar insecticides should be made during a window when CLM larvae hatch and begin feeding to maximize larval kill. In general, the earliest applications should occur between 13 and 20 days after budbreak. The duration of control may be shorter if a heavy flush occurs soon after the foliar application. It may be advantageous to target spring flush, even though CLM damage is not evident, to prevent build-up of populations that will cause damage later in the year. The value of this approach compared to spraying the second flush when CLM is likely to be present in significant numbers has not been carefully tested.

There are opportunities for combining control of CLM with that of other pests. Most recommended products for CLM control are also effective against ACP, which also develops in young flush. Well-timed late spring or early summer applications of Agri-Mek or Micromite can be useful to manage citrus rust mite. Products with the active ingredient, cyazypyr, are highly effective against CLM (as well as ACP). Several (8+) weeks of protection against CLM have been documented following soil-applied treatments of this active ingredient. It is also highly effective as a foliar treatment against CLM.

Some newly-registered insecticides for use in Florida citrus are mixtures of two modes of action that allow for simultaneous CLM and ACP control. Agri-Flex combines the active ingredient of Agri-Mek with that of Actara (a neonicotinoid) in a single formulation to target CLM larvae and ACP nymphs and adults. Cyazypyr has been combined with abamectin in another product. However, any consecutive application of an insecticide employing the same mode of action, whether it is a single or dual mode of action, increases the probability of insecticide resistance developing in the target pests. Intrepid 2F (Methoxyfenozide) +2% v/v horticultural spray oil (435) is effective against CLM larvae by interfering with the moulting hormone. Intrepid can be used alone or tank mixed with another insecticide without selecting for resistance to ACP upon which it has no effect. Application technology is as important as the choice of chemical. Timing and effective coverage are the two major considerations for proper application. Timing depends on the stage of insect development controlled by the pesticide, insect population dynamics and the residual control period after application. If many generations of an insect are produced each growing season, then all stages of development will be present. Long-term residual control from the chemical is desirable to break the cycles of overlapping generations. Timing of the first application should be just in advance of the populations reaching an economic injury threshold, but these thresholds have not been established for most pests. Equally important to timing may be development of a plant part, such as new leaves, required for egg laying (ACP) or feeding (CLM) of the pest. Populations of economic pests usually increase very rapidly during peak development. Therefore, sufficient time must be allowed for spraying all orchards before injury occurs. Proper timing of control should be based on a scouting procedure to determine when a population increase is approaching injurious levels. An attempt to follow this procedure in Florida has been based on the observation that injury from citrus rust mite feeding occurs at population densities of 70–100 mites cm^{-2} (McCoy and Albrigo, 1975). Evolution of a mite-days to injury concept is presented in the *Florida Citrus Spray Guide* (Knapp, 1992) and is present in more current versions of this management guide (see website above).

Pests with only two or three generations $year^{-1}$ need more precise timing in relation to development stage. Chemicals usually work on either the adult or larval stage and must be timed for when that stage predominates. This may be further complicated by where on the plant that stage of development can be found. Weevils provide an example of this complexity. The adults are leaf feeders, but the hatched larvae fall to the soil and burrow to roots where they feed. Neither stage is easy to spray directly and current chemicals have shorter residual times to protect the environment. This means that control lasts only a short time and many of the pests escape control from any one treatment.

The other important component of pest control is proper application. Aerial or ground spraying, granular ground spread and soil incorporation are used. Fumigants for nematodes and aldicarb were examples of soil-incorporated materials. Because of degradation and toxicity these materials were usually drilled into the ground. Granular baits for ants are usually broadcast around the base of young trees or placed under tree wraps where freeze protection is required. Other special applications include application of parasitic nematodes for weevil larvae control using the microirrigation system or a herbicide spray boom followed by irrigation (Knapp, 1992; current *Citrus Pest Management Guide*).

Most other applications use ground or aerial spray equipment. Fixed wing and helicopter application are used in many countries. These methods are limited in their ability to deliver pesticides to the interior of the tree or even the underside of leaves. They are effective on pests that inhabit the outside of the canopy and upper surface of leaves like citrus red mite, but provide poor control of scale insects that may cause heavy infestation on the trunk and limbs, and greasy spot fungus that invades through stomata on the lower leaf surface (see Chapter 7, p. 234). An important advantage of aerial applications is that they can be made quickly to large areas of trees.

In the case of insects like snow scale that primarily inhabit the trunk and main scaffolds, hand spraying of infested trees is the most effective application method (Knapp and Browning, 1989). For good canopy penetration and moderate inner wood coverage, at least some form of ground, air carrier sprayer is needed.

The general application procedure is to minimize the carrier solution (usually water) to cover more area per tank of spray and to minimize run-off that wastes material and increases the potential of soil contamination problems. This has been accomplished with traditional sprayers by using lower delivery volume nozzles and/or increasing ground speed. Newer sprayers are designed to create smaller droplets and the air carrier system is modified to carry these small droplets into the canopy to achieve good coverage with small volumes of spray (air curtain). Penetration is not as good with these sprayers. Most modern sprayers have electric eye sensors to evaluate tree height and spacing and deliver spray just to this area.

Nozzles are only turned on in response to sensing canopy at that height. Small trees or replants only receive a short burst as the sprayer passes. No spray is delivered to open spaces, which conserves pesticides and reduces environmental problems from spraying excess chemical onto the ground. Adequate dilute spray volume to cover the canopy depends on tree height and assumes full tree development in the tree row. Specific calibration is based on trees ha^{-1} (canopy volume) and therefore number of litres tree^{-1} that should be delivered.

Whichever machine or concentration is used, frequent calibration is important as with herbicide sprayers, for nozzles will wear or become clogged. It is also important to remember that most spray regulations now require that a limit of chemical per area is not exceeded (Knapp, 1992; current *Citrus Pest Management Guide*). Calibration and calculations for concentrate spraying must be done carefully so that product per area is not exceeded.

Spray equipment is expensive to buy and operate. Many of the more effective organic chemicals are also expensive. Methods that do not require or minimize spraying are desirable for cost reasons and because of environmental considerations. IPM is a realistic approach that properly maximizes use of biological control but recognizes the proper use of target-specific pesticides when needed.

Biological Control

Biological control has been practised for many years in citrus (Browning, 1992). In many cases, researchers discovered that biological control was already occurring and learned how to manage it more efficiently. Many of these natural enemies were probably imported with citrus plants just as the pests were. Purposeful importations of predators and parasites have also occurred. Most useful insects are usually found in the areas of origin of the citrus species (see Chapter 1). Significant investments were made to import and establish species of predators, which not only attack ACP and CLM, but also several other pests of citrus. Parasitoids have also been introduced and established which specifically attack ACP and CLM. For example, the parasitoids *Ageniaspis citricola* and *Tamarixia radiata* are highly specific to CLM and ACP, respectively (Qureshi and Stansly, 2009). Historically, high levels of parasitism (60–80%) by these species were observed at different times of the year, but this has been significantly impacted by increased use of insecticides for ACP management over the past several years (Qureshi *et al.*, 2009).

Following consistent releases, *T. radiata* has established in greater abundance in organic citrus production systems with parasitism averaging up to 31% (Qureshi and Stansly, 2009). Populations of this parasitoid and associated higher attack rates of ACP occur when fewer conventional insecticide sprays are applied (Qureshi *et al.*, 2009). Also, fungi such as *Hirsutella*, *Beauvaria* and *Paeciliomyces* contribute to pest control. These beneficial organisms can be conserved by limiting use of broad-spectrum insecticides to the dormant season, as supplemental border sprays, or with targeted use following unexpected flush events. Furthermore, use of selective insecticides, oils, soaps and releases of parasitoids during the growing season will contribute to development of functioning IPM.

Although predaceous mites, such as *Euseius hibisci*, and ladybird beetles, *Stethorus picipes*, attack spider mites, the current complexes in California and Florida do not usually control citrus red mite or Texas citrus mite. The same can be said for the citrus rust mite. Populations of both kinds of mites are suppressed in marginal subtropical and tropical climates during rainy periods by entomophagous fungi such as *Hirsutella thompsonii*, but citrus rust mites usually build up to injurious levels in the spring dry period before control by fungi occurs and again in the autumn after the summer rains end. In a dry Mediterranean climate, citrus rust mite control was adequate after the introduction and mass rearing of predaceous mites and the judicious use of selective miticides instead of sulfur, which greatly reduces populations of natural predators (Cohen, 1975). A similar experience occurred in California when selective chemicals were integrated with other practices (California IPM Manual Group, 1984; current IPM manual).

Citrus thrips are known to scar fruit in late spring during the second generation in California. Orchards with ground cover and minimum spray programmes that avoid organophosphates tend to have fewer thrip problems. Predaceous mites, *Euseius hibisci* and *Anystis agilis*, and the minute pirate bug prey on thrips as do natural enemies found in ground litter. A single spray at petal fall will usually control thrips if other factors are favourable for minimum thrip build-up.

In Florida, a successful biological control programme for blackfly was introduced based on work started in Mexico using two parasitic wasps (*Encarsia opulenta* and *Amitis hesperidum*). These parasites are easily killed by sprays of organophosphates or sulfur, while the blackfly is difficult to kill with insecticides. Best control is maintained by introduction of the parasites and avoidance of these chemicals (Knapp and Browning, 1989). Citrus and woolly whiteflies are controlled under Mediterranean climates by parasitic wasps such as *Cales noacki* and *Eretmocerus* spp. This control is easily disturbed by ants protecting this source of honeydew, or insecticide treatments such as sulfur that kill these wasps.

A number of parasites, predators and fungi have been observed associated with aphids in South and Central America including the brown citrus aphid.

It is believed that these might control this serious aphid to prevent economic damage. Biological control depends on low levels of a pest to sustain the control species. For an efficient virus vector like the brown citrus aphid, this level of control would probably slow down only slightly the spread of virulent CTV (see Chapter 7).

Scale insect pests of citrus are generally under some of the best biological control of any insects (Browning as reported by Smith, 1992). Many successful parasitic wasps attacking scale insects are of the genus *Aphytis*. In Florida, purple scale is controlled by *A. lepidosaphes*, a chance introduction, and Florida red scale is controlled by *A. holoxanthus DeBach*, which was purposely introduced (Bullock and Brooks, 1975). In South Africa, the predators *Cilocorus negrita* and *C. distigma* combined with *Aphytis* spp. give good control of Florida red scale. Glover, chaff and citrus snow scale are less well controlled biologically, but the first two are seldom a problem if repeated sulfur sprays are avoided and some ground cover is maintained to provide a good habitat for predators. Chaff scale is a serious pest in Brazil where clean cultivation diminishes alternate hosts and a higher humidity habitat for predators. Heavy use of sulfur sprays in Brazil kills predators and parasites, compounding the problem. *A. lingnanensis* Compere was introduced to Florida for snow scale control, but it has been largely unsuccessful. Some predators (ladybird beetle from China being tested in Australia) and parasites appear to be better biological control agents for this difficult to spray scale insect.

In California, *A. melinus* is an important parasite of the California red scale. Combined with the effects of *Comperiella bifasciata* (Howard), another effective parasite, biological control is obtained. Soft scales like brown soft scale and black scale are parasitized by *Metaphycus* spp. Ladybird beetle larvae and parasitic flies also contribute to soft scale control. The most common cause of upset of these control equilibria is the use of broad spectrum pesticides that kill predators and parasites as well as the target pest. Spray oils and specific target pesticides are essential for effective IPM programmes in controlling scale insects.

Various Lepidoptera attack citrus but usually on an inconsistent basis, thus making it difficult to develop any standardized control practices. A number of predators and parasites exist and can be promoted by sound IPM practices so that these minor pests are even less of a problem.

Area-wide Management in the Era of HLB

ACPs are able to disperse several kilometres over fallow fields, lakes and roads (Tiwari *et al.*, 2010; Martini *et al.*, 2013; Lewis-Rosenblum *et al.*, 2015). Measures for HLB management have consisted of: (i) reduction of ACP populations by intensive insecticide treatments (Qureshi *et al.*, 2014); (ii) removal of infected trees to reduce HLB inoculum (Bassanezi *et al.*, 2013);

and (iii) re-planting healthy citrus plants produced under insect-proof nurseries. Soil- and foliar-applied insecticides play a vital role in suppressing ACP populations that are infected or capable of acquiring and transmitting HLB bacteria. Among these recommended measures for HLB management, the removal of symptomatic trees is the most controversial and difficult to accomplish in the USA. First, it increases per-acre grove maintenance costs by ≈ $400 annually, and second, newly planted young trees designated to replace removed trees are more susceptible to HLB infection than mature trees (Martini *et al.*, 2015). While vector control is still a key component of HLB management in Texas, the disease incidence continues to exponentially progress in citrus groves and backyard trees, and some growers are voicing their concerns about the usefulness of vector control. Some growers in Florida are also on the brink of abandoning ACP vector management. HLB infection rates are hovering around 100% at all times, and newly established plantings are quickly becoming infected prior to reaching the bearing stage. Growers are ceasing citrus production and citrus processing plants are closing at unprecedented rates. The future industry in Florida may require a clean slate approach by re-planting with the first HLB tolerant or resistant rootstock or scion cultivars, when those are available, appropriately vetted by regulatory agencies, and meeting with public acceptance. However, even then, partially resistant trees will likely require some protection from pathogen inoculation by ACP.

Previously, grower neighbours often made independent decisions on production and pest management, and these tendencies have been overcome through outreach and demonstrations of the benefits of cooperative action. A spray programme effectively applied to a single block of citrus and not to neighbouring blocks will be ineffectual, as untreated blocks serve as sources of infected ACP and disease. Based on this knowledge, it became clear that area-wide, cooperative management of the vector is necessary. Although new area-wide ACP control programmes were initially adopted in Florida in coordinated scale and incorporated resistance management strategies, their adoption was short-lived (2–3 years) and is now in question. While growers continue to implement area-wide dormant sprays in Texas, no spray coordination is done during the active growing season. As managing citrus blocks for HLB control can add $1000 to annual cultural costs, growers may be unable to invest in such programmes (Spreen *et al.*, 2014). Therefore, a comprehensive, long-term management approach beyond only insecticides requires further investigation. By effectively optimizing spray schedules, improving the effectiveness of biological control, and affecting bacterial acquisition and transmission, the prospect of producing profitable citrus will increase.

Diseases

L.W. Timmer

There are many types of diseases including systemic, tree declines caused primarily by viruses, viroids or procaryotes, soil-borne diseases caused by fungi or nematodes, and foliar and fruit diseases caused mostly by fungi and bacteria. Reviews of citrus diseases have been published and some deal with most diseases such as the *Compendium of Citrus Diseases* (Timmer *et al.*, 2000) and *Citrus Health Management* (Timmer and Duncan, 1999) as well as a book chapter (Timmer *et al.*, 2003a), whereas one book deals with only systemic diseases (Roistacher, 1991) and one book chapter considers only fungal diseases (Timmer *et al.*, 2004). All of the above contain valuable information on the diseases dealt with below, but some may now be outdated. Disease complexes in several countries have been reported: decline diseases in India (Chadha *et al.*, 1970), Mexican lime diseases (Garza-Lopéz and Medina-Urritia, 1984) and fungal and bacterial diseases in Italy (Salerno and Cutuli, 1981).

TREE DECLINES AND DISORDERS

Bacteria, Spiroplasmas and Phytoplasmas

Huanglongbing (greening)
Huanglongbing (HLB) (yellow shoot disease) was first reported in mainland China in 1919 and it was shown to be graft-transmissible by K.H. Lin in 1956 (Bovè, 2006). The same disease was described in South Africa in 1937 as 'citrus greening'. HLB occurs in many Asian countries under various names, in southern and eastern Africa and the southwestern part of the Arabian Peninsula. It was found in Brazil in 2004 and in Florida in 2005 (Fig. 7.1). Typically, citrus trees first show symptoms in one area of the tree. Chlorotic symptoms (see leaf description below) and yellow shoots followed by dieback are typical. HLB has since been discovered in many citrus areas of the Americas, such as the Caribbean Basin, Central America and Mexico. It is a limiting factor

Fig. 7.1. Huanglongbing (HLB) examples on young tree in China (A) and Florida, USA (B). Note that the Chinese tree is pale with a thin canopy while the Florida tree has an area that is displaying yellowing, canopy thinning and excessive fruit drop.

in citrus production wherever it is found and has become a serious threat to citrus production worldwide.

Blotchy mottle – asymmetric yellow patches on a pale green background – is the most characteristic symptom of HLB (Fig. 7.2). Symptoms often begin in a particular part of the canopy and then spread to the rest of the tree. Zinc-like deficiency symptoms are commonly associated with HLB, resulting in confusion with nutritional problems. Fruit on severely affected trees are often small, lopsided with aborted seeds, and poorly coloured on the stylar end, hence the name 'greening disease'. Leaf yellowing and drop are often followed by twig dieback. Affected young trees are stunted, terminal leaves are yellowed, new leaves are small, leathery and upright, and old leaves are mottled. Since these symptoms take 4–6 months to appear, symptomless trees may be distributed from field or unprotected, and therefore affected, nurseries. The citrus species most severely affected by HLB include sweet oranges, some mandarins and their hybrids, and grapefruit. Lemons and limes may be less severely affected. Many citrus relatives are apparently more tolerant because bacterial titre levels are usually equal to species or cultivars that show more severe symptoms (Folimonova *et al.*, 2009). Although not conclusively proven as the causal agent, liberibacteria are consistently associated with this disease (Bovè, 2006). These bacteria are filamentous and polymorphic, varying in length and diameter. The bacteria have been characterized taxonomically by the nucleotide sequence of their 16S ribosomal RNA gene, and are members of a new subgroup in the alpha subdivision of the *Proteobacteria* of the Gram-negative bacteria. Three species have been recognized: *Candidatus* Liberibacter asiaticus in Asia and *Ca.* L. africanus in Africa, and Ca. L. americanus in Brazil. In addition, phytoplasmas in two different groups have been found in trees with HLB symptoms in Brazil and China, but these do not appear to be widespread. None of the bacteria associated with HLB has been grown in axenic culture.

Fig. 7.2. Leaf blotchy mottle typical of HLB (supplied by Dr Eduardo Carlos).

HLB is transmitted by grafting and by psyllid vectors. In southern Africa, the disease is transmitted by the African citrus psyllid, *Trioza erytreae* (Del Guercio). Asian and American HLB strains are transmitted by the Asian citrus psyllid, *Diaphorina citri* Kuwayama (Fig. 6.1, p. 184). The Asian psyllid, present in most of Asia for a long time, has spread in recent years throughout most of the Americas and is likely to continue to spread into new citrus areas. *Murraya paniculata* (L.) Jack, orange jasmine, and other citrus relatives are good hosts for the psyllid vectors. Both Asian and African strains of the disease and both vectors were present on Reunion Island (eastern Africa) and have been studied for their behaviour with introduction of the parasites of both vectors (Aubert *et al.*, 1984).

Polymerase chain reaction (PCR) can be used to identify the three different species of liberibacter and is also useful for diagnosis in affected trees or in the psyllid vector. For survey purposes or to identify the disease in new areas, HLB is usually identified by characteristic symptoms such as blotchy mottle followed by the use of PCR for doubtful cases.

HLB can only be controlled when coordinated efforts are made to detect the disease in the early stages and eliminate infected trees, aggressively control the vector and ensure that nurseries distribute only pathogen-free trees. This type of programme is being followed in Brazil and was used to some extent in Florida. It has been effective where large contiguous plantings are managed aggressively (Belasque, Jr. *et al.*, 2010). South Africa has shown that it is possible to live with HLB, but disease pressure is probably lower there than it is in Asia and the Americas. Most nursery trees planted in South Africa originate from budwood from nurseries located in an area with little greening. By pursuing effective biological and chemical psyllid control programmes after planting, growers can raise greening-free trees even in areas where the psyllids and HLB are endemic. Very few if any successes of bringing new plantings to bearing without significant HLB have occurred in Florida. Some growers are producing fresh citrus without HLB by using a citrus under protective screening (CUPS) system (Schumann *et al.*, 2017). An alternative method is to cover the young trees with protective screens, but this method has little research information. Another aid in getting new plantings to bearing is to use metalized reflective mulch in the tree row to disorient the psyllids. This reduces positive HLB trees by age 3, but does not completely eliminate some psyllids finding some trees and infecting them (Giles, 2018).

Biological control of the psyllids and eradication of HLB-affected trees has eliminated the disease from Reunion Island (Aubert *et al.*, 1984). However, in many countries, the effectiveness of psyllid parasites is reduced by hyperparasites. Eradication of all citrus and replanting with pathogen-free trees has been attempted in Indonesia and in coastal China as a means of controlling HLB, but these efforts have usually failed. In Guilin, Guangxi Province, China, after 13 years, this procedure appears to have succeeded where the over 4000 ha production was eradicated, replanted with clean nursery stock and coordinated psyllid control was instituted.

Because infected trees take several months to show symptoms it is not feasible to eradicate HLB once an orchard or block is infected. Small orchards cannot be maintained by eradication of symptom-affected trees unless all neighbours practise aggressive psyllid control. About as many trees are infected but don't have symptoms as do have symptoms. In Florida, some infected blocks have been maintained in production for several years by eliminating most other stresses and using an enhanced foliar nutrition programme that includes major and minor essential elements and perhaps some other important ingredients (Stansly *et al.*, 2014; Morgan *et al.*, 2018). With 'Carrizo' and 'Swingle' rootstocks, lower soil and water pH and low water bicarbonates also are important to minimize stress on the root system (Graham and Morgan, 2018).

Some recent changes in production practices in Florida that appear promising include frequent (daily) irrigation during dry periods, partial use of slow-release fertilizer plus higher rates of Mn and Zn with sulfur applied to the soil and HLB tolerant rootstocks (Handique *et al.*, 2012). 'Sugar Belle' mandarin has also been successful since it appears to have some HLB tolerance (Killiny *et al.*, 2017). Considerable effort is being directed towards breeding and finding scion and rootstock selections with HLB tolerance and hopefully resistance (see Chapters 2 and 3). A GMO version for HLB resistance using spinach bacterial resistance genes is in advanced testing (E. Mirkov as quoted by R. Santa Ana, 2012).

Citrus variegated chlorosis

Citrus variegated chlorosis (CVC) is a serious and widespread disease in Brazil, parts of Argentina and apparently in Costa Rica. Unlike HLB affected leaves, CVC leaves have small necrotic lesions (Fig. 7.3) on one or more branches and in chronic stages may be stunted with twig dieback. Leaves often have interveinal chlorosis associated with the small brown, necrotic areas on the lower surface

Fig. 7.3. Citrus variegated chlorosis (CVC) showing typical small necrotic lesions on the underside of affected leaves (photo supplied by Dr E. Carlos).

that enlarge with time. Fruits are small, hard and change colour prematurely. They are frequently sunburned and may have sunken brown areas on the surface of the rind.

CVC is caused by a strain of the bacterium *Xylella fastidiosa* that inhabits the xylem and impairs its normal function. Movement of the pathogen in xylem, especially basipetally, is slow following infection. The bacterium can be cultured from CVC-affected trees and pathogenicity has been proven (Hartung *et al.*, 1994). The strain of *X. fastidiosa* that causes CVC is different from that causing other diseases such as Pierce's disease of grapevines, plum leaf scald, phony peach and leaf scorch diseases in coffee, oaks and sycamore. It is most closely related to strains that cause coffee leaf scorch (Qin *et al.*, 2001).

Extensive spread has occurred unknowingly via propagation of infected budwood sources and by leafhopper vectors. The disease can be managed by avoiding propagation of CVC-infected budwood for new plantings, pruning out infected limbs of recently affected trees, and removal of affected trees in young plantings. Mandarins, grapefruit and lemons are less sensitive to CVC than sweet orange and are more suitable for areas that are severely affected by CVC.

Stubborn

Stubborn is an important disease of citrus in arid regions such as parts of California, North Africa and the Middle East. The disease affects most citrus cultivars and relatives and is especially severe in young plantings. Leaves are typically small and internodes are shortened, giving the tree canopy a bushy appearance. Some leaf chlorosis may be present and fruit are sparse. Fruit often have aborted seed, are acorn-shaped and the colouring of the fruit's stylar end may be delayed. Symptoms in mature trees are often less conspicuous. In contrast to many citrus diseases, symptoms are most severe under high temperature conditions.

The pathogen and the vectors have wide host ranges and citrus is probably a secondary host.

Stubborn is caused by *Spiroplasma citri* (Sagio *et al.*, 1973) a mollicute that lacks a cell wall, that is culturable. Cells of *S. citri* frequently have a filamentous form with a distinct helical morphology. The pathogen is found only in the phloem of infected plants and the symptoms reflect its pathogenic effects on that tissue. Several species of leafhopper can transmit stubborn and epidemics of stubborn are associated with large-scale periodic migrations of leafhoppers into citrus from other crops or native vegetation. Control of stubborn is based on avoiding infection in nurseries and in young plantings.

Witches' broom disease of limes
Witches' broom disease of limes (WBDL) is a lethal disease that occurs in Oman and the United Arab Emirates and has been reported recently in Iran and in other citrus areas in that region. Symptoms in the early stages of infection are one or more witches' brooms that form from repeated abnormal proliferation of axillary buds. Leaves on stems in the broom are small and paler than normal. As broom formation increases, twig and limb dieback occurs and trees decline rapidly and become unproductive.

The disease is associated with a phloem-limited phytoplasma that has not been cultured (Garnier *et al.*, 1991). It can be observed in infected trees by electron microscopy. A leafhopper, *Hishimonus phycitis* (Distant), is suspected to be the vector, although transmission from lime to lime has not been confirmed experimentally. The primary reservoir of the pathogen may be another host. Secondary spread from infected limes is suspected since the disease becomes epidemic once it enters a planting. Some spread to new areas via movement of infected plant materials has occurred. The disease is confined primarily to *C. aurantifolia* (SE Asian lime) but some other citrus species have been infected experimentally. Sweet orange is apparently resistant.

Diseases Caused by Virus and Virus-like Agents

Tristeza decline and stem pitting
Citrus tristeza virus (CTV) causes a wide range of symptoms on various species of citrus (Garnsey *et al.*, 2005; Hilf, *et al.*, 2005). Tristeza decline has resulted in the loss of over 50 million trees on sour orange rootstock since the 1930s and is a continuing threat to the more than 200 million citrus trees on this rootstock that remain worldwide. In addition to the decline that CTV causes in trees grafted on sour orange rootstock, some CTV isolates also cause serious stem pitting diseases of limes, grapefruit and sweet orange.

CTV apparently originated in Southeast Asia. It has been spread by infected budwood and aphid vectors so that it is now widespread in most citrus-growing areas. Areas where it is not as common, such as Mexico and parts of the

Mediterranean Basin, are under an increasing threat. Recently this threat in the Mediterranean was emphasized by identification of virulent strains near the region (Yahiaoui *et al.*, 2015). Differences in virulence of strains, efficiency of aphid vectors and susceptibility of different cultivars have resulted in a complex disease situation (Garnsey *et al.*, 2005). A broad range of effects has been observed ranging from minimal injury to devastating crop losses that prevent economic cultivation of sensitive species.

Foliar symptoms include vein clearing, leaf cupping and chlorosis on sensitive species such as Mexican lime. Tristeza decline occurs primarily on trees on sour orange rootstock and canopy symptoms include wilt, leaf chlorosis and premature flowering. Trees may decline gradually or collapse and die in a matter of weeks (quick decline). CTV-induced bud union phloem necrosis that produces the tree decline may appear as pinhole pitting in the bark of the sour orange rootstock just below the union (Fig. 7.4) or as a yellow to brown stain at the bud union. Trees affected by decline typically set abnormally large numbers of fruit which are small and may ripen prematurely. Nursery trees propagated on sour orange seedlings with budwood of trees infected with decline isolates are stunted and chlorotic, but rarely show a quick decline reaction.

Severe stem pitting (isolate pitting) occurs on the scion and is not affected by the rootstock on which the tree is grafted. Commercial production of grapefruit, pummelo and some sweet oranges is affected by stem pitting in many areas. The degree of pitting can vary from a few small pits that apparently have no effect on tree vigour to numerous pits that may coalesce and deform stems.

Fig. 7.4. Stem pitting associated with CTV.

In severe cases, the bark of infected plants is thickened and the wood is brittle and has a porous texture. Trees severely affected by stem pitting produce small and often misshapen fruit of poor quality. Damage from stem pitting can be greatly reduced by previous inoculation of nursery trees with mild strains of CTV (Costa and Müller, 1980). Recent work indicates that this cross-protection can be greatly enhanced by using a variant of the prevalent stem pitting strain in the local industry (Folimonova *et al.*, 2010).

CTV is a flexuous rod-shaped virus and a member of the closterovirus group. The genome is encoded in a single RNA species that is the largest plant viral RNA described to date. The sequence of nearly all isolates of CTV is similar in the 3′ portion of the genome that encodes the virion coat protein. Considerable variation present in the 5′ half of the genome suggests different origins for CTV strains, and tristeza may be a complex of different virus species.

CTV is transmitted by several species of aphids in a semi-persistent manner. The brown citrus aphid, *Toxoptera citricida* (Kirkaldy) is the most efficient vector, and tristeza decline and stem pitting problems have been most severe in areas where this aphid has become endemic. The cotton or melon aphid, *Aphis gossypii* Glover, and the spirea aphid, *A. spiraecola* Patch, are other important vectors.

Psorosis and citrus ringspot

Psorosis was the first recognized virus-like disease of citrus. Affected trees exhibit various types of transient chlorotic leaf patterns plus large bark scaling lesions on the trunks of susceptible cultivars, especially sweet orange and grapefruit. Trees with large lesions on the trunk or scaffold limbs become debilitated and fruit production is severely affected. Psorosis B is an especially severe form of psorosis. Other diseases that cause symptoms on immature leaves, such as concave gum and impietratura, but do not induce bark scaling, were included in what became known as the 'psorosis complex'. Study of the causal agents is complicated because trees are frequently co-infected by several different viruses that cannot be separated readily. Some viruses, especially those that produce chlorotic ringspot symptoms in leaves or fruit, are mechanically transmissible to non-citrus hosts, an advance which facilitated their characterization. The term 'psorosis' is now retained only for diseases previously described as psorosis A, psorosis B, citrus necrotic ringspot and naturally spread psorosis.

Leaf symptoms include various types of chlorotic patterns in young leaves that may, in some cases, persist in mature leaves. A necrotic shock reaction may occur in young shoots following graft inoculation, especially in sweet orange indicator seedlings. Ringspot patterns occur on fruit, but are not common. Most psorosis isolates produce a scaling and flaking of the bark in infected sweet orange and grapefruit trees that usually appear only in bearing trees. Callus tissue forms, bark sloughing occurs and some gum impregnation may occur in the wood under lesion areas. Rio Grande gummosis, a disease of unknown etiology, also produces lesions on the trunk and limbs that can be confused with psorosis, but gumming is more profuse and the bark usually dies around the site of the initial lesion.

The disease is caused by citrus psorosis virus (CPsV), the type member of the genus *Ophiovirus* (Sánchez de la Torre *et al.*, 2002; Martin *et al.*, 2006). The virus is tripartite and its genome consists of three ssRNAs of negative polarity. RNA 1 carries two open-reading frames: a 24-kDa polypeptide of unknown function separated by an intergenic region, and the putative RNA-dependent RNA polymerase of 280 kDa. RNA 2 carries one gene coding for a 53.6 kDa polypeptide of unknown function (the 54K protein), while RNA 3 codes for the coat protein of 48.6 kDa. The virus is mechanically transmissible to numerous herbaceous plants, including *Chenopodium quinoa* Willd. Natural spread by unknown means has been observed in several countries, most notably Argentina, where the disease remains a major problem. Incidence of psorosis in most citrus-growing areas has been reduced greatly by propagation of psorosis-free budwood.

Tatterleaf-citrange stunt

The tatterleaf-citrange stunt virus (TLV) was originally discovered in symptomless Meyer lemon trees in California (Roistacher *et al.*, 2000). TLV was found subsequently to be widespread in citrus in China where Meyer lemon originated. It has been spread to a number of countries via infected budwood. Although most citrus cultivars do not develop symptoms when infected with TLV, a severe bud union incompatibility occurs in trees that are propagated on rootstocks such as trifoliate orange or its hybrids. The extensive use of these rootstocks in modern plantings and the continued widespread use of trifoliate orange as a breeding parent for new rootstock varieties could result in widespread susceptibility.

The scion/rootstock incompatibility symptoms include formation of a deep crease in the wood of the trunk at the bud union that may also be accompanied by a brown stain or gum. The leaves become chlorotic and trees are stunted and may be easily broken off at the union. Citrange seedlings graft-inoculated with TLV develop chlorotic spotting and distortion on some leaves and stems often show a zig-zag growth habit. Diffuse chlorosis and leaf distortion also occur in recently inoculated seedlings of Mexican lime and alemow (*C. macrophylla*).

The virions are rod-shaped and the virus is a member of the capillovirus group. A very close sequence homology with viruses isolated from apples, pears and lilies suggests that they have a common origin. No vector has been described for TLV or its related viruses, but the presence of such similar viruses in distinct hosts suggests that some form of natural spread occurs. The unresolved epidemiology of TLV remains a concern in assessing its potential impact on citrus production. With clean budwood programmes, no TLV has been reported recently in Florida or California.

Leprosis

This virus is a widespread problem in many citrus areas of South America. Initial symptoms on the fruit, leaves and twigs of susceptible cultivars are chlorotic lesions that often become necrotic and gum-impregnated and show concentric patterns (Fig. 7.5). A chlorotic zone around the lesion may remain. Leaf and fruit drop occurs when infections are abundant. Leprosis is caused by citrus leprosis virus (CiLV), a bacilliform virus that is vectored by mites in the genus *Brevipalpus* (Bastianel *et al.*, 2006). The mite vectors are widespread in citrus in the world. There are two types of the virus: (i) a nuclear form (CiLV-N), the first to have been seen and thought to be a rhabdovirus; (ii) a more prevalent cytoplasmic form (CiLV-C). The cytoplasmic virus was believed to be a rhabdovirus, but sequence information and other characteristics indicated that it does not belong in that family, and CiLV-C was proposed as the type member of a new genus, *Cilevirus*, related to several (+) ssRNA viruses (Locali-Fabris *et al.*, 2006). Infections are localized and associated with feeding activity of mites that carry the causal virus (Freitas-Astúa *et al.*, 2003). The virus is also transmitted experimentally by mechanical inoculation and by grafting. The virus does not systemically infect citrus and trees do not develop symptoms on new growth after infective mites are removed. Control of the disease is primarily by management of the mite vector.

Satsuma dwarf

Satsuma dwarf was first described in Japan where it has become a significant problem. Affected trees are stunted and yields are reduced, but the disease is not lethal. Citrus mosaic, natsudaidai dwarf and navel orange infectious mottling are related diseases in Japan. Affected satsuma mandarin trees often have cupped, boat-shaped leaves and the related diseases also produce various chlorotic leaf symptoms in sensitive cultivars. Citrus mosaic may also cause fruit symptoms. Satsuma dwarf virus (SDV) has an isometric particle that has

Fig. 7.5. Leprosis lesions on fruit in Brazil (photo L.G. Albrigo).

two RNA components. Sequencing studies indicate that it is phylogenetically related to both cucumo- and nepoviruses.

Natural infections of SDV have been found in China laurestine, *Viburnum odoratissimum* Ker., and the virus can be transmitted experimentally by mechanical inoculation to a number of non-citrus hosts. Natural spread occurs in the field and the pattern of spread suggests a soil-borne vector, but none has been identified. Once an area becomes infested, SDV apparently remains in the soil since new trees replanted in these sites become infected. Long distance spread has occurred via movement of infected budwood within Japan and to other countries.

Sudden Death

Citrus sudden death (CSD) (*Morte Subita dos Citros*) has killed millions of trees in Minas Gerais and northern São Paulo State in Brazil. The disease primarily affects sweet orange on 'Rangpur' rootstock and is known only from this region. Symptoms of the disease are a rapid decline and death of trees similar to those of CTV-affected sweet orange on sour orange rootstock. CSD is graft transmissible and is likely caused by a virus. A member of the *Marafivirus* has been isolated from CSD-affected trees and is consistently associated with the disease (Román *et al.*, 2004; Yamamoto *et al.*, 2011). No vector has been identified for the disease, but spread of CSD is similar to that of tristeza and it has been suggested that it may be aphid-transmitted (Bassanezi *et al.*, 2003). Caged plants in the field are protected from infection (Yamamoto *et al.*, 2011).

Trees on many other rootstocks such as 'Cleopatra' and 'Sunki' mandarins, 'Carrizo' citrange, and 'Swingle' citrumelo are tolerant to CSD and can be used to replace 'Rangpur' (Fig. 7.6). The *Marafivirus* associated with the disease is present in species tolerant to CSD, but the trees do not decline. Seedlings of these rootstocks can also be used to inarch unaffected trees on susceptible rootstocks and prevent their decline. The disease is currently limited to this single warm area in Brazil and has not spread southward, but has the potential to be serious in other areas if introduced.

Other Virus and Virus-like Diseases

There are many other virus or virus-like diseases of citrus that are of minor or local importance. Many, such as concave gum, cristacortis and impietratura, that cause trunk deformities, are not well studied and their cause is only suspected to be viral. Others such as citrus variegation, crinkly leaf and leaf rugose are caused by well-known viruses, such as the ilarviruses, in this case, but are of little importance commercially. Chlorotic dwarf is caused by a whitefly transmitted virus and has caused significant losses in Turkey. Vein enation-woody gall is widespread and probably caused by a luteovirus, but is of

Fig. 7.6. Trees on 'Carrizo' citrange are healthy while trees on 'Rangpur' lime are all declining in the field in northern Brazil.

relatively minor importance. Yellow mosaic is important in India and is caused by a mealy-bug-transmitted badnavirus.

Diseases caused by viroids

Viroids are infectious, single-stranded, circular RNA molecules which lack a capsid protein and mRNA activity (Flores *et al.*, 2005). Viroids are the smallest plant pathogens and cause diseases in a wide range of crops. All belong to the family *Pospiviroidae* and five viroids, *Citrus exocortis viroid* (CEVd, genus *Pospiviroid*), *Hop stunt viroid* (HSVd, genus *Hostuviroid*), *Citrus bent leaf viroid* (genus *Apscaviroid*), *Citrus dwarfing viroid* (*Apscaviroid*) (former *Citrus viroid III*, CVd-III) and *Citrus bark cracking viroid* (genus *Cocadviroid*) (former *Citrus viroid IV*, CVd-IV) are recognized as distinct species. Two additional viroids, *Citrus viroid original sample*, only reported in Japan, and the recently characterized *Citrus viroid V* are still considered as tentative new species of the genus *Apscaviroid*. Each citrus viroid induces specific symptoms on Etrog citron, an indicator plant in which these viroids replicate and accumulate to high concentrations. However, only two viroids, CEVd and HSVd, have been demonstrated to cause specific diseases in citrus: exocortis and cachexia, respectively. Other citrus viroids produce more subtle effects on commercial species used as scions or rootstocks.

Exocortis is a bark-scaling disease of trifoliate orange rootstocks. CEVd is present in almost all citrus-growing regions and numerous sequence variants are known. Most scion cultivars are symptomless and disease symptoms are only expressed when infected scions are propagated on sensitive rootstocks, e.g., trifoliate orange, citranges and 'Rangpur'. Xyloporosis, a disease causing gum impregnation of the bark and pitting of the wood with corresponding bark pegs on 'Palestine' lime was described in Israel about the same time as cachexia, a disease affecting 'Orlando' tangelo, was described in the USA. Cachexia-xyloporosis also affects mandarins, mandarin hybrids, *C. macrophylla* and 'Rangpur', and is caused by HSVd. Another HSVd variant, CVd-IIa, is sometimes associated with dwarfing of trees on trifoliate orange rootstock.

Viroid-induced dwarfing of trees has limited commercial use in Australia, Israel, California and for 'Tahiti' limes in Brazil. Australian field trials indicate that inoculation of trees with viroids reduced tree size of selected sweet orange scions on trifoliate orange and citrange rootstocks without affecting tree health. Closely spaced, dwarfed trees out-yielded conventional plantings on the basis of planted area without affecting fruit quality. No effects on canopy development become apparent until 4 years after inoculation. Rapid, early canopy development before viroid-inoculation of the trees assures a high bearing volume per ha and promotes early orchard productivity, but this procedure is more time consuming than inoculating in the nursery, which has the disadvantage of over-dwarfing the scion/rootstock combination.

Citrus viroids are distributed principally by the propagation of infected scion wood and subsequently by natural root grafting and mechanical transmission on budding knives and hedging equipment, especially from lemon to lemon. Citrus viroids are not known to be seed or insect transmitted.

Blight and Other Decline Diseases

Citrus blight

Citrus blight is one of the most economically important diseases in Brazil and Florida and occurs in other citrus areas in the Americas, as well as in Australia and South Africa. Losses in Florida exceed 0.5 million trees annually and are probably much greater in Brazil. The most susceptible rootstocks are rough lemon, 'Rangpur', trifoliate orange and most of the citranges.

Blight is a wilt and decline disease of citrus trees. The disease does not occur on non-bearing trees and trees may be 8–10 years old when decline is first noted. The first symptom of the disease is usually Zn accumulation in the trunk phloem (Albrigo and Young, 1981) leading to Zn deficiency in the leaves followed by a mild wilt and greyish cast of a sector of the canopy (Fig. 7.7). These symptoms are followed by more severe wilt, leaf drop and twig dieback. Trees often decline in sequence down the row as the disease spreads tree to tree through natural root grafts. Trees with blight seldom die, unless under

Fig. 7.7. A wilting, declining left branch of a citrus tree affected with citrus blight, while the right side is relatively heathy. The small tree in front is where another blight tree was removed.

non-irrigated production, but usually become unproductive within a year or so, even if irrigated. As the canopy declines, fibrous roots are lost and eventually major roots may be rotted as well. Wilt symptoms are the result of blockage of xylem vessels with amorphous and filamentous plugging material which follows the upset of Zn distribution (Albrigo *et al.*, 1986).

The cause of blight is unknown (Derrick and Timmer, 2000; Timmer and Bhatia, 2003). Many causal agents, such as soil and nutritional factors, soil pathogens, viruses and *Xylella fastidiosa*, have been investigated, but no causal role has been confirmed for any of these agents. Blight has been transmitted by root-piece and tree-to-tree root grafts, but not by budwood or branch-to-branch grafts (Albrigo *et al.*, 1992). Thus, it is an infectious disease with a root-associated agent. However, it is not spread by soil taken from around blighted trees nor is it more common in replanted areas where blight-affected trees have been removed. Disease occurrence is initially random, but clusters of affected trees may develop or severity may become greater in certain parts of the planting. Disease increase is often linear which is not typical of infectious diseases, but is probably due to transmission through natural root grafts in close-spaced tree rows.

Trees affected by citrus blight must be replaced since there is no known cure. Before trees are removed, the disease must be accurately diagnosed to differentiate it from other decline diseases. Blight-affected trees are characterized by the failure to take up water when it is injected into the trunk with a syringe, a higher content of zinc in the bark (Albrigo and Young, 1981) than healthy trees or those that are affected by other diseases and by characteristic 12-kilodalton and larger proteins in leaves and other tissues which can be assayed serologically (Derrick *et al.*, 1990.). All of the above are useful in the identification of the disease.

Blight-affected trees should be replaced with trees on tolerant rootstocks. Sweet orange, sour orange, 'Swingle' citrumelo, and to a lesser extent 'Cleopatra' mandarin, are more tolerant to the disease (Young *et al.*, 1982). Tolerance to the disease is expressed as the age at which symptoms usually appear and as reduced incidence. Once trees on any rootstock develop the disease, they usually decline within a short time. Updates on the disease can be found in the 2018–2019 Florida Citrus Production Guide (https://crec.ifas.ufl.edu/program-areas/florida-citrus-production-guide/).

Other Decline Diseases

Dry root rot or sudden collapse is a disorder in which apparently healthy trees with a normal crop suddenly collapse and die (Timmer and Bhatia, 2003). Affected trees have blackened, rotted roots, but no gumming or pitting. The disorder is usually associated with fine-textured soils, excess moisture and poor aeration, and at times with root injury. Several fungi, such as *Fusarium* spp. and *Coprinus* sp., are associated with the disease, but are not able to induce the disorder unless trees are under severe stress. Incidence of the problem can be reduced by improved drainage, good irrigation practices and careful selection of planting sites.

Murcott or tangerine collapse and various twig diebacks affect some varieties of mandarins. Collapse occurs when trees set excessive amounts of fruit, leaving little stored carbohydrate in the leaves, twigs and particularly the roots

(Smith, 1976). Trees may recover, but produce little fruit the following year, or they may die. Cultivars such as 'Robinson' and 'Fallglo' tangerines have a genetic propensity for twig dieback. Fungi such as *Colletotrichum* or *Lasiodiplodia* spp. are readily isolated from affected twigs, but do not cause the disease if healthy twigs or branches are inoculated. Pruning out dead limbs and twigs and application of benzimidazole fungicides helped reduce the damage.

SOIL-BORNE DISEASES

Phytophthora Root Rot, Foot Rot and Brown Rot

Phytophthora spp. cause some of the most economically important diseases of citrus worldwide (Graham and Timmer, 1992). Losses occur in seedbeds from damping-off, in nurseries from root or foot rot (gummosis and collar rot) and in orchards from foot rot, fibrous root rot and brown rot of fruit. The symptoms of root rot damage are thin canopies, short, weak flushes of new growth and reduced yield. These symptoms result from deterioration of the fibrous root system and failure of the tree to replace roots rapidly enough to maintain tree health. Under favourable conditions for the fungus, lateral roots may also develop lesions depending on the susceptibility of the rootstock. On fruit, lesions are light tan to brown and firm. Under humid conditions, white mycelium appears on the fruit surface. Gummosis, foot rot or collar rot is characterized by gum exudation from the trunk, usually above the bud union since scion species are often more susceptible than rootstock species. The affected area is usually moist, sometimes having soft bark with olive to brown or black-discoloured wood underneath. Later, the bark dries and cracks and the wood beneath becomes surrounded by callus tissue at the surviving bark perimeter. The principal branches of the scaffold can be infected, producing cankers similar to lesions on the trunk.

Phytophthora citrophthora (R.E. Sm. and E.H. Sm. [Leonian]), *P. palmivora* (E.J. Butler) E.J. Butler, and *P. nicotianae* (B. de Hahn) cause root rot and gummosis. In Mediterranean climates with winter rains and dry summers, *P. citrophthora* is most important, whereas in warmer areas, *P. nicotianae* is more important. In winter rainfall areas, *P. citrophthora* is an important cause of trunk cankers. In California, where both pathogens occur, *P. citrophthora* is active in winter and *P. nicotianae* in summer. In Florida, *P. palmivora* causes a serious decline, especially of trees also attacked by root weevils, and is also an important brown rot pathogen.

Phytophthora spp. survive unfavourable periods as chlamydospores or oospores in soil or in decaying roots. Chlamydospores germinate readily when moisture is present and quickly form sporangia. Infection is usually by zoospores that are released from sporangia when free moisture is abundant. Zoospores are attracted to wounds or to the region of elongation of the root tips, where they encyst, germinate and penetrate. A wound is necessary for infection of suberized bark. All species of *Phytophthora* that affect citrus can cause brown rot. This disease is most commonly caused by *P. citrophthora* in winter rainfall areas, whereas in humid, subtropical and tropical areas, it is often caused by *P. palmivora*. Both species produce abundant sporangia on the fruit surface that can be spread through the tree by wind-blown rain. In contrast, *P. nicotianae* is mostly soil-borne and propagules must be splashed with soil and, consequently, most infections are within 1 m of the ground.

Strategies for control of foot and root rot and gummosis include improving surface drainage, installation of underground drains above impervious soil layers, improving irrigation practices, applying fungicides, or use of more tolerant rootstocks. Copper-containing fungicides, metalaxyl or mefenoxam, or fosetyl-Al or phosphorous acid may be applied as trunk paints or sprays to control foot rot or gummosis. Foliar sprays of copper products, fosetyl-Al, or phosphorous acid may be used for control of brown rot. For root rot control, any of the above products, with the exception of copper, may be applied through the irrigation system or as soil-surface sprays or drenches. However, the most effective control for all species except *P. palmivora* is the use of resistant rootstocks such as trifoliate orange and its hybrids. For control of gummosis or collar rot where the scion is usually more susceptible than the rootstock, the bud union must be kept well above soil through high budding, trees planted at the proper depth and soil kept away from the trunk. Mechanical damage to the trunk must be avoided since such wounds serve as infection points. *Phytophthora* diseases can be avoided in nurseries by collecting seed only from fruit high in the canopy, by hot water treatment (52°C for 10 minutes), or steam-air treatment (60°C for 30 minutes) of seed. Pathogen-free potting mixes, clean irrigation water and good nursery practices to avoid contamination of the mixes or nursery soils must be used. In field nurseries, soils may be fallowed or fumigated chemically to reduce populations of the pathogens.

Other Root Diseases

Mushroom root rot causes decline and death of trees in many areas of the world. The disease is often caused by species of *Armillaria*, such as *A. mellea* (Vahl:Fr.) P. Kumm and usually occurs in areas where native forests have been cleared and planted to citrus. The fungus colonizes dead wood in the soil and then spreads to living trees. Major roots are colonized and killed and characteristic mycelial fans are formed beneath the bark. The pathogen spreads slowly in the soil and affects tight clusters of trees. The disease can only be controlled by complete clearing of trees and all large roots and fallowing for several years. Rosellinia root rot, caused by *Rosellinia* spp., and Ustulinia root and collar rot, caused by *Ustulinia duesta* (Hoff.:Fr.) Lind, cause diseases with similar symptoms and epidemiology, but the pathogens are unrelated.

Damping-off of young citrus seedlings of all species is a common problem in nurseries. It may be caused by *Phytophthora* spp., *Pythium* or *Rhizoctonia solani* Kuhn. Once seedlings pass the two-leaf stage, they are no longer susceptible to this disease. Black root rot, caused by *Thielaviopsis basicola* (Berk. & Br.) Ferr., mostly affects rootstock seedlings in artificial potting mixes in greenhouse-grown nursery trees. It causes a black, dry rot of fibrous roots and is most common in cool, moist soils. Fusarium wilt, caused by *Fusarium oxysporum* (Schlect.) Snyd. & Han. f. sp. *citri* Timmer, is only a problem on Mexican limes grown in artificial soil mixes in the greenhouse.

Nematodes

Nematode problems are often expressed in a manner similar to diseases.

Slow decline
Slow decline is caused by the citrus nematode, *Tylenchulus semipenetrans* Cobb, which is cosmopolitan in citrus orchards. It is referred to as slow decline because it debilitates trees and reduces yields and fruit size. It is usually most serious in areas with fine-textured soils where trees are grown on susceptible rootstocks. Damage from citrus nematode does not produce diagnostic symptoms, but affected trees are usually weak and have thin foliage, small leaves and many dead twigs. Micronutrient deficiency symptoms are common on infested trees, and plants often appear to be slightly wilted or water-stressed. Rootlets appear coarse and soil clings to fibrous roots because of the gelatinous egg masses that are produced by the female nematode.

The citrus nematode is semi-endoparasitic and feeds in the cortex of the fibrous roots. Each female nematode produces from 75 to 100 eggs that hatch in 2–3 weeks. There are several biotypes of this nematode, all of which can attack citrus. Multiplication of citrus nematodes is favoured by fine-textured or organic soils and this nematode is not usually a problem in sandy soils. Population growth is also favoured by high salinity and the nematode increases salt damage to trees.

Resistant rootstocks can be used effectively for control wherever trifoliate orange and its hybrids can be grown. Most other commonly used citrus rootstocks are moderately to highly susceptible. The key to reducing problems with citrus nematodes is to produce nursery stock that is free of the pest and to plant orchards in areas that are free of the nematode. Native soils are usually free of nematodes that attack citrus. Old orchard soils may be freed of nematodes by chemically fumigating or fallowing for 1 to 2 years. Postplant nematicides reduce nematode populations and increase yields and fruit size. However, some of these products have caused environmental problems and microbes rapidly degrade others.

Other nematode problems
Spreading decline, caused by *Radopholus similis* (Cobb) Thorne, is a serious problem of the sandy soils of the Ridge area in Florida. This nematode is an

endoparasite that causes loss of fibrous roots in deeper soil profiles. Other endoparasites such as *Pratylenchulus coffeae* (Zimm.), Filip and Schuur., *P. vulnus* (Allen and Jensen), and *P. brachyurus* (Godfrey) Filip and Schuur. Stek. cause root rot and decline problems in many areas of the world. Sheath nematodes, *Hemicycliophorus* spp., cause localized problems in California and Australia. Sting nematode, *Belonolaimus longicaudatus* Rau, is an ectoparasite and can cause root damage and tree dwarfing in sandy soils in Florida. Many other species of nematodes parasitize citrus but do not cause serious problems in production.

DISEASES OF FRUIT AND FOLIAGE

Canker

Citrus canker is a serious bacterial disease in humid tropical and subtropical areas. The disease causes external blemishes on fruit, making them unsuitable for the fresh market and often causes fruit drop (Koizumi 1985). The disease is widespread in Asia and is spreading in southern South America and in Florida, but is generally absent from areas with Mediterranean climates.

Canker affects young leaves, stems and fruit (Fig. 7.8) of most citrus species producing water-soaked lesions of variable size. As leaf lesions age, they form a raised pustule surrounded by a chlorotic halo. Citrus canker is caused by the bacterium *Xanthomonas citri* ssp. *citri* (syn. *X. axonopodis* [Hasse] Vaut.; *X. campestris* pv. *citri* [Hasse] Dye). The Asiatic or A strain seriously affects grapefruit and early oranges and is less severe on late oranges and mandarins. The B strain in southern South America affects primarily lemons, whereas the C strain in Brazil affects Mexican lime. Citrus bacterial spot, caused by *Xanthomonas axonopodis* pv. *citrumelo* (Hasse) Vaut., causes spotting of leaves in citrus nurseries in Florida, but is not a problem in groves.

Fig. 7.8. Lesions of citrus canker (causal bacteria *Xanthomonas citri*) on leaves (A) and fruit (B) (photos supplied by Dr James H. Graham Jr).

Bacteria reproduce extensively only in young lesions on fruit and leaves. Bacteria ooze out of such lesions and are dispersed primarily by wind-blown rain. The disease is dependent on storms and wind-blown rain not only for dispersal, but also to force the bacteria into wounds and stomata (Graham *et al.*, 1992). This disease is most serious in areas with severe thunderstorms, hurricanes and typhoons in the summer. The bacterium survives for relatively short periods outside host tissue. The presence of citrus leaf miner exacerbates canker because tunnels provide entry points for the bacterium and expose additional tissue in which it can multiply (Fig. 6.2, p. 187).

Citrus canker is controlled by quarantine and eradication in countries from which it is absent or limited to certain areas (Gottwald *et al.*, 2001; Schubert *et al.*, 2001; Graham *et al.*, 2004). Diseased trees are burned in place and the area is kept free of citrus root sprouts for 6–12 months. Equipment and personnel moving from the infested area must be disinfested. In areas where the disease is endemic, windbreaks are quite effective in reducing spread of the disease and in limiting the amount of infection. Copper fungicides are effective in preventing fruit infection if applied frequently, but most other materials tested have not controlled canker well. Scheduling of copper sprays to maintain sufficient residue has been addressed in a copper spray model (Albrigo *et al.*, 2005) and further improvements have been made (http://www.agroclimate.org/tools/cudecay/).

Black Spot

This disease causes a serious external blemish of citrus fruit (Fig. 7.9) that often results in fruit drop. Black spot is widespread in the humid to semiarid citrus-growing areas in the southern hemisphere that have summer rainfall (Peres and Timmer, 2003), and was recently found in Florida (Dewdney *et al.*, 2010). The disease is more serious on lemons and late oranges than on early oranges, grapefruit and tangerines.

Fig. 7.9. Citrus black spot on fruit in Brazil (photo by L.G. Albrigo).

Black spot produces lesions on fruit varying from small brown to black spots to larger sunken lesions referred to as hard spot, virulent spot and false melanose, among others (Kotzé, 1981). Symptoms may appear in the orchard on fruit and cause premature fruit drop or the infections may remain quiescent until harvest. Lesions on leaves are uncommon, except perhaps on lemons, but quiescent infections on leaves are important in the disease cycle.

Black spot is caused by *Guignardia citricarpa* Kiely (anamorph: *Phyllosticta citricarpa* McAlp). This fungus produces pycnidia in lesions on fruit and leaves and ascocarps are formed in decomposing leaves on the orchard floor. Most infections result from ascospores that are dispersed by wind from leaf debris on the orchard floor. Infections usually occur from early to midsummer and may remain quiescent for some time. Rainfall is required for release of ascospores and moisture is essential for infection. Splash-dispersed conidia produced by pycnidia on fruit or dead twigs are important sources of inoculum in humid areas.

Fungicide applications are the primary means of control of black spot. In the more humid areas, three to six preventive sprays with copper, dithiocarbamate or strobilurin fungicides through most of the summer are needed for adequate disease control.

Pseudocercospora Leaf and Fruit Spot

This leaf and fruit spot is a devastating disease that causes considerable leaf and fruit drop and produces blemishes on fruit remaining on the tree. It was first reported in 1952 in Angola and Mozambique and now occurs through most of southern Africa and in Yemen, and is spreading to western Africa. Virtually all citrus species are affected, but damage is less severe on lemons and limes.

Lesions, 1 cm or more in diameter, that usually have light grey centres and a large chlorotic halo are formed on fruit and leaves. The centre of lesions on leaves may drop out causing a shot-hole effect. This disease is caused by *Pseudocercospora angolenis* (T. Carvalho & O. Mendes) Crous &U. Braun, comb. nov. (*synonyms: Phaeoramularia angolensis* [DeCarvalho & Mendez] P.M. Kirk; *Cercospora angolensis* Car. & Men). Hyaline, cylindrical, multiseptate conidia are produced on tufts of conidiophores on leaf lesions. Conidia are wind-dispersed and the disease is most prevalent under warm, humid conditions (Seif and Hillocks, 1997a). Preventive applications of fungicides are the only means of control and copper fungicides, chlorothalonil, flusilazole and propineb are the most effective products (Seif and Hillocks, 1997b).

Greasy Spot

Greasy spot occurs throughout the Caribbean Basin and diseases with similar symptoms have been reported in South America, Japan, Australia and other areas (Mondal and Timmer, 2006). In areas with moist, warm summers, greasy spot causes substantial leaf drop and consequent reductions in yield and fruit size. Greasy spot is more severe on grapefruit, lemons and early oranges than on late oranges and mandarins. Greasy spot rind blotch occurs on grapefruit and reduces the external quality of fruit for the fresh market.

Symptoms on leaves first appear on the underside of the leaf as yellow to tan slightly raised areas. Lesions eventually become brown to black but rarely become necrotic (Fig. 7.10A). Early leaf drop can be severe enough that alternate bearing can occur even at very low yields in tropical climates. Infections on fruit occur through stomata forming minute black lesions. Chlorophyll is retained in areas surrounding these lesions as fruit matures and an unsightly blemish results (Fig. 7.10B). The disease in the Caribbean Basin is caused by the fungus *Mycosphaerella citri* Whiteside. Evaluation of it as a problem can be determined using its behaviour in Florida as a guide (Whiteside, 1981). The other greasy spot-like diseases are probably caused by other species of *Mycosphaerella*, but no comparisons of these fungi have been made.

Most infections are caused by wind-borne ascospores produced on leaf litter on the orchard floor. Spores germinate and epiphytic growth develops on the underside of the leaf. Hyphae penetrate through stomata and spread slowly in the leaf and produce symptoms 4–6 months after infection. Conidia are produced from epiphytic mycelia on the underside of the leaf, but are considered of minor importance for infection. Rainfall is important for spore release and warm, humid nights are necessary for epiphytic growth and infection by the fungus.

The disease is controlled by fungicide applications in early to midsummer. Copper products, petroleum oil, sterol-biosynthesis-inhibiting fungicides and strobilurin fungicides are all effective. Sprays are timed to coincide with the epiphytic growth of the pathogen prior to penetration of the leaf.

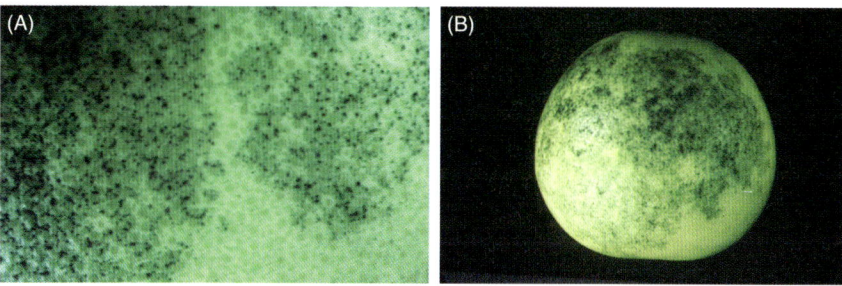

Fig. 7.10. Appearance of greasy spot fungal disease on fruit, (A) closeup on (B) whole fruit. See text for descriptions.

Melanose

External blemishes of melanose on fruit reduce its value for the fresh market, but usually do not affect yield. Melanose is an important disease of fruit produced for the fresh market in humid subtropical areas, but is not of major concern in Mediterranean climates or in high rainfall tropical areas.

Melanose appears as raised, brick-red-to-brown pustules on the leaves, twigs and fruit. Spores carried down the side of the fruit by water may cause lesions to form in a tearstain, droplet or mud cake pattern. Melanose is caused by *Diaporthe citri* Wolf (anamorph: *Phomopsis citri* Fawc.). The sexual stage contributes little to disease severity, but airborne ascospores are responsible for long-distance spread. The disease is most severe on lemons and grapefruit, but also affects oranges and mandarins. Pycnidia of the fungus are produced on dead twigs and asexual spores from those structures are dispersed by rain-splash and washed down over the fruit and leaves. Relatively long periods of wetting (12–18 hours) are required for infection even at favourable temperatures. Dead twigs decay rapidly in tropical, high rainfall areas and serve as a source of inoculum for only brief periods.

Copper fungicides are the most widely used products to control melanose because they are highly effective and have a long residual. However, they must be applied frequently when fruit growth is rapid (Albrigo *et al.*, 2005). Copper fungicides can produce necrotic lesions and darken existing blemishes when applied in hot weather. Other fungicides such as the strobilurins are effective, but may have shorter residuals, are more costly, have priority for other diseases and frequent use may lead to fungal resistance.

Postbloom Fruit Drop and Lime Anthracnose

Two diseases caused by *Colletotrichum* spp. affect citrus production in the field: postbloom fruit drop (PFD) and lime anthracnose (Timmer and Brown, 2000). In addition, postharvest anthracnose affects primarily early harvested fruit for the fresh market and is covered in the postharvest decay section. PFD affects all citrus species and causes severe crop losses, especially in humid, tropical areas. Lime anthracnose affects only Mexican (Key) lime and that disease precludes commercial Mexican lime production in humid areas. Both diseases occur primarily in the Americas, but lime anthracnose was also found in Zanzibar.

PFD produces orange-brown lesions on flower petals (Timmer *et al.*, 1994; Peres *et al.*, 2005). Fruitlets of affected flowers abscise leaving persistent buttons consisting of the peduncle, calyx and floral disc that are diagnostic for the disease. 'Tahiti' limes and lemons, as well as navel, 'Natal', and 'Valencia' oranges, are the most susceptible types of citrus. Lime anthracnose infects flowers, young leaves and fruit, and lesions range from small spots to large expanding lesions. Leaves and fruit often abscise and twigs are killed resulting in shoot dieback. PFD and lime anthracnose are caused by strains of *Colletotrichum acutatum* J.H. Simmonds that differ in pathogenicity and genetically, but are similar morphologically.

Conidia are produced in acervuli on petals in the case of PFD and on all tissues in the case of lime anthracnose. Spores are splash-dispersed and susceptible tissues are infected rapidly under moist conditions. Conidia that are splash-dispersed to mature tissues germinate to form appressoria that serve as survival structures. PFD is most severe in areas where flowering occurs more than once a year or is prolonged and rains coincide with the flowering periods.

Fungicide application may be needed during bloom to control PFD. A computer-assisted decision system, PFD-FAD (Peres *et al.*, 2002), was developed to aid in timing fungicide applications, and is widely applicable in most citrus areas. Benzimidazole (Benlate form not available) and the strobilurin fungicides are effective and other products such as captan, maneb and ferbam provide some control. Lime anthracnose is difficult to control in humid areas, but applications of benzimidazole fungicides or mancozeb are somewhat effective.

Scab Diseases

Scab diseases affect only the external quality of the fruit of susceptible citrus and are important primarily on fruit that are grown for the fresh market. Sweet orange scab primarily affects the fruit of sweet oranges and mandarins and occurs mostly in southern South America. Citrus scab affects leaves and fruit of many mandarins and their hybrids, lemons and grapefruit, and occurs in most areas where conditions are favourable.

The first symptoms of citrus scab are clear to slightly pink, water-soaked areas on leaves or fruit. These rapidly grow into raised pustules that become warty and grey with age. Lesions on fruit tend to flatten with age, especially on grapefruit, and lesions of sweet orange scab tend to be flatter than those of citrus scab. Citrus scab is caused by *Elsinoë fawcettii* Bitancourt and Jenkins, whose anamorph is *S. fawcettii* Jenkins. At least six pathotypes have been described based on host range. Sweet orange scab is caused by *E. australis* Bitancourt and Jenkins (anamorph: *Sphaceloma australis* Bitancourt and Jenkins). Another pathotype of sweet orange scab which affects natsudaidai occurs in South Korea (Hyun *et al.*, 2009).

All infections by *E. fawcettii* originate from conidia that are produced directly on scab lesions. Brief periods of wetting trigger production of hyaline conidia, which are spread by rain splash. Only a brief period of wetting (4 to 6 hours) is required for infection. Spindle-shaped, coloured conidia are also produced and can be airborne for short distances. The cycle is similar for *E. australis*, but this species does not produce spindle-shaped conidia. Ascospores may play a role since this species does not affect leaves, and there would appear to be no source of inoculum once fruit have been harvested. Fruit are susceptible to scab until they reach about 3 cm in diameter. Fungicide applications during this period are effective in controlling the disease. The most effective products include the sterol-biosynthesis-inhibiting fungicides, ferbam, benzimidazoles, strobilurins and copper materials.

Alternaria Brown Spot

Alternaria brown spot was first described in Australia in 1903. It occurs in the Mediterranean area, the Caribbean Basin, in South Africa, much of South America and probably in many other areas where susceptible cultivars are grown and environmental conditions are favourable (Timmer *et al.*, 2003b). It causes serious losses of susceptible tangerine and tangerine hybrids. A similar leaf and fruit spot affects rough lemon and 'Rangpur'.

The disease affects young leaves, twigs and fruit and produces brown-to-black lesions which vary in size from small dots to large, expanding lesions. A toxin produced by the fungus (Kohmoto *et al.*, 1991) is responsible for the symptoms and may kill large areas of leaf tissue or produce necrosis of the veins. Diseased fruit may abscise and lesions on remaining fruit vary from small, black spots to larger pockmarks (Fig. 7.11).

Alternaria brown spot is caused by a host-specific, small-spored species, *Alternaria alternata*, Fr.:Fr. (Keissler) (Akimitsu *et al.*, 2003). Isolates from tangerines and their hybrids do not affect rough lemon and most isolates from rough lemon do not affect tangerines. *A. alternata* produces thick-walled, pigmented conidia that are dispersed by wind. Conidia are produced under moist conditions on lesions on attached or recently fallen mature leaves. Moderate to high temperatures and rainfall favour disease development, but since dew is sufficient for infection, small amounts of fruit blemishes occur even in semiarid areas, such as Israel or Spain, where little rainfall occurs after flowering.

Minimizing the period of leaf wetness of the tree canopy can reduce disease incidence. Nursery trees free of the disease must be used for new plantings

Fig. 7.11. Alternaria brown spot on a 'Minneola' mandarin fruit in Florida.

and overhead irrigation must be avoided. Excessive nitrogen fertilization and irrigation that produce abundant growth flushes promote the disease. Foliar fungicide applications are needed in most affected orchards with the frequency based on disease severity. Iprodione, maneb, strobilurin and copper fungicides are the most commonly used spray products. A predictive system, the Alter-Rater (Timmer *et al.*, 2003b), has been developed in Florida and is helpful in decisions on timing of fungicide sprays.

Other Diseases of Fruit and Foliage

There are several leaf and twig blights that occur under moist conditions. Pink disease and thread blight, caused by *Corticium* spp., occurs in the humid tropics and can kill twigs, leaves and even large branches. It is most severe on cultivars with very dense canopies, which maintain higher moisture levels. Areolate leaf spot, caused by *Pellicularia filamentosa* (Pat.) Rogers, produces spreading lesions on young leaves in the humid tropics. In cool, moist climates, *Sclerotinia sclerotiorum* (Lib.) de Bary causes a twig blight of lemons. *Botrytis cinerea* Pers. ex Fr. can affect lemon flowers under cool, moist conditions leaving scars and ridges on the young fruit. Septoria spot, caused by *Septoria citri* Pass, produces small, necrotic spots on leaves and fruit. It usually occurs on tissues exposed to dews and fogs under cool temperatures in Mediterranean climates. Bacterial blast, caused by *Pseudomonas syringae* van Hall, produces leaf lesions and shoot blight under cool, moist conditions.

POSTHARVEST DECAYS

Penicillium spp. causes the most significant postharvest problems in areas with Mediterranean climates and important problems in virtually all other citrus areas. The causal species are cosmopolitan and cause millions of dollars in losses every year. Initial symptoms are soft, watery areas on the fruit surface. The spots grow and gradually become covered with white mycelium of the pathogen. If the decay is caused by green mould, the affected area is somewhat irregular and covered with green spores with a narrow, white mycelial border. With blue mould, the affected area expands more slowly, is covered by blue spores and surrounded by a broader, white margin. Symptoms of whisker mould are similar to those affected by blue mould, but develop more slowly and may have concentric circles of coremia. Penicillium decays cause fruit to become mushy and to disintegrate under humid conditions. Green mould is caused by *Penicillium digitatum* Sacc., blue mould by *P. italicum* Wehmer and whisker mould by *P. ulaiense* Hsieh, Su and Tzean. Conidia are produced in abundance in chains on brush-like conidiophores on the surface of rotted fruit. *Penicillium* spp. penetrate through wounds and decay progresses most rapidly at 22–27°C. An important step in decay control is to avoid damage to fruit during harvesting and handling. Since these fungi sporulate abundantly, sanitary measures are essential. Rotted fruit must be removed from the packinghouse area and equipment disinfested with chlorine or quaternary ammonium products. Blue mould can spread by contact in packed cartons, but is not as prevalent as green mould. Postharvest fungicides that are effective against these decays include benzimidazole compounds, imazalil, prochloraz and guazatine. Refrigeration of packed fruit is important in slowing decay and preventing spread in cartons.

There are three important stem-end rots: Diplodia stem-end rot and Phomopsis stem-end rot, which are most important in humid areas, and Alternaria black rot, which is more prevalent in drier regions. The fungal species that are involved are common in many citrus areas and disease development is restricted primarily by environmental conditions. All stem-end rots produce rather firm brown decays that usually proceed from the button end of the fruit down the centre columella and the peel. Diplodia stem-end rot progresses rapidly at the junctures between fruit segments, whereas Phomopsis stem-end rot progresses more uniformly. The decayed portion of the fruit surface is depressed with Phomopsis rot, but not with Diplodia rot. Alternaria black rot progresses down the core from the stem or stylar end, often with no external symptoms. Affected fruit develop colour prematurely and when cut reveal a dark black, rotted core. *Botrysphaeria rhodina* (Cooke) Arx (anamorph: *Lasiodiplodia theobromae* [Pat.] Griffon & Manbl.) causes Diplodia stem-end rot, and *Alternaria alternata* Ell. and Pierce causes Alternaria black rot. The causal agent of Phomopsis stem-end rot, *Diaporthe citri*, also causes melanose. All of the causal agents of these decays are common saprophytic fungi in citrus orchards. Spores of these organisms germinate and form quiescent infections on the buttons of the fruit. After the fruit are harvested and buttons senesce, the fungi grow into and rot the fruit. Application of benzimidazole fungicide prior to harvest or in the packinghouse is effective in controlling Diplodia and Phomopsis stem-end rots, but not Alternaria black rot. Application of 2,4-D to harvested fruit helps delay button senescence and slows development of black rot.

Black rot occurrence is often a problem after a freeze: Decay starts after freeze injury but yellow colouring from ethylene generated from injury is not seen for 2–4 weeks. The drop wave of this fruit occurs 3–4 weeks after freeze. Fresh fruit harvest is usually delayed for this time period until these fruit drop. Black rot often shows up after long-term storage, particularly in oranges. It is not easily detected by sight. Because of black rot, freeze damaged fruit should be harvested for processing immediately, within 2 weeks, or as for fresh fruit not harvested until black rot affected fruit has dropped from the tree. Anthracnose symptoms are often associated with bruised or injured fruit where it produces large brown to black spots. The causal agent, *Colletotrichum gloeosporioides* (Penz.), Penz & Sacc., is a common saprophyte in citrus plantings. It forms latent infections on leaf and fruit tissues and quickly colonizes senescent or injured tissues. In early harvested fruit, especially those de-greened with ethylene, the fungus invades peel, turning it first silver grey and later becoming brown to brownish black. Preharvest sprays of benzimidazole fungicide greatly reduce anthracnose incidence postharvest.

Sour rot is a common problem in most citrus areas of the world. Fully mature or overripe fruit are more frequently attacked than less mature fruit. The problem is common on stored fruit. Lesions begin as moist, light-coloured areas on the surface of the fruit. Fruit rapidly decomposes into a watery, slimy mass of peel and sections. Under moist conditions, the rotted fruit are covered with a white layer of mycelium. Sour rot is caused by *Galactomyces citri-aurantii* Butler and Peterson (anamorph: *Geotrichum citri-aurantii* [Ferraris] Butler). Arthrospores of the fungus are formed by segmentation of the mycelium of the fungus. The pathogen is a common soil inhabitant in citrus orchards that is carried with soil on fruit, harvesting bags, boxes and equipment. Since infection occurs through wounds, it is important to handle fruit carefully and not pick up any fruit that fall to the ground. Equipment and packing areas should be maintained free of soil and plant debris. None of the commonly used postharvest fungicides are highly effective for control of sour rot, but washing harvested fruit with sodium o-phenylphenate reduces disease incidence.

Phytophthora brown rot was addressed as a field problem earlier in the section on soil-borne diseases. As a postharvest problem, it occurs sporadically depending on its incidence in the field. None of the commonly used packinghouse fungicides controls the disease. Affected fruit must be eliminated prior to packing by delaying harvest to allow diseased fruit to fall and by culling in the packinghouse. Refrigeration of fruit slows brown rot development and prevents its spread to other fruit.

CONTROL MEASURES

Budwood Certification, Shoot-tip Grafting, Thermotherapy and Detection of Systemic Pathogens

One of the best means of control of systemic, graft-transmissible pathogens is the use of budwood free of pathogens. Many of these diseases are not spread naturally or are spread only slowly by their vectors. The ability to rapidly and accurately detect these pathogens is critical to most types of control efforts. Originally, identification was based on symptoms on field trees or symptoms in graft-inoculated indicator plants that were chosen for their ability to express diagnostic symptoms. Prior to biological indexing for citrus exocortis viroid on Etrog citron, it was recommended that budwood only be taken from trees >10 years old on trifoliate orange that had no symptoms of bark scaling on the rootstock or dwarfing of the tree. Biological indexing for cachexia-xyloporosis was based on the observation of stem symptoms on 'Orlando' tangelo, 'Parson's Special' mandarin, or 'Ellendale' tangor. Inoculated indicators had to be kept at 27–32°C for 3–6 months and 9–18 months for good symptom expression of exocortis and cachexia-xyloporosis, respectively. Etrog citron is widely used for rapid detection of citrus CEV and other viroids, but does not react strongly to all viroids. The use of citrus indicators is still the only method to detect uncharacterized virus-like pathogens such as the concave gum agent, and is also the the most reliable method to identify specific pathotypes of characterized pathogens such as CTV, although progress is being made toward molecular characterization of pathotypes (Hilf *et al.*, 2005; Dawson *et al.*, 2013). Electron microscopy has been used to detect CTV, and when grids are sensitized with virus-specific antibodies, can provide a rapid and sensitive, if laborious, method for identification. Inclusion bodies formed by CTV are also easily visualized by light microscopy in sections of bark or petiole phloem stained with azure A.

Purification and characterization of several citrus viruses and culture of some systemic bacteria has allowed specific antisera to be produced and serological assays to be developed. ELISA is used routinely for several citrus viruses, including CTV and SDV, and it provides a rapid and sensitive assay for viruses for which good antisera are available (Bar-Joseph *et al.*, 1979). Tissue imprint assays allow rapid and sensitive serological detection of viruses with simple equipment (Korkmaz *et al.*, 2008).

All citrus viroids replicate in Etrog citron even when plants are incubated at suboptimal temperatures for symptom expression. Consequently, citron was used as an amplification host, followed by sequential polyacrylamide gel electrophoresis (sPAGE) of RNA extracts for viroid detection. As sequence information has been obtained for different viroids, viruses, as well as for some uncultured bacteria and phytoplasma, conventional and real-time PCR assays have been developed for their detection. PCR is more sensitive than ELISA and has the potential to differentiate strains of a single pathogen that would be difficult to separate by ELISA. PCR is currently feasible for CTV, SDV, TLV and CVV, as well as several prokaryotic pathogens including HLB bacteria and citrus viroids.

Since most systemic pathogens do not invade meristems of shoots immediately after formation, it is possible to eliminate them from infected material by shoot-tip grafting *in vitro*. In this procedure, meristems of less then 0.2 mm are removed from rapidly growing shoots and grafted on a rootstock seedling. Many plants produced in this manner are pathogen-free and maintain the characteristics of the mother plant. Some viruses such as tristeza and psorosis are heat sensitive and can be eliminated by growing plants at high day and night temperatures and propagating buds from new shoots produced under these conditions. Often, it is more effective to use thermotherapy in combination with shoot-tip grafting.

Resistant Varieties

Natural plant resistance is very useful in control of citrus diseases. If rootstock resistance is helpful in controlling a particular disease, it can often be employed immediately since the marketability of the fruit is not relevant. Sour orange rootstock was widely used until decline-inducing strains of CTV invaded many citrus areas. Now, other rootstocks have been adopted. Some fruit quality may have been lost in the change to other stocks, but other rootstocks could be employed immediately. Viroids can be eliminated from budwood by shoot-tip grafting, but there are also many rootstocks that are completely resistant to these pathogens (see viroid section in this chapter).

Resistance to *Phytophthora* root rot and gummosis is considered an essential characteristic of any good rootstock. *Phytophthora* spp. have become widespread in most commercial citrus orchards and, thus, susceptible rootstocks such as sweet orange have very limited utility. Trifoliate orange has been a major source of resistance against *Phytophthora*. However, trifoliate and its hybrids are quite susceptible to *P. palmivora* and thus the rootstock selection may depend on the species present in a given location. Additionally, trifoliate rootstock resistance apparently does not operate if the rootstock is weakened from another stress factor. Feeding by root weevils and termites, or if affected by HLB, may cancel the *Phytophthora* resistance.

The availability of resistance from natural sources for most foliar fungal and bacterial diseases is limited. Many tangerine cultivars and hybrids are resistant to citrus scab and Alternaria brown spot. However, those cultivars may not have the appropriate characteristics or may not mature at the best time to meet market needs. Thus, development of cultivars for resistance to diseases of the tree canopy is more difficult than with rootstocks.

Conventional breeding is difficult with citrus because of a high degree of polyembrony and the length of time required for fruiting. However, newer techniques, such as genetic transformation, use of chemicals and radiation to induce mutations, and somaclonal selection and hybridization promise to speed up the process. However, due to the long fruiting cycle, a considerable length of time is still required to evaluate any new plants produced for maintenance of suitable horticultural characteristics.

Regulatory measures are important for preventing movement of pathogens via infected propagation materials. Most citrus-growing countries now restrict importation of citrus germplasm from other countries to avoid the introduction of new pathogens or more virulent strains. Programmes to obtain and promote propagation of pathogen-free budwood for new plantings are especially effective against pathogens that have no vectors and may also be an important component of programmes to manage those that have vectors. Disinfestation of pruning tools is also used to prevent movement of mechanically transmitted pathogens.

Eradication campaigns to eliminate introduced pathogens from other citrus production areas have had mixed results. Citrus canker has been eradicated from Australia, New Zealand and South Africa. A programme is currently underway in São Paulo State, Brazil, but it remains to be seen how successful it will be. Campaigns to eradicate CTV from Israel and Spain, after it was discovered in those areas many years ago, were unsuccessful. The success of any eradication programme depends on early detection of the disease, which is in turn dependent on effective inspection programmes and methods for rapid detection and identification of introduced pathogens.

Other regulatory measures such as the certification of nursery stock as free from foliar or soil-borne pathogens have been successful. A programme in Florida to certify stock free of nematodes was highly successful to the extent that nematodes are currently a minimal problem in the state. Programmes for *Phytophthora* in South Africa and California have also been effective in reducing the spread of these pathogens. The current trend to produce citrus nursery stock in enclosures in artificial potting mixes should facilitate the production of healthy nursery trees free of significant pathogens.

Cultural Practices

Modifications of cultural practices are of limited value in controlling most of the systemic pathogens. However, they may be of great value in reducing the impact of soil-borne and foliar pathogens. Improvements in surface and internal soil drainage can substantially reduce losses to Phytophthora foot rot and root rot to the point where chemical treatments may be unnecessary. Switching from overhead irrigation to under-the-tree sprinklers can minimize disease problems from diseases such as citrus scab and PFD. Modifications of tree spacing, proper selection of planting site, and adjustments in pruning and fertilization practices can help reduce the need for fungicide applications for control of Alternaria brown spot. Installation of windbreaks is a highly effective means of reducing citrus canker severity. Use of copper bactericides, tolerant cultivars and windbreaks has allowed continued citrus production where canker is endemic, such as in the state of Parana in southern Brazil, in Uruguay and in Argentina. Thus, changes in cultural practices can sometimes bring about substantial disease control at minimal cost.

Chemical Control

Control of most systemic pathogens with pesticides has been largely unsuccessful. Trunk injections of antibiotics are able to reduce symptoms of bacterial diseases such as HLB and CVC but are usually not economical nor eradicating.

Control of insect vectors has not proven effective with non-persistent viruses. However, where vectors transmit bacteria or spiroplasma in a persistent manner, some benefits are observed from vector control with insecticides. In Brazil, a combination of early detection, rogueing and vector control has been effective in reducing spread of CVC. Use of systemic and protectant insecticides to control the citrus psyllid are essential in any programme to control or minimize the effects of HLB.

Control of soil-borne diseases with chemicals has been useful in many situations. Formerly, many chlorinated hydrocarbon nematicides were used preplant and postplant for nematode control. They were highly effective, but their persistence created environmental problems and most are no longer used. Organophosphates and carbamates are frequently quite toxic to nematodes, but may break down in soil or resistance may develop in the pathogen. For control of Phytophthora diseases, the most commonly used materials are metalaxyl, mefenoxam, fosetyl-Al and the related phosphorous acid compounds. These are very flexible materials since they can be applied to the soil directly through the irrigation system or as trunk sprays, and are systemic. Fosetyl-Al and phosphorous acid materials can also be applied to the foliage since they are basipetally systemic.

Fungicides are commonly used for control of diseases of the tree canopy. Copper-containing products are still the mainstay of the control programme in many areas. Benomyl (Benlate) was widely used, but has been removed from the market. However, other benzimidazole fungicides, such as carbendazim and thiophanate methyl, are still available in many areas. The dithiocarbamate fungicides and the sterol-biosynthesis-inhibiting products are widely available and useful for many disease control needs. Strobilurin fungicides are registered in many countries and are highly active against most citrus diseases. Development of resistance by fungi has been a problem with the benzimidazoles and has also occurred with the strobilurins. Development of resistance management programmes, primarily by rotating the pesticide chemistry used, is essential to maintain the utility of currently used products.

Fruit Quality, Harvesting and Postharvest Technology

Postharvest handling and processing of citrus fruits are economically important final steps in transferring fruit from the orchard to the consumer. Unlike most other fruit crops, a high percentage of the worldwide citrus crop is processed into frozen concentrated or single-strength juice products (see Chapter 1). The USA has the largest per capita consumption of orange juice (about 20 l of single-strength equivalents annually per person from 1984 to 1986). Juice consumption per capita has continued to decline over the last 30 years. Western Europe and Canada are two other primary consumers of orange juice. Of interest, most of Asia and Eastern Europe still consume very small per capita quantities of juice, and remain potentially large untapped markets for citrus juices.

CHARACTERISTICS OF CITRUS FRUIT

As discussed in Chapter 2 (p. 13), citrus fruit are hesperidium berries. The hesperidium berry differs from other true berries such as tomato and grape in having a leathery peel surrounding the edible portion of the fruit. The fruit peel consists of an outer coloured exocarp (flavedo) and an inner white spongy mesocarp (albedo). The edible portion (endocarp) comprises the interior portion of the carpels which expand into segments containing juice vesicles and seeds (Fig. 8.1). Citrus fruits are unique in having juice vesicles (sacs) emanating from the carpellary membranes. The presence of the leathery rind protects the fruit from damage during handling and desiccation during storage, transport and marketing (Albrigo and Carter, 1977).

    Citrus fruits are nonclimacteric and thus lack the dramatic rise in ethylene and respiration typical of climacteric fruits such as apple, and associated with their becoming edible (ripe). Citrus fruits are also low in starch reserves and thus undergo very slow changes in internal quality during storage. Protracted storage does decrease stored acids converting them to sugars and CO_2 used in respiration. Notable exceptions are lemons (Batchelor and Bitters, 1954) and

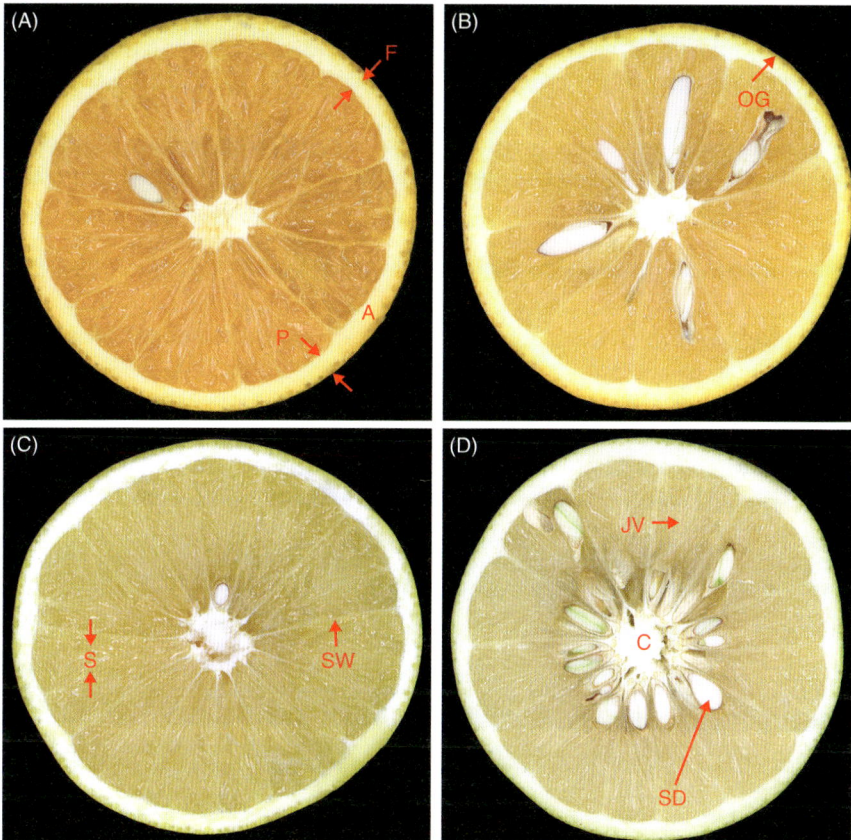

Fig. 8.1. Equatorial sections of orange (A, B) and grapefruit (C, D) and from commercial and seedless (A, C) and seedy (B, D) cultivars. A, albedo; C, core; F, flavedo; JV, juice vesicle; OG, oil gland; P, peel; S, section; SW, section wall; SD, seed.

the 'Palestine' lime (Echeverria and Ismail, 1987), which undergo considerable increases in acid content during curing.

Fruit Composition

The composition of citrus fruits varies with cultivar, climate, rootstock and cultural practices. Most citrus, like other fruits, are primarily water, but also contain over 400 other constituents including moderate levels of carbohydrates, organic acids, amino acids, ascorbic acid and minerals, and small quantities of flavonoids, carotenoids, volatiles and lipids (Erickson, 1968). Citrus fruit are low in proteins and oils. Anatomical work (Burns *et al.*, 1992) has suggested that juice vesicles contain very little oil. Further, they have less than 1 mg g^{-1} of protein on a fresh weight basis (Burns and Baldwin, 1994). Extracted juice may have higher levels coming from other tissues, but citrus fruit are a good source of pectins, vitamin C and roughage.

Besides taste, much of the reason for orange juice consumption is linked to the potential health-related benefits contained in the juice (Nagy and Attaway, 1980). However, currently consumers are interested in foods with low sugar, low fat, high mineral content and high vitamin C. The sugar content issue appears to be a significant reason for the current downturn in orange juice consumption.

Total soluble solids

Total soluble solids (TSS), which include carbohydrates, organic acids, proteins, fats and various minerals, comprise from 10–20% of the fresh weight of the fruit (Erickson, 1968). Carbohydrates account for 70–80% of the TSS in the fruit. The major groups of carbohydrates in citrus fruit include soluble monosaccharides (glucose, fructose) and oligosaccharides (sucrose), and non-soluble polysaccharides (cellulose, starch, hemicelluloses, pectins). Sucrose is the primary non-reducing sugar and is the major translocatable carbohydrate (Table 8.1). Fructose and glucose are the major reducing sugars and are present

Table 8.1. Composition of California 'Valencia' oranges. The values in the first half of the table are expressed in g (100 g^{-1}) while those in the second half are expressed in mg (100 g^{-1}) (Erickson, 1968).

	Peel	Edible portion	Juice
Acid (citric)	0.29	0.75	1.02
Ash	0.78	0.48	0.34
Fat	0.23	0.30	0.29
Moisture	72.52	85.23	87.11
Protein	1.53	1.13	1.00
Sugars			
Reducing	5.56	4.69	4.99
Sucrose	1.99	4.41	4.73
Total	7.55	9.10	9.72
TSS (total soluble solids)	15.69	13.06	12.59
Ascorbic acid	136.5	39.5	43.5
Biotin	0.005	0.001	trace
Calcium	161.0	36.7	9.5
Carotenoids	9.9	3.4	2.8
Iron	0.8	0.8	0.3
Magnesium	22.2	11.5	11.3
Phosphorus	20.8	21.8	19.5
Potassium	212.0	173.0	163.0
Sodium	3.0	1.3	0.7
Sulfur	21.0	11.5	8.5

at about a half to equal quantities of sucrose in most citrus juices. Small quantities of mannose and galactose have also been found in citrus juice. Of the polysaccharides, starch is present in small quantities, particularly as the fruit matures and only before starch is converted to sucrose, fructose and glucose. Pectin is an important polysaccharide in the cell wall matrix (see following section). TSS levels increase as fruit size increases, becoming nearly constant or only increasing slightly during stage III of development (the last 2 to 3 months except in 'Valencia' fruit, particularly growing in Mediterranean climates) (Fig. 8.2A and see Chapter 4).

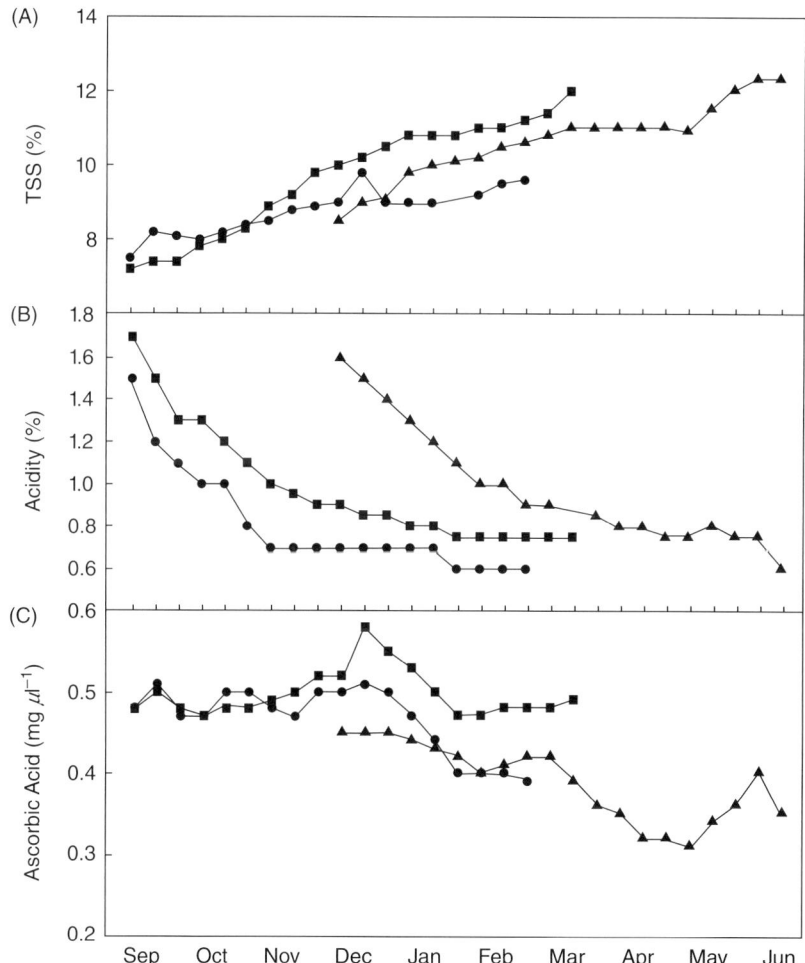

Fig. 8.2. Seasonal changes in juice quality (total soluble solids [TSS], titratable acid [TA] and ascorbic acid) of three sweet orange cultivars. ● 'Hamlin'; ■ 'Pineapple'; ▲ 'Valencia' (Harding et al., 1940).

Organic acids
Total acidity (TA) of citrus juices is an important factor in overall juice quality and in determining time of fruit harvest (Harding *et al.*, 1940). In most citrus-growing regions the ratio of TSS to TA determines whether fruit are harvestable (edible). Organic acids are the primary contributors to overall juice acidity. Citric acid is the principle organic acid (70–90% of total) followed by malic and oxalic acids with lesser amounts of succinic, malonic, quinic, lactic, tartaric and other related acids. Sweet limes have very low acid levels which are mostly malic acid (Echeverria and Ismail, 1987). Organic acid levels generally decrease seasonally as citrus fruits mature (Fig. 8.2B). The rate of decrease in acidity increases with increasing average temperatures during the latter part of fruit development. For example, acid levels decrease much more rapidly in lowland tropical than in subtropical growing regions due to higher mean temperatures (Chapter 4). Higher temperatures increase respiration rates causing less storage of acids in the vacuoles and their more rapid utilization in metabolism. The ratio of TSS:TA increases during maturation and is a good indicator of palatability (see internal quality, p. 251).

Ascorbic acid
It is known that citrus fruit are a valuable source of ascorbic acid (vitamin C). Ascorbic acid functions as a coenzyme and is an essential part of the human diet (Nagy and Attaway, 1980). Levels of ascorbic acid, however, are quite variable among citrus fruit and tend to decrease seasonally (Fig. 8.2C). Ascorbic acid levels are expressed as mg/100 ml juice and range from 18 to 20 in some tangelos to over 70 for 'Pineapple' sweet oranges. As noted in Table 8.1, ascorbic acid levels are generally higher in the peel than the extracted juice.

Limonin
Triterpene derivatives which produce bitter flavours in the juice are present in most citrus cultivars but affect palatability in only a few. Limonin was first identified as an important bitter principle in navel oranges in the late 1930s. This substance and other limonoids are present mainly in the peel and are released into the juice at juicing. Navel oranges and 'Honey' mandarins are particularly high in limonoids ranging from 9–19 and 8–25 mg l^{-1}, respectively. Generally, values of 6 are considered palatable, 9 as bitter and 24–30 as extremely bitter. The range of human detection of limonin in juice is 0.5–32 mg l^{-1} and varies considerably from person to person. Moreover, limonin becomes more difficult to detect as sugar and acid levels increase. Methods have been developed using resin exchange columns and are used commercially to reduce limonoid levels in juice, allowing processors to blend juices typically high in limonin with other juices (Kola *et al.*, 2010). Limonin levels decrease seasonally as fruit are held on the tree. In addition, limonin levels vary with rootstock selection. Lowest levels are normally found in fruit from scions on rough lemon rootstock.

Pectins

Pectins are high molecular weight carbohydrates composed of chains of anhydrogalacturonic linkages. They serve as intercellular bonding materials in many fruits and vegetables including citrus. During maturation of citrus fruit, insoluble pectins are converted to water-soluble pectins and pectinates (Nagy and Attaway, 1980). Total pectic substances decrease in the peel and pulp over the season, and water-soluble pectins increase as percentage of the total pectins. This change in pectin composition signals fruit softening or over-maturity. Pectin levels are generally low in the juice. Pectins are used in the manufacture of jellies, jams, preserves, frozen fruits, and as coatings for meats and in baking.

Internal Fruit Quality Standards

Fruit quality standards, which determine minimum levels of palatability and commercial acceptability, have been established empirically over the years for each citrus-growing region. The flavour and palatability of citrus fruit is a function of relative levels of TSS, TA and the presence or absence of various aromatic or bitter principles. In addition, the juiciness and toughness of the pulp vesicles affects the palatability of fresh citrus eaten out of the hand. Years of research and observation suggest that the ratio of TSS:TA and in particular the attainment of a certain minimum ratio is a reliable and practical means of assessing fruit quality (Harding and Fisher, 1945). Therefore, in many citrus-growing regions of the world, citrus fruit are deemed to be marketable when a minimum TSS:TA ratio is attained. This minimum ratio varies with location and local standards but generally ranges from 7–9:1 for oranges and mandarins to 5–7:1 for grapefruit. Some areas also require a minimum juice content. The TSS:TA ratio is not of importance to lemon or lime producers who harvest fresh fruit based on minimum fruit size and juice content and process lemon or lime fruit on acid and peel oil contents.

Although the TSS:TA ratio is of concern in most regions, the relative levels of TSS (°Brix) also affect palatability. Fruit or juice having high TSS:TA ratios and high Brix taste very sweet, whereas those with low ratios and Brix are tart. Fruit with high ratio and low Brix taste insipid. Many growing districts use a sliding maturity standard that allows a lower ratio (more acidity) as TSS increase.

Fruit palatability is also a function of culture and tradition. For example, the flavour of sweet oranges is accepted over much of the world, and the rich flavour and deep orange colour of mandarin fruit is prized throughout the world. Nevertheless, Asian people tend to prefer less tart mandarin juice and fruit, while Americans and Europeans tend to prefer more tart juice. In contrast, grapefruit are popular in Japan, the USA and northern Europe, whereas the Chinese and Thais prefer the lower acidity and less bitter flavour of pummelos.

Therefore, the intended market, fresh or processed, and the particular ethnic flavour preferences influence quality standard levels.

FRESH FRUIT

Production objectives and technologies differ for fresh compared with processing fruit. Although fresh citrus fruit must meet internal quality standards, emphasis is placed on external appearance and fruit size. Consumers in the largest fresh fruit markets demand oranges with bright orange peel coloration, grapefruit with bright yellow or reddish orange coloration, mandarins with orange to red-orange coloration, yellow lemons and green limes. Fruit must also be within certain guidelines for size and must be relatively free of peel blemishes. For example, small grapefruit the size of a mandarin are unmarketable, while conversely overly large mandarins and oranges are undesirable in most markets. Moderate-sized grapefruit are in demand in the Japanese market, while larger grapefruit are in demand by specialist fruit shippers in the US market. Similarly, consumers more readily accept yellow versus green lemons, although most fruit are harvested green. The European Community (EC) has come to accept only brightly coloured, relatively unblemished fruit that are characteristic of the Mediterranean countries (see Chapter 4). Segments of the US and Central American markets on the contrary will accept fruit with surface blemishes provided the internal quality is acceptable. Therefore, the most important determinants of citrus fresh fruit quality are fruit size and shape, peel colour and peel quality, i.e. the lack of surface blemishes, peel firmness and texture. Degree of seediness is also important in most markets. In general, commercially seedless fruits (averaging fewer than nine seeds per fruit) (Hensz, 1971) are more desirable than seedy fruits – perhaps a seed in every other segment is acceptable in many markets. In the past, the tangerine market in the USA accepted some seediness, while the European markets did not. Now, neither market wants seedy fruit but Brazil still accepts seedy mandarins. In these cases, availability of seedless fruit determines market acceptance. Lemon and lime markets accept moderate seediness because fruit are usually sliced and squeezed, although the presence of seeds can be annoying when preparing freshly squeezed juice. 'Persian' or 'Tahiti' lime fruit are preferred over Mexican limes due to their seedlessness (see Chapter 2).

Factors Affecting Fruit Size and Shape

Fruit size is a function of several factors including cultivar, rootstock, crop load and cultural practices such as irrigation and nutrition. For example, grapefruit are inherently larger than oranges which are generally larger than mandarins, lemons or limes, although certainly size overlap exists. Consequently, market demands will differ for each species. Most cultivars must be larger than a certain minimum size to be considered marketable. This is especially true for mandarins and mandarin-hybrids that tend to be inherently small, and for lemons and limes which are harvested primarily on fruit size and juice content. Extensive marketing strategies have been developed to sell small mandarins, typical of most seedless cultivars, particularly clementines. Names like 'Cuties', 'Pixies', etc. have been trademarked as part of their marketing strategy.

Fruit size may be improved through choice of rootstock (see Chapter 3). Generally, lemon-type rootstocks or those which impart vigour to the scion will produce larger fruit than less vigorous rootstocks such as sour orange or *Poncirus trifoliata*. This effect is more pronounced for mandarins and sweet oranges; however, in some instances mandarins become overly large with puffy peels. Fruit with loose or puffy peels ship and store poorly and are more difficult to handle and pack than those with firm peel structure.

Crop load has a significant impact on fruit size, especially for mandarins and mandarin-hybrids which tend to alternate bearing. Fruit size in the 'on' year in particular may be unacceptably small for commercial use. There are several methods of regulating crop load in mandarins: two that are most often used are growth regulators and pruning (see Chapter 5). These methods reduce the crop load in the 'on' year, thus stimulating more fruit in the 'off' year and balancing production and fruit size from year to year. From mid-flower induction period until its end, GA can be applied to revert some of the flower buds to vegetative buds (Monselise and Halevy, 1964, see Chapter 4). Growth regulators like NAA or ethephon (not registered in some locations) are applied at the beginning of the second fruit drop period after bloom (physiological drop) when fruit diameter of oranges is about 1.5–2 cm and mandarins are less than 1.5 cm (Hield *et al.*, 1962; Wheaton and Stewart, 1973). These growth regulators ideally remove about 25–30% of the crop thus allowing the remaining fruits to have higher leaf:fruit ratios and ultimately larger size. Hand-thinning can also be applied but labour is usually not available to do large numbers of trees. Some work is being done to develop mechanical thinning machines that may work on flowers or small fruit (Schupp *et al.*, 2008). The increase in economic returns occurs because fruit below minimum size standards are unmarketable. Hand-pruning or mechanical hedging and topping also removes fruit in the 'on' year and is an effective means of balancing the crop load for many mandarin cultivars. Trees are usually hedged and/or topped before they bloom but after flower bud formation, thus reducing flower number. Newly-formed shoots arising from the pruned branches are vegetative but will form flowering shoots in the following season. Alternatively, pruning can be done after the postbloom fruit drop to adjust crop load. This method also reduces leaf area, which is usually replaced by a strong vegetative flush following pruning.

Cultural practices such as irrigation, nutrition and pruning also influence fruit size but may dilute TSS (see Chapter 5). Deficient or excessive nitrogen and deficient potassium levels limit fruit size. Hand-pruning is widely used in areas specializing in fresh fruit production such as Spain, Japan and China. Shoots are removed from the centre of the tree and in some cases fruiting shoots are clipped immediately following harvest. Hand-pruning also removes deadwood which is a source of inoculum for various diseases and which may cause limb punctures of the fruit. Hand-pruning improves fruit size, colour and distribution within the tree, especially for mandarins such as satsuma and clementine cultivars.

Variation in shape is usually not of major concern for fresh fruit producers. With the exception of huanglongbing (HLB) infection, fruit shape is not influenced by seed abortion as is the case in apples where misshapen fruit are produced where seeds fail to develop. Fruit shape does vary, however, with climatic conditions. For example, grapefruit under humid subtropical or tropical conditions attain a favourable oblate shape, while those developing under arid conditions become more spherical. Off or late bloom fruit also tend to develop a pronounced neck at the stem end of the fruit termed 'sheepnosing'. Fruit with this characteristic often fail to meet consumer demands. Both climate and nutrition (K levels) appear to influence sheepnosing in grapefruit (Syvertsen et al., 2005). Similarly, mandarins produced under very vigorous growing conditions (lowland tropical) tend to have a puffy peel and often are atypical of the cultivar produced under more moderate climatic conditions.

Factors Affecting Fruit Colour

Under some circumstances, usually high growth temperatures near maturity, postharvest improvement of citrus fruit colour is necessary. From a practical standpoint, attainment of minimum peel colour standards is generally not a problem because fruit may be de-greened postharvest using ethylene gas. Fruit must have attained a minimum level of colour, however, before de-greening will be successful. For example, oranges developing in low tropical regions cannot be successfully de-greened. De-greening of citrus fruit was first observed during railcar shipment of fruit during the 1920s in the USA. Fruit shipped in railcars containing kerosene heaters were observed to become less green during shipment. It was subsequently found that acetylene produced by the heaters promoted colour development in the peel. Ethylene, which is structurally similar to acetylene, also induces the degradation of chlorophyll in the peel and in some cases at prolonged cool temperature an enhancement of carotenoid synthesis occurs (Stewart and Wheaton, 1972). Generally, de-greening produces the characteristic colour change from green to a lemon yellow in grapefruit and early oranges. After some natural colour development has occurred, the orange undercolour becomes prominent as the chlorophyll is degraded.

From these discoveries, systems have been developed to de-green fruit before sending them through the packing line (Grierson et al., 1978). The exact temperature, relative humidity (RH), ethylene concentrations and temporal conditions for de-greening vary with geographic location, cultivar and stage of colour development before de-greening. Optimum de-greening temperatures range from 25–29°C, with considerably less favourable results occurring at temperatures <20° or >35°C. At 15°C, chlorophyll degradation is slowed and at 35°C carotenoid synthesis is severely impeded. Generally, relative humidity should be as high as possible, in the range of 90–95% to prevent desiccation and peel breakdown. High humidity also permits healing of minor wounds in the peel that may have developed preharvest or due to rough handling during harvesting. A fungicide drench, usually thiabendazole at 1000 mg l^{-1} is often used before de-greening to reduce decay development (Brown *et al.*, 1988).

Since ethylene gas is a naturally occurring, endogenously produced plant growth regulator, it is effective in de-greening at relatively low concentrations, between 1 and 10 μl l^{-1}. Durations of de-greening vary with stage of fruit development and cultivars. For example, fruit which have undergone colour break de-green more rapidly than those with a dark green peel. Recommended de-greening durations vary from 24 to not longer than 72 h. Periods longer than 72 h should be avoided because of desiccation, peel injury development and increased decay. Decay incidence may increase at the longer de-greening times even if a fungicide drench is used prior to de-greening. Fruit often continue to de-green after being removed from de-greening rooms. De-greening conditions, i.e. high humidity and temperature and the presence of ethylene, are ideal for development of stem end rot caused by *Diplodia natalensis* (Brown, 1986). Therefore, in regions where this organism is a problem, duration of de-greening should be as short as possible to achieve the desired external appearance. *Colletotrichum* (anthracnose) infection in some mandarin cultivars is stimulated by ethylene de-greening (Brown and Barmore, 1976).

Peel colour is also enhanced by using dyes in some humid subtropical and tropical growing regions where fruit do not attain peel colour typical of the species. Such is the case with peel colour development of some sweet oranges in Florida. Fruit are flooded in conveyor tanks with food grade red dyes for 1.5–4 minutes at 46–48°C depending on cultivar. Colour-adding is done only on oranges, 'Temple' oranges and tangelos after de-greening, but this practice is applied by only a few packinghouses, at least in Florida, and to a limited amount of fruit.

Factors Affecting Fruit Blemishes

Fruit blemishes are of major concern to fresh fruit growers and marketers. In most regions, fresh fruit grades are based on the extent of peel damage attributable to blemishes. Moreover, considerable time and expense are spent on controlling sources or causal organisms of peel blemishes even though in some cases the internal quality of the fruit is unaffected. In many instances, control is necessary to reduce defoliation and loss of tree vigour. Consumers in most markets consistently indicate that they prefer fruit that are generally free of blemishes with colour typical of the cultivar. As many as 80 different causes for off-grade of citrus are recognized (Albrigo, 1978).

Causes of citrus fruit blemishes
There are numerous biotic and abiotic causes of peel blemishes in citrus, the severity of which varies considerably with growing conditions and geographic location. As stated previously (Chapters 4, 5, 6 and 7), the extent of environmental, production, pest and disease pressures on citrus is usually more severe in humid tropical regions than in humid subtropical regions which in turn have more pressures than cool, semiarid or arid regions. It is impractical to discuss every factor that produces peel blemishes worldwide; however, some of the most widely distributed and economically important types are discussed in the following sections.

Biotic factors
Peel blemishes are caused by a myriad of different insects, mites, fungi, bacterial and other pests (Albrigo, 1978). In some instances the pest merely discolours or disfigures the peel which affects fruit marketability for cosmetic reasons alone. Major insect pests producing blemishes on citrus fruit include scales (soft and armoured), thrips, plant bugs, coffee bean weevils, leaf-miners, crickets, grasshoppers, fruit flies and mites. Armoured scales such as red and purple scale feed on the peel and their armour (carapace) may remain on the fruit even when the insect inside has died. The fruit surface, where the insect is removed by washing during packing, may colour poorly. Thrips produce a distinctive ringing at the stem end of the fruit, but also may cause russeting similar to rust mite damage or scribbling which appears as light brown to grey patches over the entire fruit surface. Plant bugs represent a large group of insects with piercing or sucking mouthparts that puncture the peel, causing a wound response, discolouration and sometimes local de-greening from ethylene production. Often decay starts in the feeding site and leads to eventual fruit abscission. The coffee bean weevil feeds on fruit and deposits larvae in the peel. In more tropical climates, leaf-miners feed on fruit, as well as leaves. Crickets and grasshoppers will also feed on immature fruit, leaving large, deep peel depressions. Fruit flies (Mediterranean, Queensland, Caribbean and Mexican) oviposit into the peel. As the larvae develop and feed on the pulp, the fruit softens, predisposing it to fungal invasion. The growth of fruit fly larvae within the fruit make them undesirable for consumption. Fruit flies in particular have a severe economic impact on citrus production worldwide. Furthermore, areas having fruit flies cannot ship fruit to fly-free citrus-growing areas unless the fruit have received a cold or hot temperature sterilization treatment (Sharp, 1989) or have been fumigated. However, fumigation is not allowed in most instances due to environmental and health concerns. Some countries allow shipment from fly-free

zones within otherwise infested growing districts. Currently, cold sterilization with preconditioning at intermediate temperatures is commonly used (Hatton and Cubbedge, 1982). Experimental evaluations of low dose irradiation and hot water sterilization have been performed. All methods of fruit fly disinfestation can occasionally result in some fruit injury.

Mites, particularly the citrus rust mite (*Phyllocoptruta oleivora*), are a major source of peel blemishes in many citrus-growing regions. Early feeding injury leads to a sharkskin appearance. In contrast, damage in the summer leads to russeting, while late damage after complete fruit development results in dark brown, smooth injury (bronzing) (Albrigo and McCoy, 1974). Extensive rust mite damage causes increased water loss from the fruit and, in some high damage instances, fruit drop. The broad mite (*Hemitarsonemus latus*) can cause fruit injury, particularly in warm climates.

Fungi are also a significant cause of peel blemishes. Major fungal diseases include alternaria brown spot (*Alternaria citri*), scab (*Elsinoë fawcettii*), brown rot (*Phytophthora citrophthora*), greasy spot rind blotch (*Mycosphaerella citri*) and melanose (*Diaporthe citri*) (see Chapter 7). Alternaria brown spot is a particular problem on most mandarin-type fruit. The fungus enters the young fruit peel and produces small black depressions. These eventually develop into characteristic brown pustules. Scab produces raised lesions on the peel but is a problem primarily on 'Temple' orange, 'Minneola' tangelo, lemon and occasionally grapefruit. Greasy spot rind blotch is characterized by darkening of the stomata and pinkish discoloration of regions surrounding the oil glands of the peel. It is primarily a problem for grapefruit. Melanose in contrast, which is also a problem found primarily on grapefruit, produces small, dark, raised lesions on the peel. It is particularly severe where a large amount of deadwood remains in the tree on which the spore-producing pycnidia of the fungus over-winter. Brown rot is manifested as firm brown lesions on the peel. Rain or irrigation splash the spores onto the peel where they germinate; therefore, this problem most often occurs on fruit near the ground.

Citrus fruit are also blemished or misshapen by bacterial diseases, namely citrus canker (causal organism *Xanthomonas campestris* pv. *citri*), citrus variegated chlorosis (causal organism *Xylella fastidiusis* pv. *citri*) and HLB or greening (presumably caused by a fastidious phloem-limited bacterium). Citrus canker is characterized by brownish raised lesions surrounded by a yellow halo. Greening affected fruit may be small, irregularly shaped and often do not colour properly. Citrus variegated chlorosis, caused by a xylem-limited bacterium, also leads to small misshapen fruit that are overly firm and can damage juice extracting equipment (see Chapter 7).

Citrus fruit are occasionally damaged by birds and small mammals. Bird damage appears as large puncture wounds in the peel, differing substantially from much smaller insect-inflicted wounds. Seeded cultivars are prized by tropical birds. Small mammals like rodents gnaw the peel, usually causing fruit drop before the fruit reach the packinghouse.

Abiotic factors
Several abiotic factors are responsible for peel blemishes or defects as well as contributing to internal breakdown of tissues in some instances. These disorders can be further subdivided into those with environmentally related causes such as wind and hail; physiological causes such as creasing, splitting and granulation; mechanical causes such as limb rubs and plugging (rind tearing); or physical and physiological causes combined such as oleocellosis, particularly on navel oranges, and zebra skin which occurs on mandarins (Albrigo, 1978).

Wind is probably the major abiotic factor contributing to peel damage worldwide. Wind causes leaf or limb rubbing abrasions on young fruit before they develop a strong cuticle, until about 2.5 cm in diameter. The rubbing kills the outer cell layers of the peel, producing a wound periderm reaction. The peel does not colour properly as a result and develops a light to dark brown blemish. Properly located windbreaks consisting of trees or in some instances man-made structures of shade cloth or similar material serve to reduce the severity of wind scarring. Windbreaks using trees often compete with adjacent citrus trees for nutrients, water and sunlight and may cause tree stunting. Man-made windbreaks are usually cost-prohibitive and are not widely used except with very high cash value cultivars. Some experimentation is occurring in Florida using psyllid mesh cloth in windbreaks to see if they help exclude psyllids from new plantings. These may provide wind scar protection also.

Hail damage occurs in a limited number of citrus-growing regions on an irregular basis. Depending on the size of the hail, the peel becomes pitted, sometimes with deeply sunken, darkened areas. The only solution to preventing hail damage is to cover the orchards, a solution which is not cost-effective or practical in most cases, although some covered orchards exist in Japan, Korea, Morocco and Italy, often for out-of-season mandarin production as temperature control. Hail causes severe crop losses and even defoliation and limb damage in some growing seasons in many growing regions such as Gayndah, Australia (in 2013), Israel, the Mediterranean region in general and South Africa. Some growers use net covering for protection, but most growers do not.

As most physiological disorders of citrus fruit result from a myriad of environmental and physiological conditions, they are difficult to control. Fruit splitting and creasing involve separations of the flavedo or the albedo. Splitting is usually manifested as a longitudinal fissure beginning at the stylar end of the fruit where the peel is thinnest. Splitting is most severe for thin-peeled cultivars including some mandarins such as 'Ellendale' and 'Murcott' as well as sweet oranges like navels, 'Shamouti', 'Hamlin' and some 'Valencia' cultivars. Splitting incidence varies seasonally and is usually greatest where crop load is heavy, perhaps due to heavy crops resulting in fruit with thinner peel. It also appears more severe on trifoliate-type than rough lemon-type rootstocks. Splitting may result from water or nutritional stresses early in fruit development, although the fruit splitting itself occurs later in development. Splitting is more severe at high temperatures and where intensive irrigation is used or heavy rainfall occurs in late season. However, splitting is not strictly related to fruit–water relations. For example, amount of rainfall was not correlated with splitting of 'Ellendale' mandarin growing in South Africa (Rabe *et al.*, 1990). Splitting severity may be decreased by thinning the crop or by a postbloom gibberellic acid (GA_3) spray. Early season sprays of KNO_3 also have been effective in reducing splitting in some years in several countries, notably Australia, South Africa and Israel.

Creasing involves the separation of the mesocarp (albedo) from the exocarp (flavedo), resulting in sunken areas appearing in the peel. Incidence of creasing also varies seasonally, but appears related to tree and fruit stress early in the season. It is most severe on mandarins and mandarin hybrids and for trifoliate and trifoliate-hybrid rootstocks. Spray applications of GA_3 delay creasing of mandarins and reduce the severity of creasing when applied to navel and 'Valencia' oranges. NAA and GA reduced creasing in navel orange in Israel (Greenberg *et al.*, 2010). Creasing also becomes more apparent during storage for 'Robinson' tangerines. An early season K spray can reduce creasing if tree K levels are low but $(NH_4)_3PO_4$ appeared to be more beneficial than GA_3 or KNO_3 (Monselise *et al.*, 1976).

Granulation involves the hardening of juice vesicles particularly at the stem and stylar end and late in the harvest season. Mandarins, grapefruit and oranges on vigorous rootstock develop this problem. Granulation, unlike splitting and creasing, is not noticeable externally. The problem varies seasonally and is most severe in larger fruit, where a protracted bloom occurs, or following heavy summer rains. Fruit with lower than normal TSS and TA, as in cultivars on more vigorous rootstocks like rough lemon, are more prone to granulation (Erickson, 1968). Grapefruit develop the typical hardened juice vesicles but may have collapsed juice vesicles as well (Hwang *et al.*, 1988). Increases in cell wall development (Burns and Achor, 1989; Hwang *et al.*, 1990) and a higher respiration rate (Burns, 1990) are associated with the hardening of the juice vesicles.

Stylar-end breakdown occurs in 'Tahiti' limes and is associated with excessive fruit turgor. The juice cells in the stylar-end of the fruit swell and burst at high temperature with rough handling making the fruit unmarketable. Waiting until fruit have reached a minimum rind oil release pressure (RORP) and thus harvesting later in the morning helps to reduce losses due to stylar-end breakdown (Davenport *et al.*, 1976). Blossom-end clearing of grapefruit appears to be a similar problem of juice vesicle rupture. Springtime harvest at high temperature and high turgor result in the same problem in the handling of tender grapefruit (Echeverria *et al.*, 1998). Zebra skin of tangerines behaves similarly but is related to massive damage of the oil glands on the raised part of the peel rather than internal juice vesicle damage. This often occurs after running turgid fruit over stiff brushes in the packing line.

POSTHARVEST TECHNOLOGY

Harvesting and Handling

Harvesting and handling costs of citrus fruit often equal or exceed total production costs, at least if costs for HLB treatment are not needed in the production area (see Chapter 1). Improper harvesting and handling may cause extensive fruit damage and immediate decreases in pack-out percentages for fresh fruit and later increases in decay for fresh and processing fruit. Fresh fruit, especially, must be handled carefully to avoid damage or potential for future decay problems during storage or transit.

Harvesting methods
Most of the world's citrus crop is harvested manually using ladders and some type of picking container. Fruit are either snapped from the limb or clipped in some cases to avoid plugging (rind tearing). In most growing areas the picker places the fruit into a canvas or plastic sack which is then emptied into a larger wooden, plastic or metal container. For fresh fruit, particularly mandarins, smaller containers are usually used from picking to the packinghouse (Fig. 8.3A) in order to maintain quality for the final packed carton (Fig. 8.3B). The container is then moved from the orchard to a truck for transport to the packinghouse or dumped from the containers into lift trucks and then into large trailers for transport to a processing plant (Fig. 8.4A). Fruit are transported out of the orchard by these lift trucks, tractor-drawn trailers and, on steep slopes, on sleds or small vehicles on rail systems, as in Japan (Morinaga, *et al.*, 2005), or manually (China).

Pickers are usually paid on a piece-rate basis but under some conditions may work on an hourly rate. There is generally a crew foreman who oversees the harvesting and records the number of containers harvested. The foreman

Fig. 8.3. Typical mandarin orchard before picking (A). (B) Typical packed carton of mandarins for fresh market. (Image (A) courtesy of E. Carlos, Brazil).

Fig. 8.4. (A) Mechanical harvesting equipment associated with a shake and catch frame system for processing oranges. Elevators from catch frame-shakers are just visible at the bottom of image B. (B) Typical semitrailer used for transport of oranges to the processing plant are lined up on right side of photo (photos supplied by Dr. Reza Ehsani, University of California, Merced).

also inspects for improperly harvested fruit, e.g., off size, plugged or fruit harvested with a portion of the limb and leaves still attached. The limb may cause puncturing of the adjacent fruit. Plugging (rind tearing) occurs in mandarin-type fruit that are not clipped, but poor harvesting leads to plugging of all cultivars. In mandarins, economic analysis usually indicates that the higher cost for clipping is justified. Plugging predisposes the fruit to fungal infection and desiccation during transit and packing. Foremen are also responsible for ensuring that the entire crop or the properly sized or coloured portion is harvested from a particular orchard or area of an orchard.

Although most citrus cultivars have a rigid, tough peel, poor or delayed handling in the field has a significant deleterious effect on subsequent fruit quality. Reduced fresh marketing is often related to poorer handling, resulting in unreliable delivery condition of the fruit.

Two mechanical harvesting machines are available in Florida which use a rotating, vibrating tine system. One catches the fruit and the other drops it to the ground for pick-up. The equipment involved with a mechanical harvesting machine is shown in Figure 8.4B. They are not used extensively, partly because yields/hectare have decreased due to HLB disease to the point that it is not economical.

Packinghouse Procedures

Packinghouse layout and design vary from basic sorting of fruit by size or colour in the field to high-speed ultramodern sorting using a computer-controlled vision and weighing line. Nevertheless, most packinghouses contain the same basic operations, although certain specializations may be added to adjust for local conditions. The basic packinghouse is outlined in Fig. 8.5. It consists of a bin drencher and de-greening rooms (where necessary), an initial wet or dry

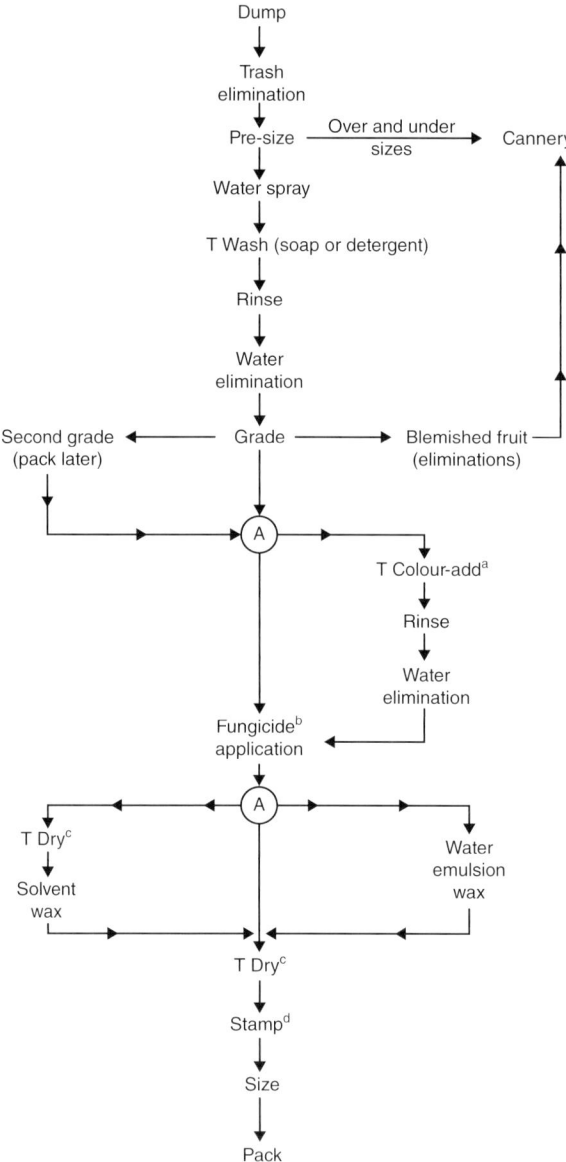

Fig. 8.5. Schematic diagram of a basic packingline layout. Aspects of this design vary with particular local conditions and requirements. (T) Time-dependent operation; (A) Alternate path (Grierson et al. 1978).
[a]Optimal, and only for oranges and tangelos.
[b]Alternative locations for application of fungicides are at the washer, in water emulsion wax or diphenyl pads placed in the packed cartons.
[c]High energy (hot air) drying is prior to solvent wax or following water emulsion wax. Drying following solvent wax is with ambient temperature air.
[d]Stamping can be done prior to a solvent wax, but not prior to a water emulsion wax.

dumping operation, pregrading, washing, grading, drying, fungicide treating, waxing-drying, sizing, packing and shipping functions (Grierson *et al.*, 1978).

De-greening with 5 mg l^{-1} ethylene is used in marginal subtropical areas for early harvested cultivars in order to remove the chlorophyll and allow full visualization of the available carotenoid pigments of the peel (see section on colour, this chapter, p. 254).

Directly from the field or following de-greening, fruit may be dumped into water that usually contains chlorine and a fungicide (wet dumping) or dumped directly onto a conveyor. In many citrus areas fruit are presorted manually to remove debris and fruit with obvious defects. Dry dumping is usually preferred to keep fungal spores from soaking into wounds. Fruit travel over a conveyor and are pre-wetted, then washed with a mild detergent and water spray rinsed as they rotate on revolving brushes. This washing removes dirt and loosely adhering mould (like sooty mould). High pressure nozzles are occasionally used to remove scale insects and strongly adhering sooty mould. Fruit may also receive a fungistatic treatment of sodium orthophenyl phenate (SOPP) at this time. This material settles in wounds and inhibits fungal growth. At this point, fruit are generally graded to eliminate off-grade fruit that will be processed before they receive fungicide and wax treatment. The fewer chemical treatments before separating small fruit and surface defect fruit for processing the better. A fungicide may be applied after this grading stage or it can be incorporated in the wax for application after drying. The fruit are thoroughly dried over absorbent rollers under high-velocity, forced air as it is essential to remove water before waxing. If fungicide is applied separately, some de-watering after washing and rinsing is advisable to avoid diluting the fungicide. Currently approved fungicides and their use are available in Palou (2014) and Ritenour *et al.* (2011). Most commonly used are TBZ and Imazalil, but some newer fungicides, Azoxystrobin, Fludioxonil, and Pyrimethanil may be used to a greater extent in the future.

During washing, some of the natural wax layers are disturbed causing an increase in weight (water) loss. Very little of the natural waxes are removed by normal washing. Wax is applied to the fruit after drying to prevent this desiccation. Many packinghouses apply natural carnauba or synthetic waxes, which are water soluble, but a few use solvent-based synthetic waxes which dry more rapidly than water-based waxes, but require careful venting of the wax drying zone because of the solvents. The wax is applied in a continuous motion via nozzles located above the conveyor or dripped onto the fruit over brushes if a water wax is used. Soft synthetic brushes may be used to polish the fruit before waxing. Wax may include a fungicide at this stage. The wax delays loss of moisture and thereby weight losses during subsequent handling and improves the appearance of the fruit. Fruit are again dried, after waxing, with heat and forced air. It is preferable that the greatest heat be applied primarily from the wet, entrance side of the dryers. The fruit then continues along the conveyor for final grading.

Most grading worldwide is still done visually. Fruit pass along belts where they are visually evaluated for defects. Usually a certain predetermined percentage of the fruit surface area must be free of blemishes. This percentage varies with grade from less than 10% to more than 50%. Although this method of grading appears quite subjective, experienced graders are very effective. Alternatively, some packinghouses have high speed photo-grading systems that can detect off-colour or physical defects. Eliminated fruit usually travel along another conveyor to a trailer which transports them to a juice plant. In dry, subtropical areas as much as 95% of the fruit may be packed if of proper size. In humid subtropical and tropical climates, packout may be as low as 50 or 60% because of blemishes and off-colour. For some markets with strict requirements as little as 30% of the fruit may be of market grade. In locations like Florida, with significant processing capabilities, fruit from probable low packout orchards are sent directly to a processing outlet.

Selected fruit are then sized for packing. The conventional methods of sizing are the roller sizer or the use of different sized openings along the conveyor. The older roller sizers consist of two tapered rollers which are positioned parallel with wider spacing as the fruit moves from the beginning to end of the rollers. Small fruit fall to initial conveyors and packing areas while larger fruit continue along the rollers until the appropriate spacing allows the fruit to drop to other conveyors. The distance between the tapered rollers can be adjusted for each type of citrus to be sized. Newer roller sizers spread the rollers as they move down a track so that the rollers become wider apart as the rollers move to the end of the machine. Fruit are delivered by conveyor to the smaller separation end and the sized fruit drop to conveyors that run perpendicular to the roller separation. With the other mechanical method of sizing, fruit are conveyed over an area with different sized openings corresponding to the various size categories. Today, more packinghouses use computer-controlled sizers that weigh each fruit individually as it travels in a cup. The computer then directs the cup to dump the fruit into the appropriate size (mass) category along the travel path. Some systems can also measure fruit optically. These systems can weight and optically measure volume in order to calculate density. Low density fruit due to freeze, over-maturity, internal drying or other damage can be eliminated.

Each fruit is usually labelled with a sticker, but recently laser printer systems have been introduced that etch the needed information into the peel of the fruit (Sood *et al.*, 2009). Iron oxide or hydroxide spray may be added to enhance the contrast (Laserfood, 2015).

Fruit are packed into several types of containers but the most common are corrugated cardboard cartons, wooden boxes or netted bags placed in cartons. The standard carton varies from region to region but usually contains between 15 and 18 kg of fruit. Netted shipping bags also vary in size from 1.8 to 6.5 kg. This method often results in having to ship excess weight to meet the minimum net weight marked on the bag (Albrigo *et al.*, 1991). Cartons are usually packed manually or sometimes mechanically using a system of suction cups which lifts a predetermined layer of fruit into the carton, placing the fruit into rows. Each carton is then stamped with the number of fruit it contains. This number is inversely related to fruit size. The carton is also stamped with information on types of fungicides used, the cultivar and the name and location of the packer. Some large market chains are now having suppliers use returnable containers to reduce waste recyclable materials. Bar codes, mostly standard today, are often used to identify other characteristics for later tracing of source information.

Storage, Shipping and Storage Related Disorders

Storage

Citrus fruits are nonclimacteric and have low respiration rates. Thus they are quite amenable to long-term storage. Nevertheless, storage conditions are cultivar dependent and fruit quality changes occur during prolonged storage. In some areas of the world citrus fruit are held in common storage during winter, or stored in caves where the temperature remains fairly constant all year round (China). This type of storage is usually for local consumption. Sweet oranges in particular may be stored for 2 months or more at 0–4°C with very little loss of fruit quality (Albrigo and Brown, 1973). Low temperatures depress fruit respiration, water loss and growth of decay organisms. Mandarins can be held at low temperatures, but usually develop too much decay and weight loss if stored for long periods as their cuticle and wax coating are thinner than most other citrus types (Albrigo, 1973). Lemons, limes and grapefruit cannot be stored at temperatures less than 10°C because they develop chilling injury (CI) which is manifested as pitted necrotic regions in the peel. Consequently, lemons are usually stored at 10–12°C and grapefruit at 10–15°C. The susceptibility to CI varies seasonally, however, and may be related to abscisic acid (ABA), proline, reducing sugar content and water loss of the peel (Purvis, 1981, 1989). In Florida, grapefruit harvested before January must be stored at 15°C to avoid CI, whereas fruit harvested after January may be stored at 10°C. ABA levels tend to decrease in the peel during this period and reducing sugar levels increase. No cause and effect relationships have been established between several of these coincident changes with CI. Squalene content in the peel has been related to reduced incidence of CI (Nordby and McDonald, 1990). Holding fruit for a week at a moderately cool temperature (10–16°C) reduces subsequent injury at temperatures as low as 1°C. Temperatures this low are not used for grapefruit, but they will store for relatively long periods if handled carefully and are protected from excessive water loss by film covering (Kawada and Albrigo, 1979; Albrigo and Ismail, 1983).

Lemon fruit are often harvested when light green and cured during storage at 12.5 to 15°C and an RH of 95%. RH should be maintained as high as possible, usually between 85 and 95%, to retard water loss by minimizing the vapour pressure gradients between the fruit and air. High humidity promotes wound healing (Ismail and Brown, 1975) but it also promotes growth of some decay organisms such as *Diplodia* stem-end rot (Brown, 1986).

Controlled atmosphere (CA) storage has modest benefits for citrus fruit storage due to the low respiration rate of citrus but is not recommended due to the limited benefit and high cost. Citrus fruit are generally not stored for as long as apples and common or refrigerated storage is more economical and widely used.

Fruit quality changes during storage and shipment
External and internal changes occur in citrus after it is harvested. Desiccation can lead to unsalable fruit when 5–10% of weight is lost. Natural waxes and the postharvest application of wax influence the rate of water loss. The faster the weight loss, the worse the appearance of the peel at a given weight loss.

Citrus fruit undergo internal quality changes during long-term shipping or storage, but these changes are a function of cultivar and storage conditions. For example, the TSS of 'Marsh' grapefruit did not change during 9 weeks of storage at 15°C, but the TSS of 'Hamlin' orange increased from 8.1 to 9.5 and 'Robinson' tangerine from 10.9 to 12.1 (Fig. 8.6A). Similarly, citric acid levels decreased for 'Hamlin' and 'Robinson' but again remained relatively stable for 'Marsh' (Fig. 8.6B). Consequently, TSS:acid ratio increased for 'Hamlin' and 'Robinson' and remained relatively unchanged for 'Marsh' (Fig 8.6C). 'Palestine' sweet lime ratios decreased, but the values were so high because of the low acidity that flavour was not affected (Fig. 8.6C) (Echeverria and Ismail, 1987).

Fruit water content also decreases during storage and shipment. Use of shrinkwrap plastic films on individual fruit is quite effective in reducing water loss, although gas exchange may also be impeded causing anaerobiosis, ethanol and aldehyde accumulation and off-flavours to develop if fruit are held at high temperatures (>30°C) for extended periods (Albrigo and Ismail, 1983). Waxed fruit also develop off-flavours if stored at high temperatures for too long because of poor gas exchange (Hagenmaier and Shaw, 1992). Poorly handled fruit often develop extensive decay in filmwraps (Albrigo and Ismail, 1983).

Contrary to the behaviour of other citrus, internal quality changes in lemons during curing are quite significant (Fig. 8.7). Juice content increases about 16% primarily from water stored in the peel. Acid content also increases significantly from 0.16 to 0.20 mgl^{-1} (24%) in 4 weeks and peel colour changes from light green to yellow (Batchelor and Bitters, 1954).

Postharvest storage diseases
Fungal diseases are the major source of postharvest fruit damage and losses. Decays usually become evident at the retail and consumer points (Burns and Echeverria, 1988) but are only controlled at the time of harvesting and packing by careful handling and proper use of fungicides. Green (caused by *Penicillium digitatum*)

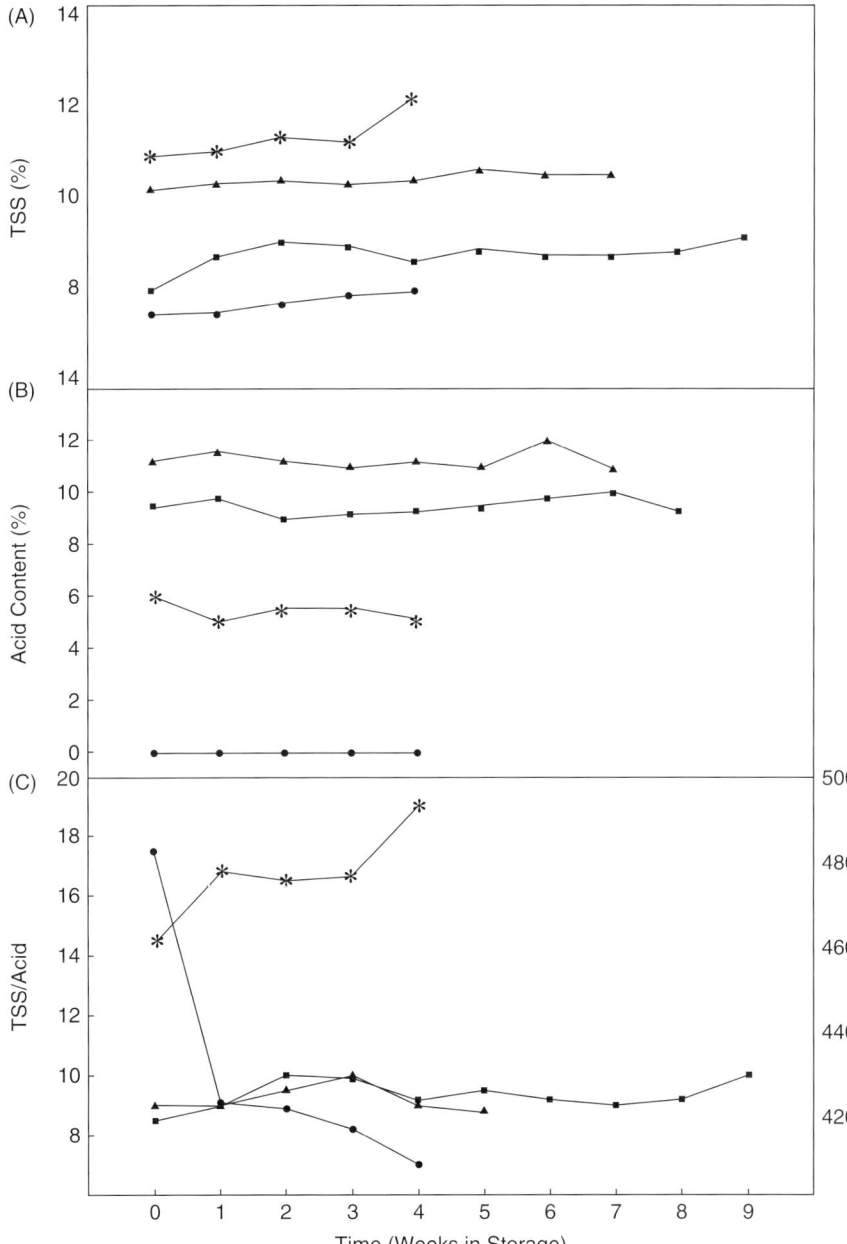

Fig. 8.6. Changes in total soluble solids (TSS) (A), citric acid; (B) and TSS/acid (C) for 'Hamlin' orange (■), 'Marsh' gapefruit (▲), 'Robinson' tangerine (✶) and 'Palestine' sweet lime (●) stored at 15°C (Echeverria and Ismail, 1987).

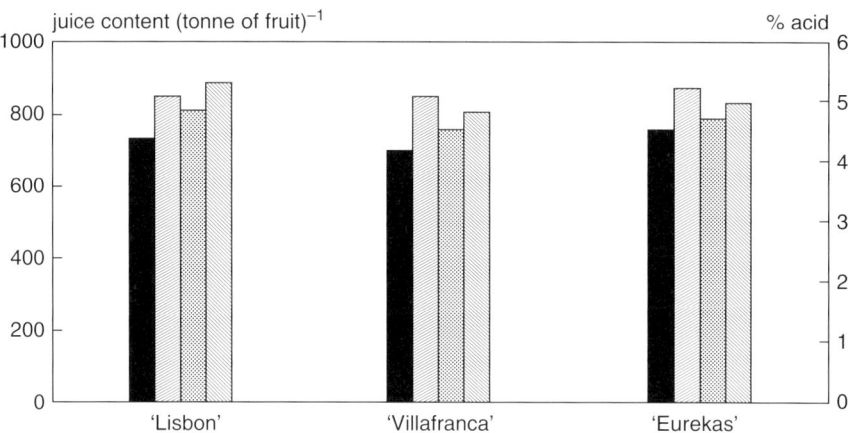

Fig. 8.7. Changes in juice content and acidity of lemons cured for 38 d (Batchelor and Bitters, 1954). ■ Juice (tonne fruit harvested)$^{-1}$; ▨ Percentage acid at harvest; ▩ Juice (tonne fruit cured)$^{-1}$; ▢ Percentage of acid cured.

and blue (caused by *Penicillium italicum*) moulds are major sources of fruit decay in the shipping carton. In the case of blue mould, spread of decay in the carton is by contact with other fruit. Eventually most of the carton becomes decayed. Green mould requires open wounds for decay development. Other common postharvest fungal diseases include stem-end decay (caused by *Physalospora rhodina* or *Diaporthe citri*), sour rot (caused by *Endomyces geotrichum*) and brown rot (caused by *Phytophthora citrophthora*). The latter two also spread by contact. These problems can be controlled by using care during picking and handling, proper fungicides and good sanitary measures in the packinghouse to keep damaged, infected fruit at a minimum. See Chapter 7 for details about citrus fruit decay organisms.

Postharvest nonpathological disorders

Some postharvest disorders are not associated with a pathogen but occur due to handling problems or physiological causes. Oleocellosis occurs when oil cells are ruptured during harvesting or during handling from the field to the packinghouse. The oil from ruptured cells inhibits degreening and discolours the peel or completely kills the tissue affected by the oil, making the fruit unmarketable. Oleocellosis is associated with low rind oil release pressure (RORP) which is a measure of the peel firmness (Eaks, 1969). Lemons and limes are especially susceptible to oleocellosis when harvested below a minimum level of RORP. Therefore, harvest may be delayed until late morning or several hours or days after rainfall until the fruit peel turgor decreases. Some fruit are also harvested based on differences between the fruit temperature and the wet bulb temperature. This method indirectly determines when the vapour pressure gradient between the fruit and air is sufficient to reduce turgor and thereby reduce oleocellosis. The fruit temperature is measured with a thermometer and the wet bulb air temperature with a sling psychrometer. When the fruit temperature exceeds the wet bulb temperature by 2–3°C, harvesting may begin and the incidence of oleocellosis is reduced, but RORP integrates these factors and is easier to use (Eaks, 1969). Careful harvesting and handling also reduces the incidence of oleocellosis. Lemons, which are very susceptible, are often left in the harvesting containers for 12–24 h or overnight before packing. Oleocellosis becomes more visible with advancement of symptoms and during grading can be removed before packing, but if fruit are moved rapidly through the packinghouse oleocellosis may not be detected until the fruit reaches the wholesale or retail dealer.

Fruit harvested late in the season and held in long-term storage even under low temperatures are subject to excessive desiccation, often causing the peel to become darkened, soft and wrinkled or furrowed around the button. This anomaly is called peel ageing (Albrigo and Brown, 1973). Another disorder, stem-end-rind breakdown (SERB), also occurs near the stem end but is characterized more by pitting of the peel except for a narrow healthy zone of peel separating the injury from the button. This healthy zone has a heavier natural wax coating and does not have stomata in its epidermis (Albrigo and Carter, 1977). Filmwraps can reduce water loss to 10 or 50% of waxed fruit while improving gas exchange (Ben-Yehoshua *et al.*, 1979), but this method has not replaced waxing primarily because of economic and application constraints (Albrigo and Ismail, 1983).

Citrus Processing

Large quantities of citrus fruit are processed (see Chapter 1). The major products are frozen concentrate orange juice (FCOJ) and not from concentrate (NFC). Other orange processed products include whole vesicles in juice, squash and canned segments. Grapefruit are processed into the same products. Some mandarins are juiced or canned as segments and lemons and limes are processed into concentrate. Essential oils and pulp are major by-products. The pulp, after drying, is used primarily as cattle feed. Pulp is also processed for pectin (Maya *et al.*, 2014).

Fruit for processing must be wholesome, having no cuts into juice segment areas or decay, and be of suitable quality for product standards. For orange processing, fruit must have TSS, TA and juice content that will produce a high yield and good balance of flavour. Internal orange colour must be sufficient to meet colour score standards or higher coloured juice must be added (blending) to meet these standards (Redd *et al.*, 1986a, 1986b, 1992).

Climate plays an important role in producing high quality juice oranges (see Chapter 4). Marginally humid subtropical climates (e.g. Florida, USA and São Paulo, Brazil) have sufficient cool autumn and winter temperatures to stimulate the development of internal orange colour while suppressing respiratory loss of acidity. But the mild autumn and winter temperatures allow photosynthesis to continue leading to higher TSS. Fruit also tend to be thin skinned and of high juice content, in excess of 50%.

The most desirable fruit for processing contain >12% TSS, a 14–16 TSS:TA ratio and a juice colour score of at least 36, which provides an A-grade FCOJ product (Redd *et al.*, 1986a, 1986b). 'Valencia' oranges easily meet these values, but early season 'Hamlin' fruit do not. New 'Valencia' and other selections mature earlier and provide better juice quality than 'Hamlin' (see Chapter 2). On the other hand, HLB disease results in poorer juice quality for processing (Bassanezi *et al.*, 2009). Grapefruit from marginally tropical climates are desirable for processing because of their high juice content, low acidity and low levels of bitter components, such as naringen.

Processing

Following harvest, fruit are transported to the processing plant where they are weighed and assigned a lot number. A random fruit sample is then taken to determine the TSS, TA and juice content of the fruit. Many citrus processors also test fruit for wholesomeness, namely, bacterial levels in the juice at this time. The percentage juice and TSS determine the kg-solids from which the producer will be paid in areas such as Florida. Alternatively, in other growing areas, producers are paid simply per tonne of fruit. Fruit are then either temporarily stored for subsequent processing or dumped directly onto conveyors. Most processors manually grade fruit, removing damaged or unwholesome fruit at the earliest stage convenient. Even though Spain is primarily a fresh fruit producer they do have some processing plants (Fig. 8.8).

After fruit are washed, juicing is accomplished by sheer press or reamer-type extractors. Reamer-type extractors contain several pairs of cups constructed of stainless steel blades. Fruit halves are lifted into the lower cup which is meshed with the upper cup hydraulically. As the cups converge, the juice and pulp are pressed through a cylindrical strainer that is forced through the centre of the fruit. The remaining peel and rag are ejected for use in cattle feed or other by-products such as pectin. The juice, pulp and oils are then transported to the finisher via pipes. The finishers control the size of pulp particles and amount. These machines combine screening and centrifugal force to separate excess large pulp from the juice and to extract essential oils. Concentrating of citrus juices is done using a TASTE evaporator (thermally accelerated short time evaporator) (Redd *et al.*, 1986a, 1986b, 1992). Many aromatic volatiles are captured during the TASTE evaporation process but are usually added back as a flavour pack to the final product. Typical evaporator capacities range from 20,000 to 400,000 l of water evaporated h^{-1}. The largest plant has over 2,000,000 l h^{-1} of evaporating capacity. This equals about 1000 tonnes of fruit h^{-1}. Juice is usually concentrated to 60–70°Brix for storage and transport.

Fig. 8.8. Fruit ready for processing at a plant in Spain. Note highly coloured fruit typical of a Mediterranean climate.

Storage tank farms are large freezer storage room(s) with 300,000–380,000 l stainless steel tanks connected by stainless steel pipe and pump systems. Smaller blending tanks are also available. Tank farms often have total capacities of 10,000,000 l of concentrate. Some storage occurs in plastic-bag lined 200 l barrels in freezer rooms. The viscous, high Brix juice is transported in these barrels or in stainless steel bulk containers on refrigerated trucks or bulk tank ships, similar to those used for crude oil transport, except for the stainless steel containers. Receiving tank farms, particularly for Brazilian orange juice companies, exist in the USA, Europe and Japan (Albrigo and Behr, 1992). Consumer- or industrial-size packages (containers) are prepared from the bulk juice concentrate out of the tank farms. In many countries, the concentrate is reconstituted to single strength juice for retail sale, blended with other juices or used as a base component for juice-based drinks.

Compared to FCOJ more and more consumers prefer NFC citrus products (FAO, 2016) that are lightly pasteurized to provide stability from microbial growth. They are stored in a low oxygen environment and have a moderate storage life compared to FCOJ. Both products allow blending to provide a more uniform product through the season.

Food Safety

For both fresh and processed citrus, food safety issues have become more important for the industry. European and US markets have both insisted on more careful handling of produce at the fresh and processed level. In the USA the Federal Food, Drug and Cosmetic Act (FFDCA) states that 'Owners, operators, or agents in charge of domestic or foreign facilities that manufacture/process, pack, or hold food for consumption in the U.S. are required to register the facility with the FDA.' The FDA Food Safety Modernization Act (FSMA) amends the FFDCA to identify standards for produce safety under Section 419. The FSMA identifies six issues which will be addressed by standards that consider 'hazards that occur naturally, may be unintentionally introduced, or may be intentionally introduced, including by acts of terrorism. The issues are: 1. soil amendments, 2. hygiene, 3. packaging, 4. temperature controls, 5. animals in the growing area, and 6. water'. Training in and maintenance of food safety standards are usually managed by following procedures in a Hazard Analysis and Critical Control Point (HACCP) programme for either juice or fresh product production. These procedures include field to consumer considerations of food safety. Many countries in Europe also require HACCP-like programmes. Training programmes are usually available at the local industry level. Worldwide, the FAO Codex is a good guide to requirements for international trade.

REFERENCES

Abbott, C.E. 1935. Blossom-bud differentiation in citrus trees. Amer. J. Bot. 22:476–485.

Abkenar, A.A., S. Isshiki, and Y. Tashiro. 2004. Phylogenetic relationships in the "true citrus fruit trees" revealed by PCR-RFLP analysis of cpDNA. Scientia Horticulturae 102(2):233–242.

Achor, D.S., H. Browning, and L.G. Albrigo. 1997. Anatomical and histochemical effects on feeding by citrus leafminer larvae (*Phyllocnistis citrella* Stainton) in citrus leaves. J. Amer. Soc. Hort. Sci. 122: 829–836.

Agusti, M., F. Garcia-Mari and J.L. Guardiola. 1982. Gibberellic acid and fruitset in sweet orange. Scientia Horticulturae 17:257–264.

Akimitsu, K., T.L. Peever, and L.W. Timmer. 2003. Molecular, ecological and evolutionary approaches to understanding Alternaria diseases of citrus. Mol. Plant Pathol. 4:435–446.

Albrigo, L.G. 1973. Some parameters influencing development of surface wax citrus fruits. Proc. Int. Soc. Citriculture 3:107–115.

Albrigo, L.G. 1977. Rootstocks affect 'Valencia' orange fruit quality and water balance. Proc. Int. Soc. Citriculture 1:62–65.

Albrigo, L.G. 1978. Occurrence and identification of preharvest fruit blemishes in Florida citrus orchards. Proc. Fla. State Hort. Soc. 91:78–81.

Albrigo, L.G. 1996. Seasonal leaf area development in citrus-relation to leafminer. Proc. Int. Conf. Managing Citrus Leafminer pg. 70.

Albrigo, L. G. 1999. Effects of foliar applications of urea or Nutriphite on flowering and yields of 'Valencia' orange trees. Proc. Fla. State Hort. Soc. 112:1–4.

Albrigo, L.G. and R.H. Behr. 1992. Distribution and consumption consequences for production and marketing of citrus from the American Continent. Proc. Int. Soc. Citriculture 1992;4:1204–1206.

Albrigo, L.G. and R.F. Brooks. 1977. Penetration of citrus cuticles and cells by citrus snow scale, *Unapsis citri* (Comst.). Proc. Int. Soc. Citriculture 2: 463–467.

Albrigo, L.G. and G. E. Brown. 1973. Storage studies with Valencia oranges. Proc. Int. Soc. Citriculture 3:361–367.

Albrigo, L.G. and R.C. Bullock. 1977. Injury to citrus fruits by leaf footed and citron plant bugs. Proc. Fla. State Hort. Soc. 90:63–67.

Albrigo, L.G. and R.R. Carrera. 2015. Managing drought stress of oranges under São Paulo-Minas Gerais, Brazil conditions to optimize flower bud induction and productivity. Acta Hortic. 1065:1251–1255.

Albrigo, L.G. and R.D. Carter. 1977. Structure of citrus fruits in relation to processing. In: Nagy, S., Shaw, P.E. and Veldhuis, M.K. (eds), Citrus Science and Technology. AVI Publishing Co., Connecticut, pp. 33–73.

Albrigo, L.G. and E.J. Chica. 2011. Citrus shoot age requirement to fulfill flowering potential. Proc. Fla. State Hort. Soc. 124:56–59.

Albrigo, L.G. and M.A. Ismail. 1983. Potential and problems of film-wrapping citrus in Florida. Proc. Fla. State Hort. Soc. 96:329–332.

Albrigo, L.G. and C.W. McCoy. 1974. Characteristic injury by citrus rust mite to orange leaves and fruit. Proc. Fla. State Hort. Soc. 87:48–55.

Albrigo, L.G. and R.V. Russ. 2002. Considerations for improving honeybee pollination of Florida citrus hybrids. Proc. Fla. State Hort. Soc. 115:27–31.

Albrigo, L.G. and J.P. Syvertsen. 2000. Foliar urea as a substitute for soil applied N during establishment of citrus trees. International Citrus Congress (9th: 2000 : Orlando, Florida), 2003. p. 417–420.

Albrigo, L.G. and R.H. Young. 1981. Phloem zinc accumulation in citrus trees affected with blight. HortScience 16:158–160.

Albrigo, L.G., C.C. Childers, and J.P. Syvertsen. 1981. Structural damage to citrus leaves from spider mite feeding. Proc. Int. Soc. Citriculture 2:649–652.

Albrigo, L.G., J.P. Syvertsen, and R.H. Young. 1986. Stress symptoms of citrus trees in successive stages of decline due to blight. J. Amer. Soc. Hort. Sci. 111(3):465–470.

Albrigo, L.G., C.W. McCoy, and D.P.H. Tucker. 1987. Observations of cultural problems with the 'Sunburst' mandarin. Proc. Fla. State Hort. Soc. 100:115–118.

Albrigo, L.G., J.K. Burns, and F.M. Hunt III. 1991. Weight loss considerations in preparing and marketing weight-fill bagged citrus. Proc. Fla. State Hort. Soc. 104:74–77.

Albrigo, L.G., K.S. Derrick, L.W. Timmer, D.P.H Tucker, and J.H. Graham. 1992. Inability to transmit citrus blight by limb grafts. Proceedings of the 12th Conference of the International Organization of Citrus Virologists, pp. 131–138.

Albrigo, L.G., D.S. Achor and R.V. Russ. 2001. Low production of Ambersweet orange in Florida related to poor pollen production and germination. Proc. Fla. State Hort. Soc. 114:127–131.

Albrigo, L.G., J.I. Valiente and H. W. Beck. 2002. Flowering expert system development for a phenology based citrus decision support system. Proc. 6th IS on Modelling in Fruit, ed. T. DeJong, Acta Hort. 584:247–254.

Albrigo, L.G., H.W. Beck, L.W. Timmer and E. Stover. 2005. Development and testing of a recommendation system to schedule copper sprays for citrus disease control. J. ASTM International 2(9), JAI12904:1–12.

Albrigo, L.G., H.W. Beck and J.I. Valiente. 2006a. Testing a flowering expert system for the 'Decision Information System for Citrus'. Acta Hort. 707:17–24.

Albrigo, L.G., J.I. Valiente, and C. Van Parys de Wit. 2006b. Influence of drought and winter temperatures on year-to-year citrus fruit set and yield variation in São Paulo, Brazil. Proc. 2004 Int. Soc. Citriculture, p. 263–267.

Alferez, F., J.K. Burns and L. Zacarias. 2004. Postharvest peel pitting of citrus is induced by changes in relative humidity. Proc. Fla. State Hort. Soc. 117:355–358.

Ali, A.G. and C.J. Lovatt. 1994. Winter application of low-biuret urea to the foliage of 'Washington' navel orange increased yield. J. Amer. Soc. Hort. Sci. 119(6): 1144–1150.

Allen, J.C. 1978. The effect of citrus rust mite damage on citrus fruit drop. J. Econ. Entomol. 71:746–750.

Alva, A.K. and J.P. Syvertsen. 1991. Soil and citrus tree nutrition are affected by salinized water. Proc. Fla. State Hort. Soc. 104:135–138.

Amar, M.H., M.K. Biswas, Z. Zhang, and W. Guo. 2011. Exploitation of SSR, SRAP and CAPS-SNP markers for genetic diversity of Citrus germplasm collection. Scientia Horticulturae 128:220–227.

Anderson, C.A. 1987. Calcium. J. Plant Nutr. 10:9–16, 1907–1916.

Anderson, C.A. and L.G. Albrigo. 1977. Seasonal changes in the relationships between macronutrients in orange (C. sinensis Osb.) leaves and soil analytical data in Florida. Proc. Int. Soc. Citriculture 1:20–25.

Anderson, J. A., L.V. Gusta, D.W. Buchanan, and M.J. Burke. 1983. Freezing of water in citrus leaves. J. Amer. Soc. Hort. Sci. 108: 397–400.

Anon. 1997. Centre for Biosafety and Sustainability. (1997, February 8). Transformation methods and genetic elements introduced into transgenic plants. Retrieved from https://www.bats.ch/bats/publikationen/1997-2_gmo/1.2_gmo_transformation.php

Anonymous. 2008. USDA-APHIS. 2008. Citrus greening national quarantine map. Retrieved from http://www.aphis.usda.gov/plant_health/plant_pest_info/citrus_greening/downloads/pdf_files/nationalquarantinemap.pdf.

Ashkenazi, S. and Y. Oren. 1988. The use of citrus exocortis virus (CEV) for tree size control in Israel - practical aspects. Proc. Int. Soc. Citriculture 2:917–919.

Aubert, B., A. Sabine, and P. Picard. 1984. Epidemiology of the greening disease in Reunion Island before and after the biological control of the African and Asian citrus psyllids. Proc. Int. Soc. Citriculture 1:440–442.

Ayalon, S. and S.P. Monselise. 1960. Flower bud induction and differentiation in the 'Shamouti' orange. Proc. Amer. Soc. Hort. Sci. 75:216–221.

Bailey, L.H. and E.Z. Bailey (eds). 1978. Hortus III. MacMillan, New York.

Bain, J.M. 1958. Morphological, anatomical and physiological changes in the developing fruit of the Valencia orange, *Citrus sinensis* (L.) Osbeck. Aust. J. Bot. 6:1–24.

Bar-Joseph, M. 1993. Citrus viroids and citrus dwarfing in Israel. Acta Hortic. 349:271–276.

Bar-Joseph, M., S.M. Garnsey, D. Gonsalves, M. Moscovitz, D.E. Purcifull, M.F. Clark and G. Loebenstein. 1979. The use of enzyme-linked immunosorbent assay for detection of citrus tristeza virus. Phytopathology 69:190–194.

Barmore, C.R. and W.S. Castle. 1979. Separation of citrus seed from fruit pulp for rootstock propagation using pectolytic enzyme. HortScience 14:526–527.

Barone, E., G. Bounous, D. Gioffre, P. Inglese, and R. Zappia. 1988. Survey and outlook of bergamot (Citrus aurantium sub. bergamia Sw.) industry in Italy. Proc. Int. Soc. Citriculture 4:1603–1611.

Barrett, H.C. 1985. Hybridization of Citrus and related genera. Fruit Varieties Journal 39:11–16.

Barrett, H.C. and A.M. Rhodes. 1976. A numerical taxonomic study of affinity relationships in cultivated Citrus and its close relatives. Syst. Botany 1:105–136.

Basiouny, F. M. and E. H. Biggs. 1974. Effects of UV-B radiation on pigment changes and quality of citrus fruits. Proc. Fla. State Hort. Soc. 87:281–289.

Bassanezi, R.B., A. Bergamin Filho, L. Amorim, N. Gimenes-Fernandes, T.R. Gottwald, and J.M. Bové. 2003. Spatial and temporal analyses of citrus sudden death as a tool to generate hypotheses concerning its etiology. Phytopathology 93, 502–512.

Bassanezi, R.B., L.H. Montesino and E.S. Stuchi. 2009. Effects of huanglongbing on fruit quality of sweet orange cultivars in Brazil. Eur. J. Plant Pathol. 125:565–572.

Bassanezi, R.B., J. Belesque, Jr. and, L.H. Montesino. 2013. Frequency of symptomatic tree removal in small citrus blocks on citrus huanglongbing epidemics. Crop Protection 52:72–77.

Bastianel, M., J. Freitas-Astúa, E.W. Kitajima, and M.A. Machado. 2006. The citrus leprosis pathosystem. Summa Phytopathologia 32, 211–220.

Batchelor, L.D. and W.P. Bitters. 1954. Juice and citric acid content of three lemon varieties. California Citrograph 39:187.

Bayer, M., T. Nawy, C. Giglione, M. Galli, T. Meinnel, and W. Lukowitz. 2009. Paternal control of embryonic patterning in *Arabidopsis thaliana*. Science 323:1485–1488.

Beavers, J. B., and A.G. Selhime. 1975. Development of Diaprepes abbreviatus on potted citrus seedlings. Florida Entomologist, 58:271–273.

Belasque Jr., J., R.B. Bassanezi, P.T. Yamamoto, A.J. Ayres, A. Tachibana, A.R. Violante, A. Tank Jr., F. Di Giorgi, F.E.A Tersi, G.M. Menezes, J. Dragone, R.H. Jank Jr., and J.M. Bové. 2010. Lessons from huanglongbing management in São Paulo State, Brazil. J. Plant Pathol. 92, 285–302.

Ben-Yehoshua, S., I. Kobiler, and B. Shapiro. 1979. Some physiological effects of delaying deterioration of citrus fruits by individual seal packaging in high density polyethylene film. J. Amer. Soc. Hort. Sci. 104:868–872.

Berhow, M.A., S. Hasegawa, K. Kwan, and R.D. Bennett (2000) Limonoids and the chemotaxonomy of Citrus and the Rutaceae family. In Citrus limonoids. Functional chemicals in agricalrue and foods. (Ed.) Berhow M.A., Hasegawa S. and Manners G.D.. ACS Symposium series 758 pp. 212–229.

Bevington, K.B. and P.E. Bacon. 1977. Effects of 'rootstocks on the response of navel orange trees to dwarfing inoculations. Proc. Int. Soc. Citriculture 2:567–569.

Bevington, K.B. and W.S. Castle. 1982. Development of the root system of young 'Valencia' orange trees on rough lemon and Carrizo citrange rootstocks. Proc. Fla. State Hort. Soc. 95:33–37.

Bevington, K.B. and W.S. Castle. 1985. Annual root growth pattern of young citrus trees in relation to shoot growth, soil temperature and soil water content. J. Amer. Soc. Hort. Sci. 110:840–845.

Bielorai, H., S. Dasberg, Y. Erner, and M. Brum. 1981. The effect of various soil moisture regimes and fertilizer levels on citrus yield response under partial wetting of the root zone. Proc. Int. Soc. Citriculture 2:585–589.

Bielorai, H., S. Dasberg, Y. Erner, and M. Brum. 1988. The effect of saline irrigation water on Shamouti orange production. Proc. Int. Soc. Citriculture 2:707–715.

Bitters, W.P., D.H. Cole, and C.D. McCarty. 1977. Citrus relatives are not irrelevant as dwarfing stocks or interstocks for citrus. Proc. Int. Soc. Citriculture 2:561–567.

Bitters, W.P., D.A. Cole, and C.D. McCarty. 1981. Effect of budding height on yield and tree size of 'Valencia' orange on two rootstocks. Proc. Int. Soc. Citriculture 1:109–110.

Blazquez, M. 2005. The right time and place for making flowers. Science 309: 1024–1025.

Borrel, M. and A. Diaz. 1981. Effects of mechanical pruning on yield of citrus trees. Proc. Int. Soc. Citriculture 1:190–194.

Borroto, C.G. and A.M. Rodríguez. 1977. The effect of water stress on flowering and fruit set of Valencia oranges in Cuba, In: W. Grierson (Ed.). Proc. Int. Soc. Citriculture 1977:1069–1073.

Boswell, S.B., L.N. Lewis, C.D. McCarty, and K.W. Hench. 1970. Tree spacing of 'Washington' navel orange. J. Amer. Soc. Hort. Sci. 95:523–528.

Boswell, S.B., C.D. McCarty, K.W. Hench, and L.N. Lewis. 1975. Effect of tree density on the first ten years of growth of 'Valencia' navel orange trees. J. Amer. Soc. Hort. Sci. 100:370–373.

Bovè, J. M. 2006. Huanglongbing: a destructive, newly-emerging, century-old disease of citrus. J. Plant Pathol. 88:7–37.

Bowman, K. 2001. Rootstocks Released by Dr. Kim Bowman. Retrieved from http://www.ars.usda.gov/Research/docs.htm?docid=22496

Bowman, K. 2014a. United States Department of Agriculture, Agricultural Research Service. (2014, September 29). Release of US-1283, Citrus rootstock. Retrieved from https://crec.ifas.ufl.edu/extension/citrus_rootstock/Rootstock_Literature/US%201283%20Ninkat%20Mandarin%20x%20Gotha%20Rd%20TF.pdf

Bowman, K. 2014b. United States Department of Agriculture, Agricultural Research Service. (2014, September 29). Release of US-1279, Citrus rootstock. Retrieved from https://crec.ifas.ufl.edu/extension/citrus_rootstock/Rootstock_Literature/US%201279%20Changsha%20x%20Gotha%20Rd%20TF.pdf

Bowman, K. D., F. G. Gmitter, Jr., G. A. Moore, and R. L. Rouseff. 1991. Citrus fruit sector chimeras as a genetic resource for cultivar improvement. J. Amer. Soc. Hort. Sci. 116:888–893.

Bowman, K. D., H. K. Wutscher, R. D. Hartman, and A.E. Lamb. 1997. Enhancing development of improved rootstocks by tissue culture propagation and field performance of selected rootstocks. Proc. Fla. State Hort. Soc. 110:10–13.

Brar, G. and T.M. Spann. 2014a. Low night temperatures affect growth of container-grown liner trees of trifoliate orange under different photoperiods. Acta Hort. 1042, 97–100.

Brar, G.R.P.S. and T.M. Spann. 2014b. Photoperiodic phytochrome-mediated vegetative growth responses of container-grown citrus nursery trees. Scientia Hort. 176:112–119.

Bredell, G.S. and Barnard. 1977. Microjets for macro-efficiency. Proc. Int. Soc. Citriculture 1:87–92.

Brewer, R.F., K. Opitz, F. Aljibury and K. Hench. 1977. The Effects of cooling by overhead sprinkling on "June Drop" of navel oranges in California. Proc. Int. Soc. Citriculture. 3:1045–1048.

Brown, G.E. 1986. Diplodia stem-end rot, a decay of citrus fruit increased by ethylene degreening treatment and its control. Proc. Fla. State Hort. Soc. 99:105–108.

Brown, G.E. and C.R. Barmore. 1976. The effect of ethylene, fruit colour and fungicides on susceptibility of 'Robinson' tangerines to anthracnose. Proc. Fla. State Hort. Soc. 89:198–200.

Brown, G.E., P. Mawk, and J.O. Craig. 1988. Pallet treatment with benomyl of citrus fruit on trucks for control of diplodia stem-end rot. Proc. Fla. State Hort. Soc. 101:187–190.

Browning, H. 1992. CREC report (75 years of excellence in entomology research and extension). Citrus Industry 73:13–14, 36–38.

Buchanan, D.W., F.S. Davies, and D.S. Harrison. 1982. High and low volume under- tree irrigation for citrus cold protection. Proc. Fla. State Hort. Soc. 95:23–26.

Bullock, R.C. and R.F. Brooks. 1975. Citrus pest control in the USA. In: Citrus. Ciba-Geigy Agrochemicals, Basel, Switzerland, pp. 35–37.

Bullock, R.C. and R.W. Miller. 1994. Suppression of *Pachnaeus litus* and *Diaprepes abbreviatus* (Coleoptera: Curculionidae) adult emergence with *Steinernema carpocapsae* (Rhabditida: Steinernematidae) soil drenches in field evaluations. Proc. Fla. State Hort. Soc. 107:90–92.

Burns, J.K. 1990. Respiratory rates and glycosidase activities of juice vesicles associated with section drying in citrus. HortScience 25:544–546.

Burns, J.K. and D.S. Achor. 1989. Cell wall changes in juice vesicles associated with 'section drying' in stored late-harvested grapefruit. J. Amer. Soc. Hort. Sci. 114:283–28 7.

Burns, J.K. and E.A. Baldwin. 1994. Glycosidase activities in grapefruit flavedo, albedo and juice vesicles during maturation and senescence. Physiol. Plantarum 90:37–44.

Burns, J.K. and E. Echeverria. 1988. Assessment of quality loss during commercial harvesting and postharvest handling of 'Hamlin' oranges. Proc. Fla. State Hort. Soc.101:76–79.

Burns, J.K., D. S. Achor, and E. Echeverria. 1992. Ultrastructural studies on the ontogeny of grapefruit juice vesicles (Citrus paradisi Macf.CV Star Ruby). Echeverria Int. J. Plant Sci. 153:14–25.

Bustan, A. and E. E. Goldschmidt. 1998. Estimating the cost of flowering in a grapefruit tree. Plant Cell and Environment 21:217–224.

Bustan, A.Y. Erner and E.E. Goldschmidt. 1995. Interactions between developing Citrus fruit and their vascular support system. Ann. Bot. 76: 657–666.

Bustan, A.Y. Erner, and E.E. Goldschmidt. 1996. Carbohydrate supply and demand during fruit development in relation to productivity of grapefruit and 'Murcott' mandarin. Acta Hort. 416:81–88.

California IPM Manual Group. 1984. Integrated Pest Management for Citrus. Statewide IPM Project, University of California at Davis, pp.144.

Camacho-B, S.E. 1981. Citrus culture in the high altitude American tropics. Proc. Int. Soc. Citriculture 1:321–325.

Cameron, J.W. and R.K. Soost. 1979. Sexual and nucellar embryony in F1 hybrids and advanced crosses of Citrus with Poncirus. J. Amer. Soc. Hort. Sci. 104:408–410.

Campos-Herrera, R., F.E. El-Borai, and L.W. Duncan. 2015. It takes a village: entomopathogenic nematode community structure and conservation biological control in Florida (US) orchards. In: Campos Herrera, R. editor, Nematode Pathogenesis of Insects and Other Pests. Springer, pp. 329–352.

Carpenter, J.B., R.M. Burns, and R.F. Sedlacek. 1981. Phytophthora resistant rootstocks for Lisbon lemons in California. Citrograph 67:287–292. I

Cary, P.R. 1981. Citrus tree density and pruning practices for the 21st century. Proc. Int. Soc. Citriculture 1:165–168.

Cassin, J., J. Bourdeaut, A. Fougue, V. Furan, J.P. Gaillard, J. LeBourdelles, G. Montagut, and C. Moreuil. 1969. The influence of climate upon the blooming of citrus in tropical areas. Proc. Int. Soc. Citriculture 1:315–324.

Castle, W.S. 1978. Citrus root systems: their structure, function, growth and relationship to tree performance. Proc. Int. Soc. Citriculture 1:62–69.

Castle, W.S. 1987. Citrus rootstocks. In: Rom, R.C. and R.F. Carlson (eds), Rootstocks for Fruit Crops. John Wiley and Sons, New York, pp. 361–399.

Castle, W.S. 2010. A career perspective on citrus rootstocks, their development, and commercialization. HortScience 45:11–15.

Castle, W.S. 2015. Citrus Rootstock Selection Guide. Retrieved from http://www.crec.ifas.ufl.edu/extension/citrus_rootstock/templates/guide/

Castle, W.S. and J.C. Baldwin. 2006. Rootstock effects on Murcott tangor trees grown in a calcareous alfisol or a spodosol. Proc. Fla. State Hort. Soc. 119:136–141.

Castle, W.S. and J.C. Baldwin. 2011. Young-tree performance of juvenile sweet orange scions on Swingle citrumelo rootstock. HortScience 46:541–552

Castle, W.S. and J.J. Ferguson. 1982. Current status of greenhouse and container production of citrus nursery trees in Florida. Proc. Fla. State Hort. Soc. 95:52–56.

Castle, W.S. and A.H. Krezdorn. 1973. Rootstock effects on root distribution and leaf mineral content of 'Orlando' tangelo trees. Proc. Fla. State Hort. Soc. 86:80–84.

Castle, W.S. and R.E. Rouse, R.E. 1990. Total nutrient content of Florida citrus nursery plants. Proc. Fla. State Hort. Soc. 103:42–44.

Castle, W.S. and C.O. Youtsey. 1977. Root system characteristics of citrus nursery trees. Proc. Fla. State Hort. Soc. 90:39–44.

Castle, W.S., R.R. Pelosi, C.O Youtsey, F.G. Gmitter Jr., R.F. Lee, C.A. Powell, and X. Hu. 1992. Rootstocks similar to sour orange for Florida citrus trees. Proc. Fla. State Hort. Soc. 105:56–60.

Castle, W.S., J. Nunnallee and J. A. Manthey. 2009. Screening citrus rootstocks and related selections in soil and solution culture for tolerance to low-iron stress. HortScience 44:638–645.

Castle, W.S., J.C. Baldwin, and R.P. Muraro. 2010a. Rootstocks and the performance and economic returns of 'Hamlin' sweet orange trees. HortScience 45(6):875–881.

Castle, W.S., J. C. Baldwin and R. P. Muraro and R. Litell. 2010b. Performance of 'Valencia' sweet orange trees on 12 rootstocks at two locations and an economic interpretation as a basis for rootstock selection. HortScience 45 (4):523–533.

Castle, W.S., K. D. Bowman, J. W. Grosser, S. H. Futch, and J. H. Graham. 2016. Florida Citrus Rootstock Selection Guide, 3rd Edition. Retrieved from http://edis.ifas.ufl.edu/hs1260

Chadha, K.L., Randhawa, N.S., Bindra, O.S., Chohan, J.S. and Knorr, L.C. 1970. Citrus Decline in India: Causes and Control. Punjab Agricultural University, Ohio State University and United States Agency for International Development, India, pp. 97.

Chaires, P. (2013, March 29). 9 New Citrus Varieties Face Trial. Retrieved from http://www.growingproduce.com/citrus/9-new-citrus-varieties-face-trial/

Chaires, P. 2014. Playing the citrus rootstock market has risks and rewards. Florida Grower 107(11):18–19.

Chaires, P. (2015, May 20). Interest Continues To Grow In Protected Citriculture. Retrieved from https://www.growingproduce.com/citrus/varieties-rootstocks/interest-continues-to-grow-in-protected-citriculture/

Chapman, H.D. 1968. The mineral nutrition of citrus. In: Reuther, W., Batchelor, L.D. and Webber, H.D. (eds), The Citrus Industry. University of California Press, California, pp. 127–289.

Chapot, H. 1975. The citrus plant. In: Citrus, Technical monograph no. 4, Ciba-Geigy Agrochemicals, Basel, Switzerland, pp. 6–13.

Chen, C., F. Gmitter, Jr., C. Huan and P. Cancalon. 2011. Quantification of furanocoumarins and acids in grapefruit hybrid populations to evaluate their relation and assist selection. Proc. Fla. State Hort. Soc. 124:111–114.

Chen, X.D. and L.L. Stelinski. 2017. Resistance management for Asian citrus psyllid, *Diaphorina citri* Kuwayama, in Florida. Insects. 8(3)103; doi:10.3390/insects8030103.

Chica, E. J. 2011. Expression patterns of flowering genes during flower induction and determination in sweet orange (*Citrus sinensis* L. Osbeck). Ph.D. University of Florida Dissertation, pp. 126.

Chica, E. J. and L.G. Albrigo. 2011. Stimulating flowering in basal buds of sweet orange summer shoots by removal of terminal buds early in the flower bud induction period. Proc. Fla. State Hort. Soc. 124:60–64.

Chica, E. J. and L.G. Albrigo. 2013a. Expression of flower promoting genes in sweet orange during floral inductive water deficits. J. Amer. Soc. Hort. Sci. 138:88–94.

Chica, E.J. and L.G. Albrigo. 2013b. Changes in CsFT transcript abundance at the onset of low-temperature floral induction in sweet orange. J. Amer. Soc. Hort. Sci. 138:184–189.

Chica, E. J. and L.G. Albrigo. 2015. Patterns of accumulation of CSFT, CSAP1 and CSLFY transcripts in leaves and buds are related to flowering gradients and cohorts in sweet orange. Acta Hortic. 1065:1257–1265.

Coggins, C.W., Jr. 1981. The influence of exogenous growth regulators on rind quality and internal quality of citrus fruits. Proc. Int. Soc. Citriculture 1:214–216.

Cohen, I. 1975. From biological to integrated control of citrus pests in Israel. In: Citrus. Ciba-Giegy Agrochemicals, Basel, Switzerland, pp. 38–41.

Cohen, M. 1968. Exocortis virus as a possible factor in producing dwarf citrus trees. Proc. Int. Soc. Citriculture 81:115–119.

Cohen, M. and H.J. Reitz. 1963. Rootstocks for Valencia orange and Ruby Red grapefruit: results of a trial initiated at Fort Pierce in 1950 on two soil types. Proc. Fla. State Hort. Soc. 76:29–34.

Constantinidou, H.A., O. Menkissaglu and O. Stergiadou. 1991. The role of ice nucleation active bacteria in supercooling of citrus tissues. Physiol. Plant. 81(4):548–554.

Cooper, W.C. and Chapot, H. (1977) Fruit production with special emphasis on fruit for processing. In: Nagy, S., Shaw, P.E. and Veldhuis, M.K. (eds) Citrus Science and Technology, 1st edn, Vol 2. AVI Publishing Co., Westport, Connecticut, pp. 1–127.

Costa, A.S. and G.W. Müller. 1980. Tristeza control by cross protection: A U.S.-Brazil cooperative success. Plant Disease. 64:538–541.

Crocker, T.E., W.P. Bell, and J.F Bartholic. 1974. Scholander pressure bomb technique to access the relative leaf water stress of 'Orlando' tangelo scion as influenced by various citrus rootstocks. HortScience 9:453–455.

Croxton S.D. and P.A. Stansly. 2014. Metalized polyethylene mulch to repel Asian citrus psyllid, slow spread of Huanglongbing and improve growth of new citrus plantings. Pest Manag. Sci. 70: 318–323.

Culbert, D. L. and H.W. Ford. 1972. The use of a multi-celled apparatus for anaerobic studies of flooded root systems. HortScience 7(1):29–31.

D'Onghia, A.M., K. Djelouah, and V. Savino. 2000. Serological detection of Citrus psorosis virus in seeds but not in seedlings of infected mandarin and sour orange. J. Plant Path. 82 (3):233–235.

Dasberg, S.A., A. Bar-Akiva, S. Spazisky, and A. Cohen. 1988. Fertigation vs. broadcasting in an orange grove. Fertilizer Research 15:147–154.

Davenport, T.L. 1990. Citrus flowering. In: Janick, J. (ed.), Horticultural Reviews. Timber Press, Portland, Oregon, pp. 349–408.

Davenport, T.L., C.W. Campbell, and P.G. Orth. 1976. Stylar-end breakdown in 'Tahiti' lime - some causes and cures. Proc. Fla. State Hort. Soc. 89:245–248.

Davies, F.S. 1986a. The navel orange. In: Janick, J. (ed.), Horticultural Reviews. AVI Publishing Co., Westport, Connecticut, pp. 129–180.

Davies, F.S. 1986b. Growth regulator improvement of postharvest quality. In: Wardowski, W.F., Nagy, S. and Grierson, W. (eds), Fresh Citrus Fruits. AVI Publishing Co., Westport, Connecticut, pp. 79–99.

Davies, F. S. and L.G. Albrigo. 1994. Citrus 1st Ed. CABI, Wallingford, UK.

Davies, F.S. and L. K. Jackson. 2009. Citrus Growing in Florida, 5th Edition. University Press of Florida, Gainesville, Fl.

Davies, F.S. and M. Maurer. 1992. Reclaimed wastewater for irrigation of citrus in Florida. HortTechnology 3:163–167.

Dawson, W.O., S.M. Garnsey, S. Tatmeni, S. Folimonova, S.J. Harper and S. Gowda. 2013. Citrus tristeza virus-host interactions. Frontiers Microbiol. 4:11–20.

De Barreda, D.G. 1977. Present status of weed control practices in Spain. Proc. Int. Soc. Citriculture 1:158–161.

Derrick, K.S. and L.W. Timmer. 2000. Citrus blight and some other diseases of recalcitrant etiology. Annu. Rev. Phytopathol. 38:181–205.

Derrick, K.S., R.F. Lee, R.H. Brlansky, L.W. Timmer, B.G. Hewitt, and G.A. Barthe. 1990. Proteins associated with citrus blight. Plant Disease 74:168–170.

DeVilliers, J.I. 1969. The effect of differential fertilization on the yield, fruit quality and leaf composition of navel oranges. Proc. Int. Soc. Citriculture 1:1661–1668.

Dewdney, M.M., N.A. Peres, M. Ritenour, T. Schubert, R. Atwood, G. England, S.H. Futch, T. Gaver, T. Hurner, C. Oswalt, and M.A. Zekrl. 2010. Citrus black spot arrives in Florida. Citrus Industry 91(7):18–21.

Downing, A.S., S.G. Erickson, and M.J. Kraus. 1991. Field evaluation of entomopathogenic nematodes against citrus root weevils (Coleoptera: Curculionidae) in Florida citrus. Florida Entomol. 74:584–586.

Duarte, A.M.M. and J.L. Guardiola. 1996. Flowering and Fruit Set of 'Fortune' Hybrid Mandarin. Effect of Girdling and Growth Regulators. Proc. Int. Soc. Citriculture 1996; 106.

Duncan, L.W. and E. Cohn. 1990. Nematode parasites of citrus. In: Luc, M., Sikora, R.A. and Bridge, J. (eds) Plant Parasitic Nematodes in Subtropical and Tropical Agriculture. CAB International, Wallingford, UK.

DuPlessis, S.F. 1984. Crop forecasting for navels in South Africa. Proc. Fla. State Hort. Soc. 96: 40–43.

DuPlessis, S.F. and T.J. Koen. 1988. The effect of N and K fertilization on yield and fruit size of Valencia. Proceedings of the Sixth International Citrus Congress, 663–672.

Dutt, M. and J.W. Grosser. 2009. Evaluation of parameters affecting Agrobacterium-mediated transformation of Citrus. Plant Cell, Tissue and Organ Culture 98:331–340.

Eaks, I. L. 1969. Rind disorders of oranges and lemons in California. Proc. First Internl. Citrus Symposium 3:1343–1354.

Echeverria, E. and M. Ismail. 1987. Changes in sugars and acids of citrus fruits during storage. Proc. Fla State Hort. Soc. 100:50–52.

Echeverria, E., J. Burns, and W. Miller. 1998. Progress on blossom-end clearing of grapefruit. Proc. Fla. State Hort. Soc. 111:255–257.

Economides, C.V. 1976. Performance of Marsh seedless grapefruit on six rootstocks in Cyprus. J. Hortic Sci. 51:393–400.

Embleton, T.W., C.K. Labanauskas, W.W. Jones, and C.B. Cree. 1963. Interrelations of leaf sampling methods and nutritional status of orange trees and their influence on macro- and micronutrient concentrations in orange leaves. Proc. Amer. Soc. Hort. Sci. 82:131–141.

Embleton, T.W., W.W. Jones, C. Pallares, and R.G. Piatt. 1978. Effects of fertilization of citrus on fruit quality and ground water nitrate-pollution potential. Proc. Int. Soc. Citriculture 2:280–285.

Engler, A. 1931. Rutaceae, pp. 187–359. In A. Engler, K. Pranti eds. Die naturlichen Pilanzenfamilien, 2nd ed., Vol. 9a, Wilhelm, Leipzig.

Erickson, L.C. 1968. The general physiology of citrus. In: Reuther, W., Batchelor, L.D. and Webber, H.J. (eds), The Citrus Industry. University of California Press, Berkeley, California, pp. 86–126.

Erickson, L.C. and B.L. Brannaman. 1960. Abscission of reproductive structures and leaves of orange trees. Proc. Amer. Soc. Hort. Sci. 75:222–229.

Erner, Y. and B. Bravdo. 1983. The importance of inflorescence leaves in fruit setting of 'Shamouti' orange. Acta Horticulturae 139:107–113.

Fallahi, E. and D.R. Rodney. 1992. Tree size, yield, fruit quality and leaf mineral nutrient concentration of 'Fairchild' mandarin on six rootstocks. J. Amer. Soc. Hort. Sci. 117:28–31.

FAO Commodities and Trade Division. 1991. Citrus fruit fresh and processed annual statistics. CCP:CI/91 Food and Agriculture Organization of the United Nations, Rome, Italy.

FAO. 2016. Citrus fruit statistics 2015. Retrieved from http://www.fao.org/3/a-i5558e.pdf

Febres, V.J., R.F. Lee and G.A. Moore. 2007. Genetic transformation of citrus for pathogen resistance I. In: I.A. Khan (Ed.). Citrus Genetics, Breeding and Biotechnology. Cambridge, MA: CABI Press. pp. 307–327.

Febres, V., L. Fisher, A. Khalaf and G.A. Moore. 2011. Citrus transformation: challenges and prospects. In Prof. MarÃa Alvarez (Ed.), Genetic Transformation, ISBN: 978–953–307–364–4.

Fennah, R.G. 1940. Observations on behaviour of citrus root-stocks in St. Lucia, Dominica and Montserrat. Tropical Agriculture. 17:72–76.

Ferdman, R.A. 2014. How America fell out of love with orange juice. Retrieved from https://qz.com/176096/how-america-fell-out-of-love-with-orange-juice/

Ferguson, J.J. and J.A. Menge. 1986. Response of citrus seedlings to various field inoculation methods with Glomus deserticola in fumigated nursery soils. J. Amer. Soc. Hort. Sci. 111:288–292.

Ferguson, L., M.A. Ismail, F.S. Davies and T.A. Wheaton. 1982. Pre- and postharvest gibberellic acid and 2, 4–dichlorophenoxyacetic acid applications for increasing storage life of grapefruit. Proc. Fla. State Hort. Soc. 95:242–245.

Flores, R., C. Hernández, A.E. Martínez de Alba, J.A. Daròs, and F. DiSerio. 2005. Viroids and viroid-host interactions. Annu. Rev. Phytopathol. 43:117–139.

FDACS, DPI (Florida Department of Agriculture and Consumer Services, Division of Plant Industry). (2013, February 14). Citrus Nursery Stock Certification Manual. Retrieved from http://freshfromflorida.s3.amazonaws.com/Citrus-Nursery-Stock-Certification-Manual-02-14-13.pdf

Folimonova, S.Y., C. J. Robertson, S. M. Garnsey, S. Gowda, and W. O. Dawson. 2009. Examination of the responses of different genotypes of citrus to Huanglongbing (citrus greening) under different conditions. Phytopathology 99(12): 1346–1354.

Folimonova, S.Y., C.J. Robertson, T. Shilts, A.S. Folimonov, M.E. Hilf, S.M. Garnsey and W.O. Dawson. 2010. Infection with strains of citrus tristeza virus does not exclude superinfection by other strains of the virus. J. Virology 84(3): 1314–1325.

Ford, H.W. and D.P.H. Tucker. 1975. Blockage of drip irrigation filters and emitters by iron-sulfur-bacterial products. HortScience 10:62–64.

Ford, S.A., R.P. Muraro, and G.F. Fairchild. 1989. Economic comparison of southern and northern citrus production areas in Florida. Proc. Fla. State Hort. Soc. 102:27–32.

Fornes, F., R. M. Belda, M. Abad, P. Noguera, R. Puchada, A. Maquierira and V. Noguera. 2003. The microstructure of coconut coir dusts for use as alternatives to peat in soilless growing media. Australian Journal of Experimental Agriculture 43(9):1171–1179.

Freitas-Astúa, J.; E.C. Locali, R. Antonioli, V. Rodrigues, E.W. Kitajima, M.A. Machado. 2003. Detection of citrus leprosis virus in citrus stems, fruits and the mite vector. Virus Reviews & Research. 8, supl.1: 196.

Frost, H.B. and P.K. Soost. 1968. Seed reproduction: development of gametes and embryos. In: Reuther, W., Batchelor, L.D. and Webber, H.J. (eds.), The Citrus Industry. University of California Press, California, pp. 290–324.

Fucik, J.E. 1977. Hedging and topping in Texas grapefruit orchards. Proc. Int. Soc. Citriculture 1:172–175.

Fucik, J.E. 1978. Sources of variability in sour orange seed germination and seedling growth. Proc. Int. Soc. Citriculture 1:141–143.

Futch, S. and C.W. McCoy. 1992. Citrus root weevils. Citrus and Vegetable Magazine 55:27, 30–33, 36–37.

Futch, S.H. and M. Singh. 2009. Weeds. In: M.E. Rogers, M.M. Dewdney, and T.M. Spann (eds). 2009 Florida citrus pest management guide. Univ. of Fla., Coop. Ext. Serv., IFAS. SP-43. p. 127–138.

Gallasch, P.T. and N.J. Ainsworth. 1988. Developments in the Australian citrus industry. Proc. Int. Soc. Citriculture 4:1613–1623.

Garcia-Luis, A., P. Santamarina, and J. L. Guardiola. 1989. Flower formation from Citrus unshiu buds cultured in vitro. Ann. Bot. 64(5):515–519.

Garcia-Luis, A., M. Kauduser, P. Santamarina, and J.L. Guardiola. 1992. Low temperature influence on flowering in Citrus: the separation of inductive and bud dormancy releasing effects. Physiol. Plant. 86:648–652.

Garcia-Luis, A., M. E. M. Oliveira, Y. Bordon, D. L. Siqueira, S. Tominga, and J. L. Guardiola. 2002. Dry matter accumulation in citrus fruit is not limited by transport capacity of the pedicel. Ann. Bot. 90(6):755–764.

Garcia-Marí, F., C. Granda, S. Zaragoza and M. Agusti. 2002. Impact of *Phyllocnistis citrella* (Lepidoptera: Gracillariidae) on leaf area development and yield of mature citrus trees in the Mediterranean area. J. Econ. Entomol. 95(5):966–974.

Gardner, F.E. and G.E. Horanic. 1961. A comparative evaluation of rootstocks for Valencia and Parson Brown oranges on Lakeland fine sand. Proc. Fla. State Hort. Soc.74:123–127.

Gardner, F.E., P.C. Reece and G.E. Horanic. 1950. The effect of 2,4–D on pre-harvest drop of citrus fruit under Florida conditions. Proc. Fla. State Hort. Soc. 63:7–11.

Garnier, M., L. Zreik, and J.M. Bove. 1991. Witches' broom, a lethal mycoplasmal disease of lime trees in the Sultanate of Oman and the United Arab Emirates. Plant Disease. 75:546–551.

Garnsey, S.M., E.L. Civerolo, D.J. Gumpf, C. Paul, M.E. Hilf, R.F. Lee, R.H. Brlansky, R.K. Yokomi, and J.S. Hartung. 2005. Biological characterization of an international collection of Citrus tristeza virus (CTV) isolates. In: Hilf, M.E., Duran-Vila, N. and Rocha-Peña, M.A., eds. Proceedings of the 16th Congress International Organization of Citrus Virologists (IOCV). IOCV, Riverside, pp. 75–93.

Garza-Lopéz, J.G. and V.M. Medina-Urrutia. 1984. Diseases of Mexican lime *Citrus aurantifolia* (Christm.) Swingle in Mexico. Proc. Int. Soc. Citriculture 1: 311–315.

Georgis, R., A.M. Koppenöfer, L.A. Lacey, G. Bélair, L.W. Duncan, P.S. Grewal, M. Samish, L. Tan, P. Torr, and R.W.H.M. van Tol. 2006. Successes and failures in the use of parasitic nematodes for pest control. Biol. Control. 38:103–123.

Giles, F. 2018. Shining a light on psyllids. Florida Grower 111(8):6–8.

Gmitter, F.G., Jr. and X. Hu. 1990. The possible role of Yunnan, China, in the origin of contemporary Citrus species (Rutaceae). Economic Botany 44:267–277.

Gmitter F.G. Jr., Chen C., Machado M.A., Alves de Souza A., Ollitrault P., Froehlicher Y. and Shimizu T. (2012) Citrus genomics. Tree Genet. Genomes 8: 611–626.

Goldberg-Moeller, R., L. Shalom, L. Shlizerman, S. Samuels, N. Zur, R. Ophir, E. Blumwald, and A. Sadka. 2013. Effects of gibberellin treatment during flowering induction period on global gene expression and the transcription of flowering-control genes in Citrus buds. Plant Science 198:46–57.

Goldschmidt, E.E. 1999. Carbohydrate supply as a critical factor for citrus fruit development and productivity. HortScience 34:1020–1024.

Goldschmidt, E.E., and D.A. Golomb. 1982. The carbohydrate balance of alternate-bearing citrus trees and the significance of reserves for flowering and fruiting. J. Amer. Soc. Hort. Sci. 107(2):206–208.

Gottwald, T.R., G. Hughes, J.H. Graham, X. Sun, and T. Riley. 2001. The citrus canker epidemic in Florida: the scientific basis of regulatory eradication policy for an invasive species. Phytopathology 91:30–34.

Gottwald T.R., R.B. Bassanezi, L. Amorim, and A. Bergamin-Filho. 2007. Spatial pattern analysis of citrus canker-infected plantings in São Paulo, Brazil, and augmentation of infection elicited by the Asian leafminer. Phytopathology 97:674–683.

Grafton-Cardwell, E., L.L. Stelinski, and P.A. Stansly. 2013. Biology and management of Asian citrus psyllid, vector of huanglongbing pathogens. Annu. Rev. Entomol. 58:413–432.

Graham J. and K. Morgan. 2018. Why bicarbonates matter in HLB management. Citrus Industry 99(9):16–24.

Graham, J.H. and J.P. Syvertsen. 1984. Influence of vesicular-arbuscular mycorrhizae on the hydraulic conductivity of roots of two citrus rootstocks. New Phytologist 97:277–284.

Graham, J.H. and J.P. Syvertsen. 1985. Host determinants of mycorrhizal dependency of citrus rootstock seedlings. New Phytologist 101:667–676.

Graham, J.H. and L.W. Timmer. 1992. Phytophthora diseases of citrus, In: Kumar, J., Chaube, H.S., Singh,U.S., and Mukhopadhyay, A.N. (eds.). Plant Diseases of International Importance. Diseases of Fruit Crops. Vol. III. Prentice Hall, Englewood Cliffs, NJ, pp. 250–269.

Graham, J.H., T.R. Gottwald, T.D. Riley and D. Achor. 1992. Penetration through stomata and strains of *Xanthomonas campestris pv citri* in citrus cultivars varying in susceptibility to bacterial diseases. Phytopathology 82:1319–1325.

Graham, J.H., C.W. McCoy, and J.S. Rogers. 1996. Insect-plant pathogen interactions: Preliminary studies of Diaprepes root weevils injuries and Phytophthora infections. Proc. Fla. State Hort. Soc. 109:57–62.

Graham, J.H., T.R. Gottwald, J. Cubero, and D. Achor. 2004. *Xanthomonas axonopodis* pv. citri: Factors affecting successful eradication of citrus canker. Mol. Plant Pathol. 5:1–15.

Graham, J.H., L.W. Timmer and M.M. Dewdney. 2013. 2014 Florida Citrus Pest Management Guide: Phytophthora Foot Rot and Root Rot. Univ. Florida IFAS Extension Publication # 156.

Greenberg, J., S. Holtzman, M. Fainzack, Y. Egozi, B. Giladi, Y. Oren, and I. Kaplan. 2010. Effects of NAA and GA3 sprays on fruit size and the incidence of creasing of 'Washington' navel orange. Acta Hortic. 884:273–279.

Grierson, W., W.M. Miller, and W.F. Wardowski. 1978. Packingline Machinery for Florida Citrus Packinghouses. Bulletin 803, University of Florida.

Grimm, G.R. 1956. Preliminary investigations on dieback of young transplanted citrus trees. Proc. Fla. State Hort. Soc. 69:31–34.

Grosser, J.W. and F.G. Gmitter Jr. 1990. Protoplast fusion and citrus improvement. In: Janick, J. (ed.), Plant Breeding Reviews. Timber Press, Portland, Oregon, pp. 339–374.

Grosser, J.W. and F.G. Gmitter. 2013. Breeding disease-resistant citrus for Florida: adjusting to the canker/HLB world – part 2: rootstocks. Citrus Industry. March 2013:10–16.

Grosser, J.W. and F.G. Gmitter Jr. 2014. The ultimate long-term HLB solution. A report on progress in breeding HLB tolerant/resistant citrus scions and rootstocks. Citrus Industry 95(11):6–8.

Grosser, J.W., F.G. Gmitter Jr., and J.L. Chandler. 1997. Development of improved sweet orange cultivars using tissue culture methods. Proc. Fla. State Hort. Soc. 110:13–16.

Grosser, J.W., J.H. Graham, C.W. McCoy, A. Hoyte, H.M. Rubio, D.B. Bright, and J.L. Chandler. 2003. Development of 'tetrazyg' rootstocks tolerant of the diaprepes/Phytophthora complex under greenhouse conditions. Proc. Fla. State Hort. Soc. 116:263–267.

Grosser, J.W., V. Medina-Urrutia, G. Ananthakrishnan and P. Serrano. 2004. Building a replacement sour orange rootstock: Somatic hybridization of selected mandarin + pummelo combinations. J. Amer. Soc. Hort. Sci. 129:530–534.

Grosser, J.W., J.L. Chandler, and L.W. Duncan. 2007a. Production of mandarin + pummulo somatic hybrid rootstocks for improved tolerance/resistance to sting nematode. Scientia Holticulturae. 113:33–36.

Grosser, J.W., X. Deng, R. Goodrich. 2007b. Somaclonal variation in sweet orange: practical applications for variety improvement and possible causes. In: I.A. Khan (Ed.). Citrus Genetics, Breeding and Biotechnology. Cambridge, MA: CABI Press. p. 219–233.

Guardiola, J.L. 1981. Flower initiation and development in citrus. Proc. Int. Soc. Citriculture 2:242–246.

Guardiola, J.L., C. Monerri, and M. Agustí. 1982. The inhibitory effect of gibberellic acid on flowering in Citrus. Physiol. Plant. 55:136–142.

Habeck, D.H. 1977. The potential of using insects for biological control of weeds in citrus. Proc. Int. Soc. Citriculture 1:146–148.

Hagenmaier, R.D. and P.E. Shaw. 1992. Gas permeability of fruit coating waxes. J. Amer. Soc. Hort. Sci. 117:105–109.

Halbert, S. E., and K. L. Manjunath. 2004. Asian citrus psyllids (Sternorrhyncha: Psyllidae) and greening disease of citrus: A literature review and assessment of risk in Florida. Fla. Entomol. 87:330–353.

Halim, H, G.R. Edwards, B.G. Coombe, and D. Aspinall. 1988. The dormancy of buds of *Citrus sinensis* (L.) Osbeck inserted into rootstock stems: factors intrinsic to the inserted bud. Ann. Bot. 61:525–529.

Hall, D.G., T.R. Gottwald, and C.H. Bock. 2010. Exacerbation of citrus canker by citrus leafminer *Phyllocnistis citrella* in Florida. Fla Entomol. 93: 558–566.

Handique, U., R.C. Ebel and K.T. Morgan. 2012. Influence of soil applied fertilizer on greening development in new growth flushes of sweet orange. Proc. Fla. State Hort. Soc. 125:36–39.

Harding, P.L. and D.F. Fisher. 1945. Seasonal Changes in Florida Grapefruit. USDA Technical Bulletin, no. 886, Washington, D.C., pp. 100.

Harding, P.L., J.R. Winston, and D.F. Fisher. 1940. Seasonal Changes in Florida Oranges. USDA Technical Bulletin, no. 753, Washington, D.C.

Hardy, S., G. Sanderson, P. Barkley and N. Donovan. (2007, September). Dwarfing citrus trees using viroids. Retrieved from http://citeseerx.ist.psu.edu/viewdoc/download?doi=10.1.1.424.4416&rep=rep1&type=pdf

Hare, J.D. and P.A. Phillips. 1992. Economic effects of the citrus red mite (Acari, Tetranychidae) on southern California coastal lemons. J. Econ. Entomol. 85: 1926–1932.

Hartung, J. S., J. Beretta, R.H. Brlansky, J. Spisso, and R.F. Lee. 1994. Citrus variegated chlorosis bacterium: axenic culture, pathogenicity, and serological relationships with other strains of Xylella fastidiosa. Phytopathology 84:591–597.

Hatton, T.T. and R.H. Cubbedge. 1982. Conditioning Florida grapefruit to reduce chilling injury during low-temperature storage. J. Amer. Soc. Hort. Sci.107:57–60.

Hearn, C.J. 1984. Development of seedless orange and grapefruit cultivars through seed irradiation. J. Amer. Soc. Hort. Sci. 109:270–273.

Hearn, C.J. 1985. Citrus scion improvement program. Fruit Varieties Journal 39:34–37.

Hearn, C.J. 1986. Development of seedless grapefruit cultivars through budwood irradiation. J. Amer. Soc. Hort. Sci. 111:304–306.

Hearn, C.J. and D.J. Hutchison. 1977. The influence of 'Robinson' and 'Page' citrus hybrids on 10 rootstocks. Proc. Fla. State Hort. Soc. 90:44–47.

Henrick, C.A. 1982. Juvenile hormone analogs: structure-activity relationship. In: Coats, J.R. (ed.), Insecticide Mode of Action. Academic Press, New York, pp. 315–402.

Hensz, R.A. 1971. 'Star Ruby', a new deep-red-fleshed grapefruit variety with distinct tree characteristics, Journal of the Rio Grande Valley Horticultural Society 25:54–58.
Heppner, J.B. and T.R. Fasulo. 2010. Citrus leafminer, *Phyllocnistis citrella* Stainton (Insecta: Lepidoptera: Phyllocnistinae). Retrieved from http://edis.ifas.ufl.edu./in165
Hield, H. Z., R. M. Burns and C. W. Coggins. 1962. Some fruit thinning effects of napthaleneacetic acid on Wilking mandarin. Proc. Amer. Soc. Hort. Sci. 81:218–222.
Hilf, M.E., V.A. Mavrodieva, and S.M. Garnsey. 2005. Genetic marker analysis of a global collection of isolates of Citrus tristeza virus: Characterization and distribution of CTV genotypes and association with symptoms. Phytopathology 95:909–917.
Hilgeman, R.H. 1977. Response of citrus trees to water stress in Arizona. Proc. Int. Soc. Citriculture 1:70–74.
Hodgson, R.W. 1967. Horticultural varieties of citrus. In: Reuther, W., Batchelor, L.D. and Webber, H.D. (eds) The Citrus Industry. University of California Press, California, pp. 431–591.
Hooker, J.D. 1875. The Flora of British India. 7 Volumes, Reeve & Co., London.
Hutchison, D.J. 1974. Swingle citrumelo - a promising rootstock hybrid. Proc. Fla. State Hort. Soc. 87: 89–91.
Hutchison, D.J. 1977. Influence of rootstock on the performance of 'Valencia' sweet orange. Proc. Int. Soc. Citriculture 2:523–525.
Hutchison, D.J. 1985. Rootstock development screening and selection for disease tolerance and horticultural characteristics. Fruit Varieties Journal 39:21–25.
Hwang, Y.-S., L.G. Albrigo, and D.J. Huber. 1988. Juice vesicle disorders and in-fruit seed germination in grapefruit, Proc. Fla. State Hort. Soc. 101:161–165.
Hwang, Y.-S., D.J. Huber, and L.G. Albrigo. 1990. Comparison of cell wall components in normal and disordered juice vesicles of grapefruit. J. Amer. Soc. Hort. Sci. 115:281–287.
Hyun, J.W., S.H. Yi, S.J. MacKenzie, L.W. Timmer, K.S. Kim, S.K. Kang, H.M. Kwon, and H. C. Lim. 2009. Pathotypes and genetic relationship of worldwide collections of *Elsinoë* spp. causing scab diseases of citrus. Phytopathology 99:721–728.
Iglesias, D.J., F.R. Tadeo, E. Primo-Miller and M. Talon. 2003. Fruit set dependence on carbohydrate availability in citrus trees. Tree Physiology 23:199–204.
Iglesias, D.J., M. Cercos, J.M. Colmennero-Flores, M.A Haranjo, G. Rios, E. Carrera, O. Ruiz-Rivero, I. Lliso, R. Morillon, F.R. Tadeo, and M. Talon. 2007. Physiology of citrus fruiting. Braz. J. Plant Physiol. 19:333–362.
Inch, S., E. Stover, R. Driggers and R. F. Lee. 2014. Freeze response of citrus and citrus-related genotypes in a Florida field planting. HortScience. 49(8):1010–1016.
Inoue, H. 1990. Effects of temperature on bud dormancy and flower bud differentiation in satsuma mandarin. Journal of the Japanese Society of Horticultural Science 58:919–926.
Ismail, M. A. and G. E. Brown. 1975. Phenolic content during healing of 'Valencia' orange peel under high humidity. J. Amer. Soc. Hort. Sci. 100:249–251.
Ito, M., K. Ueki, and K. Ito. 1981. Approaches to weed management in citrus from the aspect of weed science. Proc. Int. Soc. of Citriculture 2:483–485.

Iwagaki, I. 1981. Tree configuration and, pruning of satsuma mandarin in Japan. Proc. Int. Soc. Citriculture 1:169–172.

Jackson, L.K. 1991. Citrus Growing in Florida. University of Florida Press, Florida.

Jackson, L.K. and F.S. Davies. 1984. Mulches and slow-release fertilizers in a citrus young tree care program. Proc. Fla. State Hort. Soc.97:37–39.

Jackson, L.K. and D.P.H. Tucker. 1992. Citrus tree planting and establishment practices. Citrus Industry 73:38–43.

Jahn, O.L. 1973. Inflorescence types and fruiting patterns in 'Hamlin' and 'Valencia' oranges and 'Marsh' grapefruit. Am. J. Bot. 60:663–670.

Jahn, O.L. 1979. Penetration of photosynthetically active radiation as a measurement of canopy density of citrus trees. J. Amer. Soc. Hort. Sci. 104:557–560.

Johnson, E.G., J. Wu, B.D. Bright, and J.H. Graham. 2014. Association of Candidatus Liberibacter asiaticus' root infection, but not phloem plugging with root loss on huanglongbing-affected trees prior to appearance of foliar symptoms. Plant Pathology 63: 290–298.

Jones, W.W. and C.B. Cree. 1965. Environmental factors related to fruiting of 'Washington' navel oranges over a 38–year period. Proc. Amer. Soc. Hort. Sci. 86:267–271.

Jones, W.W. and Embleton, T.W. (1967) Yield and fruit quality of 'Washington' navel orange trees as related to leaf nitrogen and nitrogen fertilization. Proc. Amer. Soc. Hort. Sci. 91, 138–142.

Jones, W.W., T. W. Embleton, M.L. Steinnecker and C.B. Cree. 1964. The effect of time of harvest on fruiting and carbohydrate supply of Valencia orange. Proc. Amer. Soc. Hort. Sci. 84:152–157.

Jordan, L.S. 1981. Weeds affect citrus growth, physiology, yield, fruit quality. Proc. Int. Soc. Citriculture 2:481–483.

Kadyampakeni, D.M., K. T. Morgan, A.W. Schumann, P. Nkedi-Kizza. and K. Mahmoud 2013b. Phosphorus and potassium distribution and adsorption on two Florida sandy soils. Soil Sci. Soc. Am. J. 78:325–334.

Kadyampakeni, D.M., K. T. Morgan, and A.W. Schumann. 2013a. Water and nutrient uptake in citrus open hydroponic systems. Proc. Fla. State Hort. Soc. 126, 58–61.

Kato, T. 1986. Nitrogen metabolism and utilization in citrus. In: Janick, J. (ed.), Horticultural Reviews. AVI Publishing Co., Westport, Connecticut, pp. 181–216.

Kato, T., S. Kubota, and S. Rambang. 1982. Uptake and utilization of nitrogen by satsuma mandarin trees in low temperature season. Bulletin of the Shikoka - Agricultural Experimental Station 40: 1–5.

Kaufmann, M.R. 1977. Citrus - a case study of environmental effects on plant water relations. Proc. Int. Soc. Citriculture 1:57–62.

Kawada, K. and L.G. Albrigo. 1979. Effects of film packaging, in-carton air filters, and storage temperatures on the keeping quality of Florida grapefruit. Proc. Fla. State Hort. Soc. 92:209–212.

Kaya, M., A.S. Sousa, M.J. Crepéau, S.O. Sorensen and M-C. Raleti. 2014. Characterization of citrus pectin samples extracted under different conditions: Influence of acid tye and pH of extraction. Ann. Bot. 114:1319–1326.

Kesinger, M.C. 2012. 2012 Ann. Rept. Bureau of Citrus Budwood Registration. Fla. Dept. Agric. and Cons. Services.

Ketchie, D. O. 1969. The effect of high temperature on citrus. International Citrus Symposium (1st : 1969 : University of California, Riverside. 1: 267–270).

Khan, I.A. 2007. Citrus Genetics, Breeding and Biotechnology. Khan, I.A. (Ed). Cambridge, MA: CABI Press. ISBN: 9780851990194.

Killiny, N., N.F. Valim, S.E. Jones, A.A. Omar, F. Hijaz, F.G. Gmitter, Jr. and J.W. Grosser. 2017. Metabolically speaking: Possible reasons behind the tolerance of 'Sugar Bell' mandarin hybrid to huanglongbing. Plant Physiol. Biochem. 116:36–47.

Kirkpatrick, J.D. and W.P. Bitters. 1968. Physiological and morphological response of various citrus rootstocks to salinity. Proc. Int. Soc. Citriculture 1:391–400.

Kitagawa, H., T. Matsui, and K. Kawada. 1988. Some problems in marketing citrus fruits in Japan. Proc. Int. Soc. Citriculture 4:1581–1587.

Kleinschmidt, G.D. and J.W. Gerdeman. 1972. Stunting of citrus seedlings in fumigated nursery soils related to absence of endomycorrhizae. Phytopathology 62:1447–1453.

Knapp, J.L. 1992. Florida Citrus Spray Guide. SP43 Florida Cooperative Extension Service, University of Florida, Florida, pp. 46.

Knapp, J.L. and H.W. Browning. 1989. Citrus blackfly: management in commercial orchards. The Citrus Industry 70:50–51.

Koch, K.E. and C.R. Johnson. 1984. Photosynthate partitioning in split-root citrus seedlings with mycorrhizal and non-mycorrhizal root systems. Plant Physiology 75:26–30.

Kohmoto, K., K. Akimitsu and H. Otani. 1991. Correlation of resistance and susceptibility of citrus to Alternaria alternate with sensitivity to host specific toxins. Phytopathology 81:719–722.

Koizumi, M. 1985. Citrus canker, the world situation. In: L.W. Timmer, ed. Citrus Canker: an International Perspective. University of Florida, Lake Alfred, pp. 2–7.

Kola, O., C. Kaya, H. Duran and A. Altan. 2010. Removal of limonin bitterness by treatment of ion exchange and adsorbent resins. Food Sci. Biotechnol. 19:411–416.

Koo, R.C.J. 1963. Effects of frequency of irrigation on yield of orange and grapefruit. Proc. Fla. State Hort. Soc. 76:1–5.

Koo, R.C.J. 1985. Response of 'Marsh' grapefruit trees to drip, under tree sprays and sprinkler irrigation. Proc. Fla. State Hort. Soc. 98:29–32.

Koo, R.C.J. and R.P. Muraro. 1982. Effect of tree spacing on fruit production and net returns of 'Pineapple' oranges. Proc. Fla. State Hort. Soc. 95:29–33.

Koo, R.C.J., C.A. Anderson, D.A. Calvert, I. Stewart, D.P.H. Tucker, and H.K. Wutscher. 1984. Recommended fertilizers and nutritional sprays for citrus. Bulletin 536–D, University of Florida Agricultural Experiment Station, Florida.

Korkmaz, S., B. Cevik, S. Onder, K. Koc. and O. Bozan. 2008. Detection of citrus tristeza virus (CTV) from satsuma mandarins (*Citrus unshiu*) by direct tissue blot immunoassay (DTBIA), DAS-ELISA and biological indexing. New Zealand J. Crop and Hort. Sci. 36:239–246.

Kotzé, J. M. 1981. Epidemiology and control of citrus black spot in South Africa. Plant Disease. 65:945–950.

Krezdorn, A.H. 1969. The use of growth regulators to improve fruit set in citrus. Proc. 1st Intel. Citrus Symp. 3:1113–1119.

Krezdorn, A.H. and M. Cohen. 1962. The influence of chemical fruit-set sprays on yield and quality of citrus. Proc. Fla. State Hort. Soc. 75:53–60.

Kriedmann, P.E. and H.D. Barrs. 1981. Citrus orchards. In: Kozlowski, T.T. (ed.), Water Deficits and Plant Growth. Academic Press, New York, pp. 325–417.

Langdon, K.W., R. Schumann, L.L. Stelinski, and M.E. Rogers. 2018a. Spatial and temporal distribution of soil-applied neonicotinoids in citrus tree foliage. J. Econ. Entomol. 111: 1788–1798.

Langdon, K.W., R. Schumann, L.L Stelinski, and M.E. Rogers. 2018b. Influence of tree size and application rate on expression of thiamethoxam in citrus and its efficacy against *Diaphorina citri* (Hemiptera: Leviidae). J. Econ. Entomol. 111: 770–779.

Langeland, K., M. Netherland, and W. Haller. 2006. Efficacy of Herbicide Active Ingredients Against Aquatic Weeds. Univ. Florida IFAS Extension Publication #SS-AGR-44.

Lapointe, S.L. 2000. History and importance of Diaprepes to agriculture in the Caribbean region. Diaprepes short course. 8–12. Citrus Research and Education Center, Lake Alfred, FL.

Lapointe S.L., D.G. Hall, Y. Murata, A.L. Parra-Pedrazzoli, J.M.S. Bento, E.F. Vilela, and W.S. Leal. 2006. Field evaluation of a synthetic female sex pheromone for the leafmining moth *Phyllocnistis citrella* (Lepidoptera: Gracillariidae) in Florida citrus. Fla Entomol. 89: 274–276.

Lapointe, S.L., L.L. Stelinski, T.J. Evens, R.P. Niedz, D.G. Hall, and A. Mafra-Neto. 2009. Sensory imbalance as mechanism of orientation disruption in the leafminer *Phyllocnistis citrella*: elucidation by multivariate geometric designs and response surface models. J. Chem. Ecol. 35: 896–903.

Lapointe, S.L., C.P. Keathley, L.L. Stelinski, W.H. Urrutia, and A. Mafra-Neto. 2015. Disruption of the leafminer *Phyllocnistis citrella* (Lepidoptera: Gracillariidae) in citrus: effect of blend and placement height, longevity of disruption and emission profile of a new dispenser. Fla Entomol. 98: 743–748.

Laserfood Inc. 2015. Tattoo etching of fruit with chemical enhancement. Retrieved from http://www.laserfood.es/index.php/en/

Lea-Cox, J.D., J.P. Syvertsen and D.A.Graetz. 2001. Springtime 15nitrogen uptake, partitioning, and leaching losses from young bearing Citrus trees of differing nitrogen status. J. Amer. Soc. Hort. Sci. 126:242–251.

Lee, J. A., S. E. Halbert, W.O. Dawson, C. J. Robertson, J.E. Keesling, and B. H. Singer. 2015. Asymptomatic spread of huanglongbing and implications for disease control. Proc. Nat. Acad. Sciences USA. 112: 7605–7610.

Lee, R.F., C.N. Roistacher, C.L. Niblett, M. Rocha-Pena, S.M. Gamsey, R.K. Yokomi, D.G. Gumpf, and J.A. Dobbs. 1992. Presence of *Toxoptera citricidus* in Central America, a threat to citrus in Florida and the United States. The Citrus Industry 73:8.

Legaz, F., R. Ibanez, D.G. de Barreda, and E. Primo Millo. 1981. Influence of irrigation and fertilization on productivity of 'Navelate' sweet orange. Proc. Int. Soc. Citriculture 2:591–595.

Lenz, F., 1969. Effect of day length and temperature on the vegetative and reproductive growth of 'Washington' navel orange. Proc. First Int. Citrus Symp. I:333–338.

Leonard, C.D., I. Stewart, and I.W. Wander. 1961. A comparison of ten nitrogen sources for 'Valencia' oranges. Proc. Fla. State Hort. Soc.74:79–86.

Lewis-Rosenblum, H., X. Martini, S. Tiwari, and L.L. Stelinski. 2015. Seasonal movement patterns and long-range dispersal of Asian citrus psyllid in Florida citrus. J. Econ. Entomol. 108: 3–10.

Lim, U.T. and M.A. Hoy. 2006. Overwintering of the citrus leafminer, *Phyllocnistis citrella* (Lepidoptera: Gracillariidae), without diapause in Florida. Fla Entomol. 89: 361–366.

Lima, J.E.O. 1982. Observations on citrus blight in São Paulo, Brazil. Proc. Fla. State Hort. Soc. 95:72–75.

Lima, J.E.O. and F.S. Davies. 1984. Growth regulators, fruit drop, yield and quality of navel oranges in Florida. J. Amer. Soc. Hort. Sci. 109:81–84.

Liu, Y-Z. and X.X. Deng. 2007. Citrus breeding and genetics in China. The Asian and Austral. J. Plant Sci. and Biotech. Global Sci Books 2007, 27 pp.

Lo Giudice, V. 1981a. Present status of citrus nematode control in the Mediterranean area. Proc. Int. Soc. Citriculture 2:384–387.

Lo Giudice, V. 1981b. Present status of citrus weed control in Italy. Proc. Int. Soc. Citriculture 2:485–487.

Locali-Fabris, E.C., J. Freitas-Astuas, A.A. Souza, M.A. Takita, G. Astua-Monge, R. Antonioli-Luizon, V. Rodrigues, M.L.P.N. Tangon, and M.A. Machado. 2006. Complete nucleotide sequence, genomic organization and phylogenetic analysis of Citrus leprosis virus cytoplasmic type. J. Gen. Virol. 87:2721–2729.

Lord, E.M. and M.J. Eckard. 1985. Shoot development in *Citrus sinensis* L. ('Washington' navel orange). I. Floral and inflorescence ontogeny. Bot. Gaz. 146:320–326.

Lord, E.M. and K.J. Eckard. 1987. Shoot development in *Citrus sinensis* (Washington Navel orange). II. Alternation of developmental fate of flowering shoots after GA3 treatment. Bot. Gaz. 148:17–22.

Lovatt, C.J., S.M. Streeter, T.C. Minter, N.V. O'Connell, D.L. Flaherty, M.W. Freeman, and P.B. Goodall. 1984. Phenology of flowering in *Citrus sinensis* (L.) Osbeck, cv. 'Washington' navel orange. Proc. Int. Soc. Citriculture 1:186–190.

Lovatt, C.J., Y. Zheng, and K.D. Hake. 1988. Demonstration of a change in nitrogen metabolism influencing flower initiation in Citrus. Israel Journal of Botany 37:181–188.

Lovatt, C.J., O. Sagee, and A.G. Ali. 1992. Ammonia and/or its metabolites influence flowering, fruit set, and yield of the 'Washington' navel orange. Proc. Int. Soc. Citriculture, 412–416.

Marler, T.E. and F.S. Davies. 1987. Growth of barerooted and container-grown 'Hamlin' orange trees in the field. Proc. Fla. State Hort. Soc. 100:89–93.

Marler, T.E. and F.S. Davies. 1988. Soil water content and leaf gas exchange of young field-grown 'Hamlin' orange trees. Proceedings of the Interamerican Society of Tropical Horticulture 32:51–64.

Marler, T.E. and F.S. Davies. 1990. Microsprinkler irrigation and growth of young 'Hamlin' orange trees. J. Amer. Soc. Hort. Sci. 115:45–51.

Marler, T.E., J.J. Ferguson, and F.S. Davies. 1987. Growth of young 'Hamlin' orange trees using standard and controlled-release fertilizers. Proc. Fla. State Hort. Soc.100:61–64.

Martin, S., M.L. Garcia, A. Troisi, L. Rubio, G. Legarreta, O. Grau, D. Alioto, P. Moreno, and J. Guerri. 2006. Genetic variation of populations of Citrus psorosis virus. J. Gen. Virol. 87:3097–3102.

Martini, X., T. Addison, B. Fleming, I. Jackson, K. Pelz-Stelinski, and L.L. Stelinski. 2013. Occurrence of *Diaphorina citri* (Hemiptera: Liviidae) in an unexpected ecosystem: The Lake Kissimmee state park forest, Florida. Fla Entomol. 96: 658–660.

Martini X., M. Hoffmann, M.R. Coy, L.L. Stelinski, and K.S. Pelz-Stelinski 2015. Infection of an insect vector with a bacterial plant pathogen increases its propensity for dispersal. PLoS ONE 10:e0129373 10.1371/journal.pone.0129373

Matthews, G.A. 1992. Pesticide Application Methods. Longman: Harlow, UK, 336 pp.

Maust, B. 1992. Nutrition of citrus trees in the nursery. MSc thesis, University of Florida, Gainesville.

Maxwell, N.P. and R.E. Rouse. 1984. Growth and yield comparison of ten-year-old red grapefruit trees from field- and container-grown nursery stock. Journal of the Rio Grande Valley Horticultural Society 37:71–73.

McCoy, C.W., 1981. Pest control by the fungus *Hirsutella thompsonii*. In: Burges, H.D. (Ed.), Microbial Control of Pests and Plant Diseases 1970–1980. Academic press, London, pp. 499–512.

McCoy, C.W. 1985. Current status of biological control in Florida. In: Hoy, M.A. and Herzog, D.C. (eds), Biological Control in Agricultural IPM Systems. Academic Press, New York, pp. 481–499.

McCoy, C.W. 1988. The biology of the citrus rust mite and its effects on fruit quality. In: Ferguson, J.J. and Wardowski, W.F. (eds), Citrus Short Course Proceedings - Factors Affecting Fruit Quality. Florida Cooperative Extension Service, Florida, pp. 54–68.

McCoy, C.W. and L.G. Albrigo. 1975. Feeding behavior and nature of injury to the orange caused by citrus rust mite. *Phyllocoptruta oleivora* (Prostigmata, Eriophyoidea). Annals of the Entomological Society of America 68:289–297.

McCutchan H. and K.A. Shackel. 1992. Stem-water potential as a sensitive indicator of water stress in prune trees (*Prunus domestica* L. cv. French). J. Amer. Soc. Hort. Sci. 117:607–611.

Mendel, K. 1969. The influence of temperature and light on the vegetative development of citrus trees. Proceedings of the First International Citrus Symposium 1:259–265.

Menge, J. A., H. Lembright, and E.L.V. Johnson. 1977. Utilization of mycorrhizal fungi in citrus nurseries. Proc. Int. Soc. Citriculture 1:129–132.

Meredith, F.I. and R.H. Young. 1969. Effect of temperature on pigment development in 'Red Blush' grapefruit and 'Ruby' blood orange. Proc. First Int. Citrus Symp. I:271–275.

Miller, W.M. and J.K. Burns. 1992. Grade lowering defects and grading practices for Indian River grapefruit. Proc. Fla. State Hort. Soc. 105:129–130.

Mink, G.I. 1993. Pollen- and seed-transmitted viruses and viroids. Ann. Rev. Phytopathol. 31:375–402.

Mirecki, R.M. and A.H. Teramura. 1984. Effects of Ultraviolet-B irradiance on soybean. Plant Physiol. 74:475–480.

Mondal, S.N. and L.W. Timmer. 2006. Greasy spot, a serious endemic problem for citrus production in the Caribbean Basin. Plant Disease. 90:532–538.

Monselise, S.P. 1947. The growth of citrus roots and shoots under different cultural conditions. Palest. J. Bot. 6:43–54.

Monselise, S.P. and A.H. Halevy. 1964. Chemical inhibition and promotion of citrus flower bud induction. Proc. Am. Soc. Hortic. Sci. 84:141–146.

Monselise, S.P., M. Weiser, N. Shafir, R. Goren, and E.E. Goldschmidt. 1976. Creasing of orange peel - physiology and control. J. Hortic Sci. 51:341–351.

Monzo, C., and P.A. Stansly. 2017. Economic injury levels for Asian citrus psyllid control in process oranges from mature trees with high incidence of huanglongbing. Plos One. 12(4): e0175333.

Morgan, K. 2016. Citrus MicroSprinkler Irrigation Scheduler. Retrieved from http://fawn.ifas.ufl.edu/tools/irrigation/citrus/scheduler

Morgan, K.T., R.E. Rouse and R.C. Ebel. 2018. Foliar applications of essential nutrients on growth and yield of 'Valencia' sweet orange infected with huanglongbing. HortScience 51:1482–1493.

Morinaga, K., O. Sumikawa, O. Kawamoto, H. Yoshikawa, S. Nakao, M. Shimazaki, S. Kusaba, and N. Hoshi. 2005. New technologies and systems for high quality citrus fruit production, labor-saving and orchard construction in mountain areas of Japan. Journal of Mountain Science 2:59–67.

Morris, A., R.P. Muraro, and W.S. Castle. 2011. Optimal grove replanting to mitigate endemic HLB. Citrus Ind. 12–16

Moss, G. I. 1969. Influence of temperature and photoperiod on flower induction and inflorescence development in sweet orange (*Citrus sinensis* L. Osbeck). J. Hort. Sci. 44(4):311–320.

Moss, G.I. 1970. The influence of temperature on fruit-set in cuttings of sweet orange (*Citrus sinensis* (L) Osbeck). Hort. Res. 10:97–107.

Moss, G. I. 1973. The influence of temperature during flower development on subsequent fruit set of sweet orange (*Citrus sinensis*) cv. 'Washington Navel'. Hort. Res. 13:65–73.

Moss, G.I. 1978. Propagation of citrus for future plantings. Proc. Int. Soc. Citriculture 1:132–135.

Mungomery, W.V., K.R. Jorgensen, and J.A. Barnes. 1978. Rate and timing of nitrogen application to navel oranges: effects on yield and fruit quality. Proc. Int. Soc. Citriculture 1:285–288.

Nagy, S. and J.A. Attaway (eds) .1980. Citrus Nutrition and Quality. American Chemical Society, Washington, D.C.

Navarro L and J. Juárez J (2007) Shoot-tip grafting in vitro. In: Khan IA (ed) Citrus genetics. Breeding and biotechnology. CAB International, Wallingford, UK, pp. 353–364.

Nemec, S. 1978. Response of six citrus rootstocks to three species of Glomus, a mycorrhizal fungus. Proc. Fla. State Hort. Soc. 91:10–14.

Newcomb, D.A. 1977. Citrus seed production. Proc. Int. Soc. Citriculture 1:124–126.

Nicolosi E. (2007) Origin and taxonomy. In: Khan I.A. (ed.) Citrus Genetics, Breeding and Biotechnology. CAB International, Wallingford, UK, pp. 19–44.

Nicolosi, E., Z.N. Deng, A. Gentile, S. La Malfa, G. Continella, and E.Tribulato. 2000. Citrus phylogeny and genetic origin of important species as investigated by molecular markers. Theor. Appl. Genet.100:1155–1166.

Nishikawa, F., T. Endo, T. Shimada, H. Fujii, T. Shimizu, M. Omura, and Y. Ikoma. 2007. Increased CiFT abundance in the stem correlates with floral induction by low temperature in Satsuma mandarin (*Citrus unshiu* Marc.) J. Expt. Bot. 58:3915–3927.

Nordby, H.E. and R.E. McDonald. 1990. Squalene in grapefruit wax as a possible natural protectant against chilling. Lipids 25:807–810.

O'Bannon, J.H. and H.W. Ford. 1977. Resistance in citrus rootstocks to *Radopholus simihs* and *Tylenchulus semipenetrans* (Nematoda). Proc. Int. Soc. Citriculture 2:544–549.

O'Bannon, J.H., V. Chew, and A.T. Tomerlin. 1977. Comparison of five populations of *Tylenchulus semipenetrans* to *Citrus*, *Poncirus* and their hybrids. J. Nematol. 9:162–165.

Oberholzer, P.C.J. 1969. Citrus production in southern Africa. Proceedings of the First International Citrus Symposium 1:111–120.

Oren, Y. and E. Israeli. 1977. Herbicide application through irrigation systems (herbigation) in citrus. Proc. Int. Soc. Citriculture 1:152–154.

Ortuño, A., A. Baidez, P. Gomez, H.C. Axcas, T. Pottas, A. Garcia-Lidón, and J.A. Del Rio. 2006. *Citrus paradisi* and *Citrus sinensis* flavonoids: Their influence in the defence mechanism against *Penicillium digitatum*. Food Chemistry 98:351–358.

Oslund, C.R. and T.L. Davenport. 1987. Seasonal enhancement of flower development in 'Tahiti' limes by marcottage. HortScience 22:498–501.

Palou, L. 2014. Chapter 2: *Penicillium digitatum, Penicillium italicum* (Green mold, Blue mold). In: Bautista-Banos, S. (Ed.), Postharvest decay control strategies. Elsevier Academic Press, London; pp. 45–102.

Parsons, L.R., T.A. Wheaton, N.D. Faryna, and J.L. Jackson. 1991. Improve citrus freeze protection with elevated microsprinklers. Proc. Fla. State Hort. Soc. 104:144–147.

Parsons, L.R., T.A. Wheaton and W. S. Castle. 2001. High application rates of reclaimed water benefit citrus tree growth and fruit production. HortScience. 36(7):1273–1277.

Passos, D.S., A.P. Cunha, Y.S. Coelho, and E.M. Rodriques. 1977. Behavior of orange trees under three spacings in the state of Bahia, Brazil. Proc. Int. Soc. Citriculture 1:169–171.

Pelz-Stelinski, K.S., R.H. Brlansky, T.A. Ebert, and M.E. Rogers. 2010. Transmission parameters for Candidatus Liberibacter asiaticus by Asian citrus psyllid (Hemiptera: Psyllidae). J. Econ. Entomol. 103: 1531–1541.

Peng, T., X. Zhu, N. Duan, J.H. Liu. 2014. PtrBAM1, a β-amylase-coding gene of *Poncirus trifoliata*, is a CBF regulon member with function in cold tolerance by modulating soluble sugar levels. Plant Cell Environ. 37(12):2754–67.

Peres, N.A., L.W. Timmer, J.E. Adaskaveg, and J.C. Correll. 2005. Life styles of *Colletotrichum acutatum*. Plant Disease. 89:784–796.

Peres, N.A.R., and L.W. Timmer. 2003. Citrus black spot caused by *Guignardia citricarpa*. CABI Crop Protection Compendium. CAB International, Wallingford, UK.

Peres, N.A.R., S. Kim, H.W. Beck, N.L. Souza, and L.W. Timmer. 2002. A fungicide application decision (FAD) support system for postbloom fruit drop of citrus (PFD). Online. Plant Health Progress DOI:10.1094 PHP-2002–0731–01–RV.

Phillips, R.L. 1974. Performance of 'Pineapple' orange trees at three spacings. Proc. Fla. State Hort. Soc. 87:81–84.

Phillips, R.L. 1980. Hedging and topping practices for Florida citrus. Citrus Industry 61:5–10.

Phung, H.T. and E.B. Knipling. 1976. Photosynthesis and transpiration of citrus seedlings under flooded conditions. HortScience 11:131–133.

Pillitteri, L.J., C. J. Lovatt, and L. L. Walling. 2004. Isolation and characterization of a TERMINAL FLOWER homolog and its correlation with juvenility in Citrus. Plant Physiol. 135(3): 1540–1551.

Poinar, G. O. 1990. Taxonomy and biology of Steinernematidae and Heterorhabditidae. Entomopathogenic nematodes in biological control. In R. Gaugler and H. K. Kaya (eds.). Boca Raton, CRC.

Purvis, A.C. 1981. Free proline in peel of grapefruit and resistance to chilling injury during cold storage. HortScience 6:160–161.

Purvis, A.C. 1989. Soluble sugars and respiration of flavedo tissue of grapefruit stored at low temperatures. HortScience 24:320–322.

Qin, X., V.S. Miranda, M. Machado, M.E. Lemos, and J.S. Hartung. 2001. An evaluation of the genetic diversity of *Xylella fastidiosa* isolated from diseased citrus and coffee. Phytopathol. 91:599–605.

Qureshi, J.A. and P.A. Stansly. 2007. Integrated approaches for managing the Asian citrus psyllid *Diaphorina citri* (Homoptera: Psyllidae) in Florida. Proc. Fla. State Hort. Soc. 120: 110–115.

Qureshi, J.A., and P.A. Stansly. 2009. Exclusion techniques reveal significant biotic mortality suffered by Asian citrus psyllid *Diaphorina citri* (Hemiptera: Psyllidae) populations in Florida citrus. Biol. Control. 50: 129–136.

Qureshi, J.A. and P.A. Stansly. 2010. Dormant season foliar sprays of broad spectrum insecticides: An effective component of integrated management for *Diaphorina citri* (Hemiptera: Psyllidae) in citrus orchards. Crop Prot. 29: 860–866.

Qureshi, J.A., M.E. Rogers, D.G. Hall, and P.A. Stansly. 2009. Incidence of invasive *Diaphorina citri* (Hemiptera: Psyllidae) and its introduced parasitoid *Tamarixia radiata* (Hymenoptera: Eulophidae) in Florida citrus. J. Econ. Entomol. 102: 247–256.

Qureshi, J.A., B.C Kostyk, and P.A. Stansly. 2014. Insecticidal suppression of Asian citrus psyllid *Diaphorina citri* (Hemiptera: Liviidae) vector of huanglongbing pathogens. PloS One. 9:e112331.

Rabe, E. (1991) Bench-rooted citrus nursery trees. *The Citrus Industry* 72, 52–53.

Rabe, E. and H.P. van der Walt. 1992. Effect of urea sprays, girdling and paclobutrazol on yield in 'Shamouti' sweet orange trees in a subtropical climate. Journal of the South African Society of Horticultural Sciences 2:77–81.

Rabe, E., P. H. van Rensburg, H. van der Walt and J. Bower. 1990. Factors influencing preharvest fruit splitting in Ellendale. HortScience 25(9):135.

Raghuvanshi, S.S, (1968) Cytological evidence bearing on evolution in citrus. Proceedings of the First International Citrus Symposium 1, 207–214.

Rasmussen, G.F. and P.F. Smith. 1961. Evaluation of fertilizer practices for young orange trees. Proc. Fla. State Hort. Soc. 74:90–95.

Ray, P.K. 2002. Breeding Tropical and Subtropical Fruits. Springer-Verlag, Narosa Publ. House, New Delhi.

Redd, J.B., C.M. Hendrix Jr. and D.L. Hendnx. 1986a. Quality Control Manual for Citrus Processing Plants, Vol. I. Regulation, Citrus Methodology, Microbiology, Conversion Charts, Tables, Other. Intercit Incorporated, Safety Harbor, Florida. 250 pp.

Redd, J.B., P.E. Shaw, C.M. Hendrix Jr. and D.L. Hendnx. 1986b. Quality Control Manual for Citrus Processing Plants, Vol.III. General, Systems, Important Volatiles, Shelf-life, Specialty and By-Products, Relationships – Raw to the Processed Product, Miscellaneous Conversion Charts and Tables. AgSciences, Inc. Auburndale, Florida. 335 pp.

Redd, J.B., D.L. Hendrix, and C.M. Hendrix Jr. 1992. Quality Control Manual for Citrus Processing Plants, Vol. II. Processing and Operating Procedures, Blending Techniques, Formulating, Citrus Mathematics and Costs. AgScience Incorporated, Auburndale, Florida 337 pp.

Reuther, W. and D. Rios-Castano. 1969. Comparison of growth, maturation and composition of citrus fruits in subtropical California and tropical Colombia. Proceedings of the First International Citrus Symposium 1:277–300.

Richardson, M.L., C.J. Westbrook, D.G. Hall, E. Stover and Y.P. Duan. 2011. Abundance of citrus leafminer larvae on Citrus and Citrus-related germplasm. HortScience. 46:1260–1264.

Rieger, M. 1989. Freeze protection of horticultural crops. In: Janick, J. (ed.), Horticultural Reviews, Vol. XI. Timber Press, Oregon, pp. 46–109.

Ritenour, M.A., J. Zhang and M. Dewdney. 2011. Postharvest decay control recommendations for Florida citrus fruit. Univ. Fla. IFAS Extension CIR359A; pp. 6.

Rodriquez, O. and S. Moreira. 1969. Citrus nutrition - 20 years of experimental results in the state of São Paulo, Brazil. Proceedings of the First International Citrus Symposium 3:1579–1586.

Rogers, J.S. and J.F. Bartholic. 1976. Estimated evapotranspiration and irrigation requirements for citrus. Proceedings of the Soil and Crop Science Society of Florida 35:111–117.

Rogers, M.E., M.M. Dewdney and T. Vashisth (eds). 2018. 2017–18 Florida Citrus Production Guide, Institute of Food and Agricultural Sciences, Florida Cooperative Extension Service, University of Florida, Gainesville

Roistacher, C.N. 1991. Graft-transmissible Diseases of Citrus. Handbook for Detection and Diagnosis. FAO, Rome.

Roistacher, C.N. 1996. The economics of living with citrus disease: Huanglongbing (greening) in Thailand. In J.V. da Graça, P. Moreno and R.K. Yokomi (eds.), Proceedings of the 13th Conference of the International Organization of Citrus Virologists (IOVC), pp. 279–285, Riverside, CA.

Roistacker, C.N., J. Bash and D.J. Gumpf. 2000. Continued attempts over a 22–year period to separate components of the citrus tatter leaf-citrange stunt virus complex. Proceedings Conference IOCV, Riverside (US) 14:179–184.

Román, M. P., M. Cambra, J. Juárez, P. Moreno, N. Duran-Vila, F.A.O. Tanaka, E. Alves, E.W. Kitajima, P.T. Yamamoto, R.B. Bassanezi, D.C. Teixeira, W.C. Jesus Junior, A.J. Ayres, N. Gimenes-Fernandes, F. Rabenstein, L.F. Girotto, and J.F. Bové. 2004. Sudden death of citrus in Brazil: a graft-transmissible, bud union disease. Plant Disease. 88:453–467.

Roose, M.L. 1988. Isozymes and DNA restriction fragment length polymorphisms in Citrus breeding and systematics. Proc. Int. Soc. Citriculture 1:57–67.

Roth, R.L., D.R. Rodney, and B.R. Gardner. 1974. Comparison of irrigation methods, rootstocks and fertilizer elements on 'Valencia' orange trees. Proceedings of the Second International Drip Irrigation Congress 2:103–108.

Rouse, R.E. 1982. Evaluation of 5 media and 5 fertilizer treatments for container citrus. Journal of the Rio Grande Valley Horticultural Society 35:167–171.

Rouse, R.E. 1988. Bud forcing method affects budbreak and scion growth of citrus grown in containers. Journal of the Rio Grande Valley Horticultural Society 41:69–73.

Rouse, B., P. Roberts, M. Irey, M. Boyd, and T. Willis. 2010. Monitoring trees infected with huanglongbing in a commercial grove receiving nutritional/SAR foliar sprays in southwest Florida. Proc. Fla. State Hort. Soc. 123:118–120.

Ruiz, R., A., García-Luis, C. Monerri and J.L. Guardiola. 2001. Carbohydrate availability in relation to fruitlet abscission in Citrus. Ann. Bot. 87:805–812.

Russo, F. 1981. Present situation and future prospect of the citrus industry in Italy. Proc. Int. Soc. Citriculture 2:969–973.

Sagio, P., M. Lhospital, D. Lafléche, G. Dupont, J.M. Bové, J.G. Tully and E. A. Freundt. 1973. Spiroplasm citri Gen. and sp.nn.: a mycoplasm-like organism associated with 'Stubborn' disease of citrus. Int. J. Systemic Bacteriol. 23:191–204.

Sahin-Cevik, M. and G.A. Moore. 2011. Quantitative trait loci analysis of morphological traits in Citrus. Plant Biotech. Reports 6:47–57.

Salerno, M. and G. Cutuli. 1981. The management of fungal and bacterial diseases of citrus in Italy. Proc. Int. Soc. Citriculture 1:360–362.

Sánchez de la Torre, M.E., C. López, O. Grau and M.L. García. 2002. RNA 2 of Citrus psorosis virus is of negative polarity and has a single open reading frame in its complementary strand. Journal of General Virology 83:1777–1781

Santa Ana, R. 2012. Spinach genes may stop deadly citrus disease. http://todayagrilife.org/2012/03/26/transgenic-citrus-trees/

Sanz, A., C. Monerri, J. Gonzalez-Ferrer and J.L. Guardiola. 1987. Changes in carbohydrates and mineral elements in Citrus' leaves during flowering and fruit set. Physiol. Plantarum 69:93–98.

Saunt, J. 1990. Citrus Varieties of the World. Sinclair International, UK.

Savage, E.M. and F.E. Gardner. 1965. The origin and history of Troyer and Carrizo citranges. The Citrus Industry 46:4–7.

Schaffer, B., J.E. Peña, A.M. Colls, A. Hunsberger. 1997. Citrus leafminer (Lepidoptera Gracillariidee) in lime. Assesment of leaf damage and effects on photosynthesis. Crop Protection 16:337–343.

Schneider, H., R.G. Piatt, and W.P. Bitters. 1978. Diseases and incompatibilities that cause decline in lemons. Citrograph 63:219–221.

Schneider, W., R. Avijit, J. Hartung, J. Shao and R. Brlansky. 2015. Citrus blight research. Citrus Industry 96(12):12–14.

Schroeder, W.J. 1992. Entomopathogenic nematodes for control of root weevils of citrus. Fla Entomol. 73:563–567.

Schubert, T., S.A. Rizvi, X. Sun, T.R. Gottwald, J.H. Graham, and W. Dixon. 2001. Meeting the challenge of eradication of citrus canker in Florida – again. Plant Disease. 85:340–356.

Schumann, A., K. Morgan, B. Castle, and J. Schumann. 2016. Advanced Citrus Production Systems in Florida. http://treetpee.com/wp-content/uploads/2016/02/citrusindustrymag-article-pdf

Schumann, A., L. Waldo, A. Wright and R. Ferearezi. 2017. Update on citrus undercover production systems research. Citrus Industry 98(11):16, 18–20.

Schupp, J.R., T. Auxt Baugher, S.S. Miller, R.M. Harsh and K.M. Lesser. 2008. Mechanical thinning of peach and apple trees reduces labor input and increases fruit size. HortTechnology 18:660–670.

Scora, R.W. 1988. Biochemistry, taxonomy and evolution of modern cultivated Citrus. Proc. Int. Soc. Citriculture 1:2 77–289.

Scora, R.W., J. Kumamoto, R.K. Soost and E.M. Nauer. 1982. Contribution to the Origin of the Grapefruit, *Citrus paradisi* (Rutaceae). Syst. Botany 7(2):170–177.

Seif, A. A. and R.J. Hillocks. 1997a. Some factors affecting infection of citrus by *Phaeoramularia angolensis*. J. Phytopathol.146:385–391.

Seif, A. A. and R.J. Hillocks. 1997b. Chemical control of Phaeoramularia fruit and leaf spot of citrus in Kenya. Crop Protection. 16:141–145.

Shapiro-Ilan, D.I., and C.W. McCoy. 2000a. Virulence of entomopathogenic nematodes to *Diaprepes abbreviatus* (Coleoptera: Curculionidae) in the laboratory. J. Econ. Entomol. 93:1090–1095.

Shapiro-Ilan, D.I., and C.W. McCoy. 2000b. Susceptibility of *Diaprepes abbreviatus* (Coleoptera: Curculionidae) larvae to different rates of entomopathogenic nematodes in the greenhouse. Fla Entomol. 83:1–9.

Shapiro-Ilan, D.I., D.H. Gouge, and A.M. Koppenhöfer. 2002. Factors affecting commercial success: case studies in cotton, turf and citrus. In: Gaugler, R. (Ed.), Entomopathogenic Nematology. CAB International, Wallingford, UK, pp. 333–355.

Sharp, J.L. 1989. Preliminary investigation using hot air to disinfect grapefruit of Caribbean fruit fly immatures. Proc. Fla. State Hort. Soc. 102:157–159.

Sharpies, G.C. and R.H. Hilgeman. 1969. Influence of differential nitrogen fertilization on production, tree growth, fruit size and quality, and foliage composition of 'Valencia' orange trees in central Arizona. Proceedings of the First International Citrus Symposium 3:1569–1578.

Simpson, S.E., H.N. Nigg, N.C. Coile, and R.A. Adair. 1996. *Diaprepes abbreviatus* (Coleoptera: Curculionidae): host plant associations. Environ. Entomol. 25:333–349.

Simpson, S.E., H.N. Nigg, and J.L. Knapp. 2000. Host plants of Diaprepes root weevil and their implications to the regulatory process. Diaprepes short course. 19–37. Citrus Research and Education Center Lake Alfred, FL.

Singh, M., R.V. Tamma, and H.N. Nigg. 1989. HPLC identification of allelopathic compounds from *Lantana camara*. Journal of Chemical Ecology 15:81–89.

Smajstrla, A.G. and R.C.J. Koo. 1984. Effects of trickle irrigation methods and amounts of water applied on citrus yields. Proc. Fla. State Hort. Soc. 97:3–7.

Smajstrla, A.G., L.R. Parsons, K. Anbi, and G. Velledis. 1985. Responses of young citrus trees to irrigation. Proc. Fla. State Hort. Soc. 98:25–28.

Smith, P.F. 1966a. Citrus nutrition. In: Childers, N.F. (ed.), Fruit Nutrition. Horticultural Publications, New Jersey, pp. 174–207.

Smith, P.F. 1966b. Leaf analysis of citrus. In: Childers, N.F. (ed.), Fruit Nutrition. Horticultural Publications, New Jersey, pp. 208–228.

Smith, P.F. 1969. Effects of nitrogen rates and timing of application on 'Marsh' grapefruit in Florida. Proceedings of the First International Citrus Symposium 3:1559–1567.

Smith, P. F. 1976. Collapse of 'Murcott' tangerine trees. Proc. Amer. Soc. Hort. Sci. 101:23–25.

Smith, P.F. and W. Reuther. 1949. Observations on boron deficiency in citrus. Proc. Fla. State Hort. Soc. 62:31–37.

Smith, P.F. and W. Reuther. 1953. Mineral content of oranges in relation to fruit age and some fertilization practices. Proc. Fla. State Hort. Soc. 66:80–85.

Snowball, A.N., E.A. Halligan, and M.G. Mullins. 1988. Studies on juvenility in Citrus. Proc. Int. Soc. Citriculture 1:467–473.

Snowball A.M., I.J. Warrington, E.A. Halligan, and M.G. Mullins. 1994. Phase change in citrus: the effects of main stem node number, branch habit and paclobutrazol application on flowering in citrus seedlings. J. Hort. Sci. 69: 149–160.

Sood, P., C. Ference, J. Narciso, and E. Etxeberria. 2009. Laser etching: a novel technology to label Florida grapefruit. HortTechnology 19(3):504–510.

Soost, R.K. and M. Roose. 1996. Citrus, pp. 257–328. In J. Jules and J.N. Moore (eds). Fruit Breeding: Tree and Tropical Fruits. Wiley, New York.

Southwick, S.M. and T.L. Davenport. 1986. Characterization of water stress and low temperature effects on floral induction in citrus. Plant Physiology 81:26–29.

Spann, T.M., A. Schumann, B. Rouse and B. Abel. 2011. Foliar nutrition for HLB. Citrus Industry 92(6):6–10.

Spiegal-Roy, P. and E.E. Goldschmidt. 1996. The Biology of Citrus. Cambridge University Press, UK. 230 pp

Spreen, T.H., J.-P. Baldwin, and S.H. Futch. 2014. An economic assessment of the impact of Huanglongbing on citrus tree plantings in Florida. HortScience 49:1052–1055.

Spurling, M.B. 1969. Citrus in the Pacific area. Proceedings of the First International Citrus Symposium 1:93–101.

Stansly, P.A., R.E. Rouse, R.J. McGovern, and S.B. Davenport. 1991. Chemical deterrents to girdling of young citrus by subterranean termites. Proc. Fla. State Hort. Soc.104:156–159.

Stansly, P.A., H.A. Arevalo, J.A. Qureshi, M.M. Jones, K. Hendricks, P.D. Roberts, and F.M. Roka. 2014. Vector control and foliar nutrition to maintain economic sustainability of bearing citrus in Florida groves affected by huanglongbing. Pest management science. 70: 415–426.

Stansly, P.A., S. Croxton and J. Sherrod. 2016. Big boost in young tree growth and yield from insecticides and metalized mulch. Citrus Industry 97(3):20–22.

Stelinski, L.L., J.R. Miller, and M.E. Rogers. 2008. Mating disruption of citrus leafminer mediated by a noncompetitive mechanism at a remarkably low pheromone release rate. J. Chem. Ecol. 34: 1107–1113.

Stelinski, L.L., S.L. Lapointe, and W.L. Meyer. 2010. Season-long mating disruption of citrus leafminer, *Phyllocnistis citrella* Stainton, with an emulsified wax formulation of pheromone. J. Appl. Entomol. 134: 512–520.

Stewart, I. and T.A. Wheaton. 1972. Carotenoids in citrus: Their accumulation induced by ethylene. J Agric. Food Chem. 20:448 449.

Stover, E., S. Ciliento, M. Ritenour and C. Counter. 2002. NAA thinning of 'Murcott': Comparison of small plot and commercial harvest data. Proc. Fla. State Hort. Soc. 115:287–291.

Stover, E., T. McCollum, J. Chaparro and M. Ritenour. 2013. Under severe HLB and canker pressure 'Triumph' and 'Jackson' perform better than 'Flame' or 'Marsh'. Proc. Fla. State Hort. Soc. 125:40–46.

Suzuki, K. 1981. Weeds in citrus orchards and their control in Japan. Proc. Int. Soc. Citriculture 2:489–492.

Swietlik, D. 1992. Yield, growth and mineral nutrition of young 'Ray Ruby' grapefruit trees under trickle or flood irrigation and various nitrogen rates. J. Amer. Soc. Hort. Sci. 117:22–27.

Swingle, W.T. 1948. Botany of citrus and its wild relatives of the orange subfamily. In: Webber, H.J. and Batchelor, L.D. (eds), The Citrus Industry. University of California Press, California, pp. 129–174.

Swingle, W.T. and P.C. Reece. 1967. The botany of citrus and its wild relatives. In: Reuther, W., Batchelor, L.D. and Webber, H.J. (eds), The Citrus Industry. University of California Press, California, pp. 190–340.

Syvertsen, J.P. 1981. Hydraulic conductivity of four commercial citrus rootstocks. J. Amer. Soc. Hort. Sci. 106:378–381.

Syvertsen, J.P. 1982. Minimum leaf water potential and stomatal closure in citrus leaves of different ages. Ann. Bot. 49:827–834.

Syvertsen, J.P. 1984. Light acclimation in citrus leaves. II. CO2 assimilation and light, water and nitrogen use efficiency. J. Amer. Soc. Hort. Sci. 109:812–817.

Syvertsen, J.P. and J.H. Graham. 1985. Hydraulic conductivity of roots, mineral nutrition and leaf gas exchange of citrus rootstocks. J. Amer. Soc. Hort. Sci. 110:865–869.

Syvertsen, J.P., and M.L. Smith. 1996. Nitrogen uptake efficiency and leaching losses from lysimeter-grown Citrus trees fertilized at three nitrogen rates. J. Amer. Soc. Hort. Sci. 121:57–62.

Syvertsen, J.P., R.M. Zablotowicz, and M.L. Smith. 1983. Soil temperature and flooding effects on two species of citrus. I. Plant growth and hydraulic conductivity. Plant and Soil 72:3–12.

Syvertsen, J.P., C. Goñi and A. Otero. 2003. Fruit load and canopy shading affect leaf characteristics and net gas exchange of 'Spring' navel orange trees. Tree Physiology 23:899–906.

Syvertsen, J.P., L.G. Albrigo, J.M. Dunlop, M.A. Ritenour and R.C. Vachon. 2005. Growth conditions, crop load and fruit size affect sheepnosing in grapefruit. Proc. Fla. State Hort. Soc. 118:28–34.

Tachibana, S. and S. Nakai. 1989. Relation between yield and leaf area index in different planting densities under different cultural treatments in 'Satsuma' mandarin (*Citrus unshiu* Marc. var. praecox) tree. Journal of the Japanese Society of Horticultural Science 57:561–567.

Talhouk, A.S. 1975. Citrus pests throughout the world. In: Citrus. Ciba-Geigy Agrochemicals, Basel, Switzerland, pp. 21–23.

Talon M. and Gmitter F.G. Jr. (2008) Citrus genomics. Intl. J. Plant Genomics. Article ID 528361.

Talon, M., F.R. Tadeo, W. Ben-Cheikh, A. Gomez-Cadenas, J. Mehouachi, J. Pérez-Botella, and E. Primo-Millo. 1998. Hormonal regulation of fruit set and abscission in citrus: classical concepts and new evidence. Acta Hort. 463:209–218.

Tanaka, T. 1977. Fundamental discussion of Citrus classification. Studia Citrologica 14:1–6.

Taylor, C.A. and J.R. Furr. 1937. Use of soil moisture and fruit growth records for checking irrigation practice in citrus orchards. U.S. Dept. Agr. Cir. 426:1–23.

Taylor, K.C., L.G. Albrigo and C.D. Chase. 1996. Purification of a Zn-binding phloem protein with sequence identity to chitin-binding proteins. Plant Physiol. 110:657–664.

Thomson, W.W., L.N. Lewis and C.W. Coggins. 1967. The reversion of chromoplasts to chloroplasts in 'Valencia' oranges. Cytologia 32:117–124.

Timmer, L.W. and H.N. Benatena. 1977. Comparison of psorosis and other viruses causing leaf flecking in citrus. Proc ISC 930–935.

Timmer, L.W. and A. Bhatia 2003. Diseases of Unknown Etiology. In: Stacey, G. and N. T. Keen, N.T. (eds.). PlantMicrobe Interactions. Vol. 6. APS Press, Inc., St. Paul, MN, pp. 309–329.

Timmer, L.W. and G.E. Brown. 2000. Biology and control of anthracnose diseases of citrus, In: Prusky, D., Freeman, S. and Dickman, M.B. eds. Host Specificity, Pathology, and Host-Pathogen Interactions of Colletotrichum. APS Press, Inc., pp. 300–316.

Timmer, L.W. and L.W. Duncan, eds. 1999. Citrus Health Management, APS Press, Inc., St. Paul, MN. 197 pp.

Timmer, L. W., J. P. Agostini, J. H. Graham and W. S. Castle. 1991. Relationship of citrus rootstock to phytophthora root rot and populations of *Phytophthora parasitica*. Proc. Fla. State Hort. Soc. 104:173–178.

Timmer, L.W., J.P. Agostini, S.E. Zitko, and M. Zulfiqar. 1994. Postbloom fruit drop, an increasingly prevalent disease of citrus in the Americas. Plant Disease. 78:329–334.

Timmer, L.W., S.M. Garnsey, and J.H. Graham, eds. 2000. Compendium of Citrus Diseases, 2nd. ed. APS Press, Inc., St. Paul, MN.

Timmer, L.W., S.M. Garnsey, and P. Broadbent. 2003a. Diseases of Citrus. In: Ploetz, R.C., ed. Diseases of Tropical and Subtropical Crops. CAB International, Wallingford, UK, pp. 163–195

Timmer, L.W., T.L. Peever, Z. Solel, and K. Akimitsu. 2003b. Alternaria diseases of citrus–novel pathosystems. Phytopathologia Mediterranea 42, 99–112.

Timmer, L.W., S.N. Mondal, N.A.R. Peres, and A. Bhatia. 2004. Fungal diseases of fruit and foliage of citrus trees. In: Mujerki, K.G. and Navqi, S.A.M.H, eds. Disease Management of Fruits and Vegetables. Kluwer Academic. The Netherlands, pp. 191–227.

Tiwari, S., H. Lewis-Rosenblum, K. Pelz-Stelinski, and L.L. Stelinski. 2010. Incidence of Candidatus Liberibacter asiaticus infection in abandoned citrus occurring in proximity to commercially managed groves. J. Econ. Entomol. 103: 1972–1978.

Tiwari, S., R.S. Mann, M.E. Rogers, and L.L. Stelinski. 2011. Insecticide resistance in field populations of Asian citrus psyllid in Florida. Pest Management Science. 67: 1258–1268.

Tiwari, S., L.L. Stelinski, and M.E. Rogers. 2012. Biochemical basis of organophosphate and carbamate resistance in Asian citrus psyllid. J. Econ. Entomol. 105: 540–548.

Tolkowsky, S. 1938. Hesperides: a history of the culture and use of citrus fruits. John Bales, Sons & Curnow, 371 pp.

Torres, A.M., R.K. Soost, and T. Mau-Lastovicka. 1978. Citrus isozymes. Journal of Heredity 73:335–339.

Tucker, D.P.H. and M. Singh. 1992. Weeds. In: Knapp, J.L. (ed.), Florida Citrus Spray Guide (SP 43), University of Florida, Gainesville, pp. 16–23.

Turrell, F.M. 1961. Growth of the photosynthetic area of Citrus. Bot. Gaz. 122:285–298.

Tylor, H. L., L. F. W. Roesch, S. Gowda, W. O. Dawson, and E. W. Triplett. 2009. Confirmation of the sequence of 'Candidatus Liberibacter asiaticus' and assessment of microbial diversity in Huanglongbing-infected citrus phloem using a metagenomic approach. Molec. Plant-Microbe Interactions 22:1624–1634.

Uzun, A., T. Yesiloglu, Y. Aka-Kacar, O. Tuzcu, and O. Gulsen. 2009. Genetic diversity and relationships within Citrus and related genera based on sequence related amplified polymorphism markers (SRAPs). Sci. Hortic. 121:306–312.

Valiente, J.I. and L.G. Albrigo. 2002. Modeling flowering date of sweet orange [*Citrus sinensis* (L.) Osbeck] trees in Central Florida based on historical weather records. Proc. Intl. Soc. Citriculture. 2000, 1:296–299.

Valiente, J.I. and L.G. Albrigo. 2004. Flower bud induction of sweet orange trees [*Citrus sinensis* (L.) Osbeck]: Effect of low temperatures, crop load, and bud age. J.Amer. Soc. Hort. Sci. 129(2):158–164.

Van Bavel, C.H., M. Newman, and R.H. Hilgeman. 1967. Climate and estimated water use by an orange orchard. Agricultural Meteorology 4:27–37.

Vandiver, V.V. 1992a. Ditch bank, emerged and floating weeds. In: Knapp, J.L. (ed.), Florida Citrus Spray Guide (SP 43). University of Florida, Gainesville, pp. 32–42.

Vandiver, V.V. 1992b. Submerged aquatic weeds. In: Knapp, J.L. (ed.), Florida Citrus Spray Guide (SP 43). University of Florida, Gainesville, pp. 24–31.

Vidalakis, G., J. V. da Graça, W. N. Dixon, D. Ferrin, M. Kesinger, R. R. Krueger, R. F. Lee, M. J. Melzer, J. Olive, M. Polek, P. J. Sieburth, L. L. Williams, and G. C. Wright. 2010. Citrus quarantine sanitary and certification programs in the USA, Prevention of introduction and distribution of citrus diseases. Part 1 – Citrus quarantine and introduction programs. Citrograph pp. 26–39.

Von Staden, P.F.A. and P.C.J. Oberholzer. 1977. The performance of nucellar citrus lines on several rootstocks in South Africa. Proc. Int. Soc. Citriculture 2:532–534.

Wang, N., L.L. Stelinski, K. Pelz-Stelinski, J.H. Graham, and Y. Zhang. 2017. Tale of the huanglongbing disease pyramid in the context of the citrus microbiome. Phytopathology 107: 380–387.

Webber, J.H., W. Reuther, and H.W. Lawton. 1967. History and development of the citrus industry. In: Reuther, W., Webber, H.J. and Batchelor, L.D. (eds), The Citrus Industry. University of California Press, Riverside, pp. 1–39.

Weissling, T. J., J.E. Peña, R.M.J.R. Giblin-Davis, and J.L. Knapp. 2002. Sugarcane rootstock borer weevil, *Diaprepes abbreviatus* (L.). Featured creatures, Univ. of Florida, Gainesville, FL.

Wethern, M. 1991. Citrus debittering with ultracentrifugation/absorption combined technology. Transactions of the Citrus Engineering Conference, American Society of Agricultural Engineers, pp. 48–66.

Wheaton, T.A. 1981. Fruit thinning of Florida mandarin using plant growth regulators. Proc. Int. Soc. Citriculture 1:263–268.

Wheaton, T.A. 1986. Alternate bearing. In: Citrus flowering, fruit set, and development. Univ. Fla. Citrus Short Course. In Futch, S. and Kender, W. (eds). Pupl. CREC, University Florida, pp. 67–72.

Wheaton, T.A. 1997. Alternate bearing of citrus in Florida. In: Citrus Flowering & Fruiting Short Course pp. 87–92.

Wheaton, T.A. and I. Stewart. 1973. Fruit thinning of tangerines with naphthalene-acetic acid. Proc. Fla. State Hort. Soc. 86:48–52.

Wheaton, T.A., W.S. Castle, D.P.H. Tucker, and J.P. Whitney. 1978. Higher density plantings for Florida citrus: concepts. Proc. Fla. State Hort. Soc. 91:27–33.

Wheaton, T.A., W.S. Castle, J.D. Whitney, D.P.H. Tucker, and R.P. Muraro. 1990. A high density citrus planting. Proc. Fla. State Hort. Soc. 103:55–59.

Whiteside, J.O. 1981. Diagnosis of greasy spot based on experiences with this disease in Florida. Proc. Int. Soc. Citriculture 1:336–340.

Wilcox, D.A. and F.S. Davies. 1981. Temperature-dependent and diurnal root conductivities in two citrus rootstocks. HortScience 16:303–305.

Williamson, J.G., W.S. Castle, and K.E. Koch. 1992. Growth and 14C-photosynthate allocation in citrus nursery trees subjected to one of three bud forcing methods. J. Amer. Soc. Hort. Sci. 117:37–40.

Willis, L.E., F.S. Davies, and D.A. Graetz. 1990. Fertilization, nitrogen leaching and growth of young 'Hamlin' orange trees on two rootstocks. Proc. Fla. State Hort. Soc. 103:30–37.

Willis, L.E., F.S. Davies, and D.A. Graetz. 1991. Fertigation and growth of young 'Hamlin' orange trees in Florida. HortScience 26:106–109.

Wilson, W.C. 1983. The use of exogenous plant growth regulators on Citrus. In: Nickell, L.G. (ed.), Plant Growth Regulating Chemicals. CRC Press, Florida, pp. 207–232.

Wiltbank, W.J., R.E. Rouse and L.N. Khoi. 1995. Influence of temperature on citrus rootstock seed emergence. Proc. Fla. State Hort. Soc. 108:137–139.

Winston, R.L., M. Schwarzländer, H.L. Hinz, M.D. Day, M.J.W. Cock and M.H. Julien. (Eds.) 2014. Biological Control of Weeds: A World Catalogue of Agents and Their Target Weeds, 5th edition. USDA Forest Service, Forest Health Technology Enterprise Team, Morgantown, West Virginia. FHTET-2014–04. 838 pp.

Wondimagegnehu, M. and M. Singh. 1989. Benefits and problems of chemical weed control in citrus. Review of Weed Science 4:59–70.

Wu, Guohong Albert, Terol, J., Ibanez, V., López-García, A., Pérez-Román, E., Borredá, C., Domingo, C., Tadeo, F.R., Carbonell-Caballero, J., Alonso, R., Curk, F., Du, D., Ollitrault, P., Roose, M.L., Dopazo, J., Gmitter, F.G., Rokhsar, D.S. and Talon, M. 2018. Genomics of the origin and evolution of Citrus. Nature 554: 311–316

Wutscher, H.K. (1977) The influence of rootstocks on yield and quality of red grapefruit in Texas. Proc. Int. Soc. Citriculture 2:526–529.

Wutscher, H.K. 1979. Citrus rootstocks. In: Janick, J. (ed.), Horticultural Reviews. AVI Publishing Co., Westport, Connecticut, pp. 230–269.

Wutscher, H.K. 1989. Alteration of fruit tree nutrition through rootstocks. HortScience 24:578–584.

Wutscher, H.K. and D. Dube. 1977. Performance of young grapefruit on 20 rootstocks. J. Amer. Soc. Hort. Sci. 102:267–270.

Wutscher, H.K. and A.V. Shull. 1975. Yield, fruit quality, growth and leaf nutrient levels of 14–year-old grapefruit, *Citrus paradisi* Macf., tree on 21 rootstocks. J. Amer. Soc. Hort. Sci. 100:290–294.

Wutscher, H.K. and T.A. Obreza. 1987. The effect of withholding Fe, Zn and Mn sprays on leaf nutrient levels, growth rate and yield of young 'Pineapple' orange trees. Proc. Fla. State Hort. Soc. 100:71–74.

Xiang, C. and M.L. Roose. 1988. Frequency and characteristics of nucellar and zygotic seedlings in 12 citrus rootstocks. Scientia Horticulturae 37:47–59.

Xu, C. F., Y. H. Xia, K. B. Li, and C. Ke. 1988. Further study of the transmission of citrus huanglongbing by a psyllid, *Diaphorina citri* Kuwayama, In. L.W. Timmer, S.M. Garnsey, and L. Navarro (eds.), Proc 10th Conference of the International Organization of Citrus Virologists, Riverside, CA.

Yager, E. 1977. Drip irrigation in citrus orchards. Proc. Int. Soc. Citriculture 1:110–117.

Yahiaoui, D., A. M. D'Onghia, K. Djelouah, and A. Catara. 2015. Genetic evidence of potential virulent Citrus tristeza virus isolates in mediterranean areas. J. Plant Pathol. 97(2):243–248.

Yamamoto, P.T., R.B. Bassanezi, N.A. Wulff, M.A. Santos, A.L. Sanches, S. Rodrigo, R.S. Toloy, N. Gimenes-Fernandes, A.J. Ayres, W.C. Jesus Junior, T. Francisco, A.O. Tanaka, F.A.O. Kitajima, and J.M. Bové. 2011. Citrus sudden death is transmitted by graft-inoculation and natural transmission is prevented by individual insect-proof cages. Plant Disease. 95:104–112.

Yelenosky, G. 1985. Cold hardiness in citrus. In: Janick, J. (ed.), Horticultural Reviews, Vol. VII. AVI Publishing Co., Westport, Connecticut, pp. 201–238.

Yelenosky, G. and C.J. Hearn. 1967. Cold damage to young mandarin-hybrid trees on different rootstocks in flatwoods soil. Proc. Fla. State Hort. Soc. 80:53–56.

Yelenosky, G. and R. Young. 1977. Cold hardiness of orange and grapefruit trees on different rootstocks during the 1977 freeze. Proc. Fla. State Hort. Soc. 90:49–53.

Young, R. 1969. Effect of freezing on the photosynthetic system in citrus. In: Chapman, H.D. (ed.), Proceedings of the First International Citrus Symposium. University of California Press, Berkeley, pp. 553–558.

Young, R.H. 1977. The effect of rootstocks on citrus cold hardiness. Proc. Int. Soc. Citriculture 2:518–522.

Young, R.H. and O. Jahn. 1972. Ethylene induced carotenoid accumulation in citrus fruit rind. J. Amer. Soc. Hort. Sci. 97:258–261.

Young, R.H., L.G. Albrigo, M. Cohen, and W.S. Castle. 1982. Rates of blight incidence in trees on Carrizo citrange and other rootstocks. Proc. Fla. State Hort. Soc. 95:76–78.

Zaragoza, S. and E. Alfonso. 1981. Citrus pruning in Spain. Proc. Int. Soc. Citriculture 1:172–175.

Zhang, W.C. 1981. Development and outlook of citrus industry in China. Proc. Int. Soc. Citriculture 2:987–990.

INDEX

abscission
  fruit 101
  leaves 152
acidity of fruit (TA) 112–114, 250
  TSS:TA ratio 114, 251, 266
ACP *see* Asian citrus psyllid
'Afourer' mandarin 32
age of orchards 106
air-layering (marcotting) 125
alemow (*Citrus macrophylla*) 58
Alternaria black rot 240
Alternaria brown spot 238–239, 257
alternate bearing 98, 105, 172–173, 253
altitude 78–79, 80, 115–116
'Ambersweet' mandarin hybrid 34–35
1-aminocyclopropane-1-carboxylic acid (ACC) 91–92
ammonium ions (NH_4^+) 99, 154
anthesis 95–98, 104
anthracnose
  lime 236–237
  postharvest 240–241
antibiotics 244
ants 196–197, 201–202
aphids 195, 198, 211–212
  CTV vectors 198, 221
apomixis 13
area-wide management 212–213
areolate leaf spot 239
Argentina 10, 11
arid conditions 81, 91, 108, 128
arthropod pests *see* pests
ascorbic acid 250
asexual reproduction 13
Asia
  climate 81
  HLB 214, 217

origin of citrus 1–4
production 5, 10, 107–108, 129
Asian citrus psyllid (ACP, *Diaphorina citri*) 184–185, 216
  area-wide management 212–213
  biological control 210–211, 217
  insecticides 185–186, 205–207, 208
  screens 217
Aurantioideae subfamily 13
Australia 9
  indigenous species 14–16
Australian sour 61

bacterial diseases
  control 217, 244
  fruit & foliage 232–233, 239, 257
  tree declines 214–219
  see also huanglongbing
'Baianinha' navel orange 26
bark scaling 222, 226
bees 100
beetles 188
bergamot 8, 12
biological control 184, 198, 200, 210–212
  of root weevil 189–190
  of weeds 181–182
biotechnology
  in breeding programmes 49–51, 77, 217
  pathogen identification 242
birds 257
bitterness 250
'Bittersweet' sour orange 61
black root rot 231
black rot (Alternaria) 240
black spot 233–234
blackfly 198, 211

blight 226–228, 239
blood orange 8, 27, 111
boron (B) 158
Botrytis cinerea 239
Brazil
  climate 81, 93
  production 5, 10–11, 107
breeding programmes
  difficulties 45
  objectives 46–47, 243
  rootstock 47, 48–49, 70, 77, 119
  scion 47, 48
  techniques 48–53
  zygotic seedlings 20, 33, 46
Brevipalpus phoenicis 192, 223
Brix 251
brown citrus aphid (*Toxoptera citricida*) 198, 211–212, 221
brown rot (*Phytophthora*) 145, 229–230, 241, 243, 245, 257
brown spot (Alternaria) 238–239, 257
buckhorned trees 153, 169
budding 85, 123–125
  failure due to TLV 222
burrowing nematodes 203–204
butterflies 202

cachexia 226, 241–242
calcium (Ca) 157, 166
California 7, 9
California red scale (*Aonidiella aurantii*) 200, 212
Candidatus Liberibacter spp. 215
  see also huanglongbing
canker 183, 232–233, 257
canned products 12
'Cara Cara' navel orange 27
carbohydrates
  control of flowering 98
  in fruit 112, 247–248, 251
  fruit drop 104–105
carbon (CO_2) assimilation 86–88, 90
  water stress 137–138
carotenoids 111
'Carrizo' citrange 65–66
caterpillars 202
chaff scale (*Parlatoria pergandii*) 200, 212
chilling injury 265
chimeras 52
China
  HLB 214, 217
  production 5, 107–108, 129
chlorophyll, in peel 111

citrange (*C. sinensis* × *P. trifoliata*) 44, 65–67
citric acid 112–114, 250
citron (*Citrus medica*) 1–3, 242
citrumelo (*C.paradisi* × *P. trifoliata*) 44, 67–68
Citrus aurantifolia see lime
Citrus aurantium (sour orange) 3, 59–61
citrus blight 226–228
citrus bud mite (*Aceria sheldoni*) 190–191
citrus canker 183, 232–233, 257
Citrus deliciosa (Mediterranean mandarin) 4, 30
Citrus grandis (pummelo) 3, 20, 42–43
citrus greening disease *see* huanglongbing
Citrus jambhiri (rough lemon) 56–57, 84
citrus leafminer (CLM, *Diaphorina citri*) 183, 186–187, 206, 207–208
citrus leprosis virus (CiLV) 223
Citrus limettioides ('Palestine' sweet lime) 59, 247
Citrus limon see lemon
Citrus macrophylla (alemow) 58
Citrus medica (citron) 1–3, 242
citrus nematode (*Tylenchulus semipenetrans*) 203, 231
Citrus paradisi see grapefruit
citrus psorosis virus (CPsV) 221–222
citrus red mite (*Panonuchus citri*) 192–193
Citrus reticulata (common mandarin) 28–30
  see also mandarins
citrus rust mite (*Phyllocoptruta oleivora*) 191–192, 211, 257
Citrus sinensis see orange
citrus sudden death (CSD) 224
citrus tristeza virus (CTV) 198, 219–221, 242
Citrus unshiu (satsuma mandarin) 10, 27–28, 47
citrus variegated chlorosis (CVC) 217–218, 257
Citrus volkameriana 57
classification *see* taxonomy
clean cultivation 149
'Clementine' mandarin 29
'Cleopatra' mandarin 55, 61–63
climate 78–82
  change 115–116
  orchard site selection 127
  see also temperature
CLM (citrus leafminer) 183, 186–187, 206, 207–208
Clymenia spp. 16–17
cold *see* freezing
Colletotrichum spp. 236–237

colour
   of flesh 247
      blood oranges 27
      grapefruit 10, 36–39
   of peel 111–112, 252, 254–255
common mandarin (*Citrus reticulata*) 28–30
   see also mandarins
consumer preferences
   fresh fruit 8–9, 10, 252, 256
   juice 248, 251
container-grown trees 123, 135–136
controlled atmosphere storage 266
controlled-release fertilizer 163–164
copper (Cu) 158
   fungicides 230, 236
cottony cushion scale (*Icarya purchasi*) 199
creasing 259
CTV (citrus tristeza virus) 198, 219–221, 242
CVC (citrus variegated chlorosis) 217–218, 257
cyazypyr 208

damping-off 145, 231
'Dancy' tangerine (mandarin) 29–30
de-greening of fruit 254–255, 263
Diaphorina citri see Asian citrus psyllid
2,4-dichlorophenoxyacetic acid (2,4-D) 172, 173
Diplodia stem-end rot 240
diseases 214–245
   bacterial 214–219, 232–233, 239, 244, 257
      see also huanglongbing
   control 241–245
      chemical 244–245
      cultural practices 230, 244
      in nurseries 117–118, 122
      pathogen identification 241–242
      regulations 243–244
      resistance 47, 243
   of fruit & foliage 229–230, 232–239, 245, 257
   fungal 145, 229–231, 232–241, 245, 257
   herbicide injury 179–180
   nematodes 203–204, 231–232, 245
   physiological
      copper toxicity 158
      mineral deficiencies 155, 156, 157, 158–159
      Murcott collapse 31–32, 228–229
   postharvest 239–241, 266–268
   of roots 145, 229–232, 245
   tree declines 214–229
   viral 219–225, 242

   viroidal 225–226, 241–242
   see also pests
distribution of citrus spp. 4–11
drip irrigation 107
drought 81, 91, 104, 136
   induction of flowering 93, 107
dry root rot 228
'Duncan' grapefruit 36
dwarfing
   rootstocks 64, 69, 70–71, 133
   viroid-induced 226

economics
   and planting density 131–133
   ROI 129
   seed prices 118–119
ELISA assays 242
environmental conditions 78–116
eradication campaigns 217, 244
Eremocitrus 14–16
Eriophyidae spp. (mites) 190–192
ethephon 172
ethylene 246, 254, 255
'Eureka' lemon 42
Europe 1, 3, 7–8
evapotranspiration (ET) 140–141
exocortis 226, 241–242
export 7–11, 12

'Fallglo' mandarin hybrid 34
farm size 6
Femminello lemon 41
fertigation 163
fertilization 162–168
   and fruit quality 166–168
   interactions between elements 166
   mature trees 162, 165–166
   in nurseries 120, 122, 126
   and planting density 134
   young trees 162–164
field nurseries 120–122, 123
filmwrapped fruit 266, 269
fire ants 201–202
'Flame' grapefruit 39
flood irrigation 142, 151
flooding 128, 144–146
Florida
   flowering 93
   freezes 81, 127, 147
   HLB 7, 213
   production 6–7, 8–9, 11, 107

Florida red scale (*Chrysomphalus aonidum*) 200, 212
flowers/flowering
　environmental factors 92–94, 106–107
　growth and development 94–98
　morphology 17, 101
　physiological factors 98–100
　pollination 100–101
'Flying Dragon' trifoliate orange 70
fog 79
food safety 272
foot rot (*Phytophthora*) 145, 229–230, 243, 245
Fortunella (kumquat) 4, 14, 43
'Foster' grapefruit 37
freezing
　care of damaged trees 151–153
　damage caused by 81, 146–147, 148, 152, 240
　freeze-hardiness 47, 147–149
　protection methods 149–151
　and rootstock variety 56, 58, 60, 62, 64, 66, 68
frost 147
frozen concentrate orange juice (FCOJ) 11, 270
fruit
　alternate bearing 98, 105, 172–173, 253
　appearance 17, 246, 254
　blemishes 81, 112, 255–259
　and climatic factors 101–115
　colour *see* colour
　composition 112–114, 247–251
　diseases
　　postharvest 239–241, 266–269
　　on the tree 223, 229–230, 232–239
　drop 101, 104–105, 173
　freeze damage 152, 240
　growth 109–111, 112
　hail damage 258
　harvesting 134, 260–261
　packinghouse operations 240, 241, 261–265
　pest damage 191–192, 193, 256–257
　physiological disorders 156, 258–259, 268–269
　quality
　　external 81, 111–112, 140, 252–259
　　and fertilization 166–168
　　internal 112–115, 140, 251–252, 266
　　for juicing 269
　ripening 109–110, 246
　and rootstocks 57, 58, 60, 62, 65, 68

　set 101–104, 172
　size 110, 252–254, 264
　storage 246–247, 265–266
　water management 104, 112, 137
　wind damage 81, 258
　yield 106–109
fruit flies 197, 202–203, 256–257
fumigation
　fruit 256
　soil 120–121
fungi
　biological control agents 211
　mycorrhizae 76, 121
　pathogenic
　　fruit & leaves 229–230, 233–239, 257
　　postharvest 239–241, 266–268
　　root 145, 229–231, 243
fungicides
　postharvest 263
　seed treatment 119
　on the tree 230, 234, 235, 236, 237, 238, 239, 240, 245

genetic engineering 49–51, 77, 217
genetics 20, 45–46, 49
gibberellic acid (GA_3)
　flowers 93, 98, 172
　fruit 101, 171–172, 173, 259
girdling 98
global warming 115–116
glyphosate 180
grading of fruit 264
granulation 259
grapefruit (*Citrus paradisi*) 35–39
　HLB 39
　hybrids 32–33
　origin 3–4
　peel colour 111
　production 5, 10
　as rootstock 69
　sheepnosed 112, 254
　storage 265
grasshoppers 202
greasy spot 234–235, 257
greenhouses
　commercial 150
　nurseries 118, 119–120, 122–123
greening *see* huanglongbing
growth regulators
　flowering 93, 98, 172
　fruit 101, 171–173, 253, 259
gummosis 221–222, 229, 230, 243

hail 258
'Hamlin' sweet orange 22–23
handling of fruit 260–265
harvesting 134, 260–261
    see also postharvest period
heading-back cuts 168
heat units (hu) 78–79, 80–81
heaters (in orchards) 150
hedging 170–171, 253
height of trees 131, 168–171
'Henderson' grapefruit 39
herbicides 175–181
history 1–4, 11–12
HLB *see* huanglongbing
hormones
    flowering 93, 98, 172
    fruit 101, 171–173, 253, 259
huanglongbing (HLB) 214–217
    control 212–213, 216–217
        of the vector 185–186,
            205–207, 208
    fertilization 165, 217
    grapefruit 39
    siting of orchards 129
    soil pH 160
    in the USA 6–7, 184–185, 213
    vectors 184, 216
'Hudson' grapefruit 37
humidity 79, 81
    postharvest 255, 265–266
hybridization 18, 20
    intergeneric 43–45, 51
    man-made 32–35, 46, 51, 172
    natural 30–32, 45

ice-nucleating agents 147–148
importing countries 11
India 10
insect pests 184–190, 193–203, 256–257
insecticides 185–186, 186–187,
        205–210, 245
integrated pest management (IPM)
        204, 216–217
intercropping 6, 178
interstocks 71
iron (Fe)
    deficiency 159
    reduction of Fe^{3+} 75
irradiation of seeds 51–52
irrigation 139–144
    fertigation 163
    field nurseries 121, 126

freeze protection 150–151
herbigation 181
and root depth 72
water availability/quality 128, 144
and yield 107, 140–141
isozyme analysis 46, 49
Italy 1, 3, 8

Japan 10, 81
juice *see* orange juice
juvenile trees 48, 54, 83–85

'Kinnow' mandarin 30
kumquat (*Fortunella*) 4, 14, 43

'Lane Late' navel orange 26
Lantana camara 181–182
leaf
    diseases
        CVC 217–218
        fungal 232–239, 245
        HLB 216
    freeze damage 148, 152
    growth 89–90
    heat stress 86–88
    herbicide injury 180
    mineral content 73–75, 160–162
    morphology 17
    pests
        citrus leafminer 186–187
        root weevil 189
        spider mites 192–193
    phyllotaxis 85
    water conservation 136–137
leaf and fruit spot
        (*Pseudocercospora*) 234
leaf water potential 72–73, 142
leaf-cutting ants 196–197, 201
leafhoppers 219
lemon (*Citrus limon*) 40–42
    origin 3
    postharvest changes 246–247, 266
    production 5, 10, 11
    storage 265–266
'Leng' navel orange 26
Lepidoptera 197, 202
    see also citrus leafminer
leprosis 223
light intensity 79, 86–88
lime anthracnose 236–237

lime (*Citrus aurantifolia*) 39–40
   marcotting 125
   origin 3
   production 5, 10
   WBDL 219
lime, sweet (*Citrus limettoides*) 59, 247
liming of soil 160
limonin 25, 250
'Lisbon' lemon 42
lycopene 111

macronutrients 154–157
   fertilization 162–168
   leaf content 73–75, 160–162
magnesium (Mg) 156–157, 166
mandarins
   alternate bearing 98, 105, 172–173, 253
   hybrids 30–35, 45, 172
   Murcott collapse 31–32, 228–229
   natural varieties 27–30
   origin 4
   picking 260
   pollination 100–101, 172
   production 5, 8–9, 10, 12
   reproduction 20
   rootstocks 59, 61–63, 69
   seedless 8, 9, 35, 101, 252
   size 253
manganese (Mn) 157–158
marcotting 125
'Marrs' navel orange 26
'Marsh' grapefruit 36–37
maturation season
   mandarin 28
   sweet orange 22
mealybugs 195, 199
Mediterranean mandarin (*Citrus deliciosa*) 4, 30
melanose 236, 257
Mexico 10
Microcitrus spp. 16
microirrigation 142–143, 151
mineral nutrients 153–154
   flowering 99
   interactions 166
   leaf analysis 73–75, 160–162
   macronutrients 154–157
   micronutrients 157–159
   soil analysis 159–160
   uptake 73–76
   see also fertilization
'Minneola' tangelo 33

mites
   biological control agents 211
   pests 190–193, 211, 257
   vectors 223
miticides 193
molybdenum (Mo) 158–159
Morocco 8
morphology 17
   flowers 17, 101
   grapefruit 36
   kumquat 43
   lemon 40–41
   lime 40
   mandarin/mandarin hybrids 28, 101, 253
   orange 23, 24, 25
   pummelo 42–43
   rootstocks 71–72
moulds 239–240, 268
Murcott collapse 31–32, 228–229
'Murcott' mandarin hybrid 31–32, 98
mushroom root rot 230
mutation 51–52
mycorrhizae 76, 121

naphthaleneacetic acid (NAA) 105, 172
'Natal' sweet orange 24
navel orange 3, 25–27
'Navelate' navel orange 26
'Navelina' navel orange 26
nematodes
   biological control agents 189–190
   pests 203–204, 231–232, 245
neonicotinoids 185, 207
'Newhall' navel orange 26
nitrate (NO_3^-) 154, 164
nitrogen (N) 154–155
   control of flowering 99
   leaf content 73–75, 160–162
   see also fertilization
nucellar embryony 13, 19, 46, 82
nursery operations 117–127
nutrients *see* fertilization; mineral nutrients

oleocellosis 140, 268–269
orange (*Citrus sinensis*) 21–27
   colour 111
   hybrids 30–32
   origin 3
   production 5, 8, 9
   rootstock 63
   see also sour orange; trifoliate orange

orange juice 11, 12, 246
  composition 247–249
  processing 25, 269–271
orangeries 3, 149
orchards
  fertilization 134, 162–168
  freeze protection 149–151
  longevity 106
  planting design/density 130–134, 168, 190
  planting methods 134–136
  pruning 153, 168–171, 253–254
  site selection 127–129, 149
  weed control 173–182
organophosphates 206
origin 1–4
'Orlando' tangelo 32

packaging 264–265
  shrinkwrapping 266, 269
packinghouses 240, 241, 261–265
'Palestine' sweet lime (*Citrus limettoides*) 59, 247
'Palmer' navel orange 26
parasitoid wasps 210–211, 211, 212
parthenocarpy 100
pectins 251
peel
  ageing 269
  blemishes 81, 112, 255–259
  colour 111–112, 252, 254–255
  oleocellosis 140, 268–269
Penicillium spp. 239–240, 268
'Pera' sweet orange 24
pests 183–213
  control
    area-wide management 212–213
    biological 184, 189–190, 198, 200, 210–212, 217
    chemical 185–186, 186–187, 193, 205–210, 245
    IPM 204, 216–217
    monitoring 204–205
    pheromones 187
    screens 205, 217
  insects 184–190, 193–203, 256–257
  mites 190–193, 211, 257
  nematodes 203–204, 231–232, 245
  in nurseries 120, 122
  see also diseases
pH of soil 128, 160
pheromones 187

Phomopsis stem-end rot 240
phosphorus (P) 76, 156, 165, 166
photoperiod 92
physiology 72–75, 91, 112, 136–139, 145
Phytophthora spp. (rot) 145, 229–230, 241, 243, 245, 257
'Pineapple' sweet orange 24
pink disease 239
plant bugs 201, 256
planting design/density 130–134, 168
  root weevil control 190
planting methods 134–136
plugging 261
pollination 100–101, 172
polyembryony 19–20, 46, 82
Poncirus spp. 14
  P. trifoliata see trifoliate orange
'Ponkan' mandarin 30
postbloom fruit drop 236–237
postharvest period
  disorders 239–241, 266–269
  packinghouses 240, 241, 261–265
  storage 246–247, 265–266
potassium (K) 156, 165, 166, 167–168
processed citrus 12, 269
  economics 129
  orange juice 11, 12, 25, 246, 269–270
production 4–11, 12
propagation 85, 123–125
  failure due to TLV 222
  in vitro grafting 242
protoplast fusion 51, 77
pruning 168–171, 253–254
  after freeze damage 153
pruning paints 169
Pseudocercospora leaf and fruit spot 234
psorosis 221–222
psyllids *see* Asian citrus psyllid
pummelo (*Citrus grandis*) 3, 20, 42–43
pyrethroids 206
Pythium spp. (damping-off) 145, 231

quality of fruit
  external 81, 111–112, 140, 252–259
  and fertilization 166–168
  internal 112–115, 140, 251–252, 266
  for juicing 269

rainfall 79, 81, 140
'Rangpur' mandarin 59, 69
'Rangpur' × 'Troyer' rootstock 69

'Ray Ruby' grapefruit 39
re-greening 111
rectangularity of orchards 130–131
regulations
    food safety 272
    pest/disease control 243–244, 256–257
rejuvenation pruning 169
ringspot 221–222
Rio Grande gummosis 221–222
'Rio Red' grapefruit 39
ripening 109–110, 246
'Robinson' mandarin hybrid 33–34
'robyn' navel orange 26
root
    anatomy 71
    fungal diseases 145, 229–231, 243
    growth 82, 90–91
    nematode pests 203–204, 231–232, 245
    percentage irrigated 143
root hydraulic conductivity (Lp) 73, 91, 136
root rot (*Phytophthora*) 145, 229–230, 243, 245
root weevil complex (*Diaprepes abbreviatus*) 187–190, 209
rootstocks 54–77
    breeding 47, 48–49, 70, 77, 119
    choosing 47, 54–55, 243, 253
    descriptions 56–71
    flooding tolerance 145
    morphology 71–72
    mycorrhizae 76, 121
    physiology 72–75
rough lemon (*Citrus jambhiri*) 56–57, 84
round orange 22–25
'Ruby Red' grapefruit 39
rust mite (*Phyllocoptruta oleivora*) 191–192, 211, 257
Rutaceae family 13

salt tolerance 75, 144
satsuma dwarf virus (SDV) 223–224
satsuma mandarin (*Citrus unshiu*) 10, 27–28, 47
scab disease 237–238, 257
scale insects 195–196, 199–201, 212, 256
scion
    breeding programmes 47, 48
    grafting onto rootstock 85, 123–125, 242
    production of disease-free material 117–118
screens 205, 217
seed
    germination 82, 121
    irradiation 51–52
    in the nursery 118–119
        field planting 121
        greenhouse planting 119–120
seediness/seedlessness 247, 252
    mandarins 8, 9, 35, 101
    sweet oranges 22
seedlings 82–85, 122–123
    damping-off 145, 231
Septoria spot 239
serological assays 242
shaddock (pummelo, *Citrus grandis*) 3, 20, 42–43
shading 89
'Shamouti' sweet orange 24
sheepnosed grapefruit 112, 254
shelter 149–150
shoot growth 88–89
shrinkwrapped fruit 266, 269
site selection
    nurseries 118
    orchards 127–129, 149
size of trees 131, 168–171
sizing of fruit 264
slow decline (citrus nematode) 203, 231
'Smooth Flat Seville' 61
soil
    fumigation 120–121
    mineral levels 154, 159–160
    moisture content 91, 107, 110, 138
        and irrigation scheduling 141
    and orchard site 127–128
    pH 128, 160
    residual herbicides 175–178, 179, 180
    and root morphology 71–72
    waterlogging 128, 144–146
soil-borne diseases 145, 229–232, 245
sooty mould 198
sour orange (*Citrus aurantium*) 3, 59–61
sour rot 241
South Africa 9, 216
South America 5, 9, 10–11
Spain 7–8
spider mites 192–193, 211
Spiroplasma citri 219
splitting 258–259
spraying
    herbicides 180–181
    insecticides 209–210
spreading decline 231–232
'Star Ruby' grapefruit 37
stem pitting 220–221
stem tap sampling 205
stem-end rots 240

stem-end-rind breakdown 269
stomata 136–137
storage
    of fruit 246–247, 265–269
    of juice 271
    of seed 119
stress response 91–92
stubborn 218–219
stunted growth 67, 76
stylar-end breakdown 259
subtropics 80–82
    climate change 116
    yield 106–108
sudden collapse 228
sudden death 224
'Sugar Belle' mandarin hybrid 35
sugars, in fruit 247–248
sulfur (S) 157
'Summerfield' navel orange 26
'Sun Chu Sha' mandarin 69
'Sunburst' mandarin hybrid 34
supercooling 147
sweet orange see orange (*Citrus sinensis*)
sweet orange scab 237
'Swingle' citrumelo 67–68

TA (total acidity) 112–114, 250
    TSS:TA ratio 114, 251, 266
'Tahiti' lime 40
Tamarixia radiata 210–211
tangelo (mandarin x grapefruit hybrid) 32–33
tangerine 27, 29–30
    see also mandarins
tangor (mandarin x orange hybrid) 30–32
tatterleaf-citrange stunt virus (TLV) 222
taxonomy 13
    Citrus 21–45
    different systems 13–21
    intergeneric hybrids 43–45
temperature
    climate 78, 81, 115
    de-greening 255
    flower induction 92–94, 106–107
    fruit growth 110
    fruit set/drop 104
    and orchard site 127
    storage 265
    vegetative growth 86–88
    see also freezing
'Temple' orange (or mandarin) 30–31
termites 202
terraced orchards 131

Texas citrus mite (*Eotetranychus banksi*) 192–193, 211
thermotherapy 242
thinning-out cuts 168
'Thompson' grapefruit 37–38
thorns 17, 94
thread blight 239
thrips 193, 194, 198, 211, 256
tissue culture 47, 53, 119
topping 171, 253
transpiration 134, 137, 145
transport
    logistics 129
    of nursery trees 126–127
trickle irrigation 142–143
trifoliate orange (*Poncirus trifoliata*) 14
    cold tolerance 147
    hybridisation with *Citrus* 44
    as rootstock 63–65, 70–71, 72, 243
tristeza decline 219–221
tropics 78–80
    climate change 115–116
    yield 106, 108
'Troyer' citrange 65–66
TSS (total soluble solids) 112, 247–249
    and rootstock 57, 60
    TSS:TA ratio 114, 251, 266
twig dieback 229
2,4–D 172, 173

ultraviolet light 79, 88
urea sprays 97, 99
USA
    climate 80, 93
    freezes 81, 127, 147
    HLB 7, 184–185, 213
    production 5, 6–7, 8–9, 10, 12
uses 11–12

'Valencia' sweet orange 23–24
vegetative growth 82–91
'Verna' lemon 41
'Vernia' sweet orange 25
vines (weeds) 179
viroids 225–226, 241–242
viruses
    CiLV 223
    CPsV 221–222
    CSD 224
    CTV 198, 219–221, 242
    identification 242

other 224–225
   SDV 223–224
   TLV 222
   vectors 198, 221, 223
vitamin C (ascorbic acid) 250
'Volkamer' lemon 57

'Washington' navel orange 3, 25
water
   flowering induction 93, 107
   and fruit 104, 112, 137
   plant physiology 72–73, 91, 112, 136–139, 145
   and planting density 134
   and siting of orchards 128
   stress 93, 104, 137–139
   and yield 108–109
   see also irrigation
waterlogging 128, 144–146
watersoaked leaves 148, 152
wax on leaves 136
waxing of fruit 263, 266
weeds
   biological control 181–182
   herbicides 175–181
   mechanical control 174–175
   types of 173–174, 176–177

weevils (root weevil) 187–190, 209
'West Indian' lime 40
whiteflies 194, 198, 211
'Willowleaf' mandarin (*Citrus deliciosa*) 4, 30
wind damage
   fruit 81, 258
   in nurseries 122
wind machines 150
windbreaks 149, 244, 258
winter chlorosis 91, 154–155
witches' broom disease of limes (WBDL) 219
wood, freeze damage 152–153
wood borers 201

Xanthomonas spp. 232–233
Xylella fastidiosa 218
xyloporosis 226, 241–242

yellow shoot disease *see* huanglongbing
yield
   climatic factors 78, 106–109
   planting density 131
zebra skin 140, 259
zinc (Zn) 158, 226

CABI – who we are and what we do

This book is published by **CABI**, an international not-for-profit organisation that improves people's lives worldwide by providing information and applying scientific expertise to solve problems in agriculture and the environment.

CABI is also a global publisher producing key scientific publications, including world renowned databases, as well as compendia, books, ebooks and full text electronic resources. We publish content in a wide range of subject areas including: agriculture and crop science / animal and veterinary sciences / ecology and conservation / environmental science / horticulture and plant sciences / human health, food science and nutrition / international development / leisure and tourism.

The profits from CABI's publishing activities enable us to work with farming communities around the world, supporting them as they battle with poor soil, invasive species and pests and diseases, to improve their livelihoods and help provide food for an ever growing population.

CABI is an international intergovernmental organisation, and we gratefully acknowledge the core financial support from our member countries (and lead agencies) including:

Discover more

To read more about CABI's work, please visit: **www.cabi.org**

Browse our books at: **www.cabi.org/bookshop**,
or explore our online products at: **www.cabi.org/publishing-products**

Interested in writing for CABI? Find our author guidelines here:
www.cabi.org/publishing-products/information-for-authors/